Principles
of
Biochemical Toxicology

Second Edition

John A. Timbrell
Reader in
Biochemical Toxicology
School of Pharmacy
University of London

Taylor & Francis
London • Washington DC

UK   Taylor & Francis Ltd, 4 John St., London WC1N 2ET

USA   Taylor & Francis Inc., 1900 Frost Rd., Suite 101, Bristol,
PA 19007

Reprinted 1992 and 1994

British Library Cataloguing in Publication Data

A catalogue record for this book is available from the British Library

ISBN 0-85066-829.8
ISBN 0-85066-832.8 (pbk.)

Typeset in 10 on 12 point Times Roman by
Chapterhouse, The Cloisters, Halsall Lane, Formby

Printed in Great Britain by Burgess Science Press, Basingstoke, Hants

Principles of Biochemical Toxicology

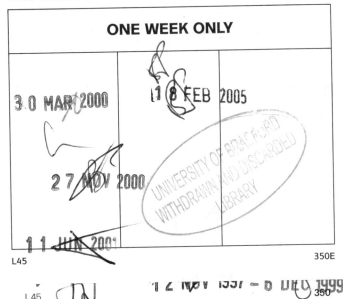

For
Anna, Becky and Cathy

Contents

'...and she had never forgotten that, if you drink much from a bottle marked "poison", it is almost certain to disagree with you, sooner or later.'

From *Alice's Adventures in Wonderland, by Lewis Carroll.*

Preface to the second edition

The first edition of this book was written as a result of my involvement with the organization of a MSc course in Experimental Pathology (Toxicology). This revised second edition has similarly benefited from my involvement with the organization and teaching of an undergraduate course in Toxicology and Pharmacology and teaching toxicology in a Pharmacy degree course. Also, there have been many advances in toxicology in the last ten years partly as a result of an increasingly mechanistic approach to the subject. For these reasons this edition of the book is considerably larger than the first but the original format has been retained as I feel it has been a successful one. It is my hope that the revisions and additions will make the book useful to a wider range of students than before.

I am grateful to Professor Sandy Florence and Professor Norman Bowery for allowing me a few months sabbatical time without which I would still be at the beginning of the task. I must thank also Dr Dennis Smith and Dr Andrew Hutt for the information they supplied, and Karen Henderson in the School of Pharmacy library for her help with the bibliography.

All of the new diagrams were prepared using the software Fig P (Biosoft) except for figure 6.15 for which I am grateful to Cathy.

Finally, I reserve special thanks for my family for their forebearance over the last year. Without their support it would have been a much more difficult task. This time, thanks to a word processor the preparation of the manuscript was an easier, but more solitary, occupation.

London, May 1991

Chapter 1

Introduction

Background

Toxicology is the subject concerned with the study of the noxious effects of chemical substances on living systems. It is a multi-disciplinary subject, as it embraces areas of pharmacology, biochemistry, chemistry, physiology and pathology, and although it has sometimes been considered as a subdivision of some of these other subjects, it is truly a scientific discipline in itself.

Toxicology may be regarded as the science of poisons; in this context it has been studied and practised since antiquity, and a large body of knowledge has been amassed. The ancient Greeks used hemlock and various other poisons, and **Dioscorides** attempted a classification of poisons. However, the scientific foundations of toxicology were laid by **Paracelsus** (1493–1541) and this approach was continued by **Orfila** (1787–1853).

Development of toxicology as a separate science has been slow, however, particularly in comparison with subjects such as pharmacology and biochemistry, and toxicology has a much more limited academic base. This may in part reflect the nature of the subject, which has evolved as a practical art, and also the fact that many practitioners were mainly interested in descriptive studies for screening purposes or to satisfy legislation.

Scope

The interest in and scope of toxicology continues to grow rapidly and the subject is of profound importance to human and animal health. The increasing numbers (currently around 100 000) of foreign chemicals (xenobiotics) to which humans and other organisms in the environment are exposed underlies this growth. These include drugs, pesticides, environmental pollutants, industrial chemicals and food additives about which we need to know much, particularly concerning their safety. Of particular importance, therefore, is the ability to predict toxicity and this requires a sound mechanistic base to be successful. It is this mechanistic base that comes within the scope of biochemical toxicology, which forms the basis for almost all of the various branches of toxicology.

The development of toxicology has been hampered by the requirements of regulatory agencies which have encouraged the 'black box' approach of empiricism as discussed by Goldberg (see Bibliography). This routine gathering of data on toxicology, preferably of a negative nature, required by the various regulatory bodies of the industrial nations, has tended to constrain and regulate toxicology.

Furthermore, to paraphrase Zbinden, misuse of toxicological data and adverse regulatory action in this climate of opinion has discouraged innovative approaches to toxicological research and has become an obstacle to the application of basic concepts in toxicology. However the emphasis on and content of basic science at a recent International Congress of Toxicology is testimony to the progress that has taken place in toxicology in the decade since Goldberg and Zbinden wrote their articles (see Volans *et al.*, Bibliography).

Ideally, basic studies of a biochemical nature should be carried out if possible before, but at least simultaneously with, toxicity testing, and a bridge between the biochemical and morphological aspects of the toxicology of a compound should be built. It is apparent that there are many gaps in our knowledge concerning this connection between biochemical events and subsequent gross pathological changes. Without an understanding of these connections, which will require a much greater commitment to basic toxicological research, our ability to predict toxicity and assess risk from the measurement of various biological responses will remain inadequate.

Thus, any foreign compound which comes into contact with a biological system will cause certain perturbations in that system. These biological responses, such as the inhibition of enzymes, and interaction with receptors, macromolecules or organelles, may not be toxicologically relevant. This point is particularly important when assessing *in vitro* data, and involves the concept of a dose threshold, or the lack of such a threshold, in the 'one molecule, one hit' type of theory of toxicity.

Biochemical aspects of toxicology

Biochemical toxicology is concerned with the mechanisms underlying toxicity, particularly the events at the molecular level and the factors which determine and affect toxicity.

The interaction of a foreign compound with a biological system is two-fold: there is the effect of the organism on the compound and the effect of the compound on the organism. It is necessary to appreciate both for a mechanistic view of toxicology. The first of these includes the absorption, distribution, metabolism and excretion of xenobiotics, which are all factors of importance in the toxic process and which have a biochemical basis in many instances. The mode of action of toxic compounds in the interaction with cellular components, and at the molecular level with structural proteins and other macromolecules, enzymes and receptors, and the types of toxic response produced, are included in the second category of interaction. However, a biological system is a *dynamic* one and therefore a series of events may follow the initial response. For instance, a toxic compound may cause liver or kidney damage and thereby limits its own metabolism or excretion.

The anatomy and physiology of the organism affect all the types of interaction given above, as may the site of exposure and entry of the foreign compound into the organism. Thus, the gut bacteria and conditions in the gastrointestinal tract convert the naturally occurring compound **cycasin**, methylazoxymethanol glycoside, into the potent carcinogen **methylazoxy-methanol** (figure 1.1). Administered by other routes, cycasin is not carcinogenic.

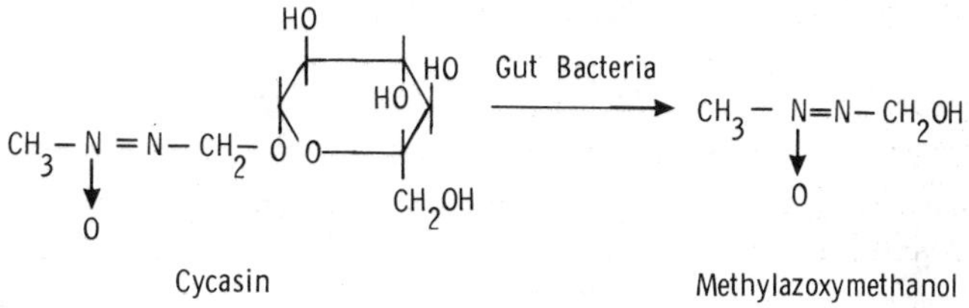

FIGURE 1.1. Bacterial hydrolysis of cycasin.

The distribution of a foreign compound and its rate of entry determine the concentration at a particular site and the number and types of cells exposed. The plasma concentration depends on many factors, not least of which is the metabolic activity of the particular organism. This metabolism may be a major factor in determining toxicity, as the compound may be more or less toxic than its metabolites.

The excretion of a foreign substance may also be a major factor in its toxicity and a determinant of the plasma and tissue levels. All of these considerations are modified by species differences, genetic effects and other factors. The response of the organism to the toxic insult is influenced by similar factors. The route of administration of a foreign compound may determine the effect, whether systemic or local. For instance, tetraethyl lead causes a local effect on the skin if applied topically, but a systemic effect on the blood if it gains entry into the body. Normally only the tissues *exposed* to a toxic substance are affected unless there is an indirect effect involving a physiological mechanism such as an immune response. The distribution of a toxic compound may determine its target-organ specificity, as does the susceptibility of the particular tissue and its constituent cells. Therefore, the effect of a foreign compound on a biological system depends on numerous factors, and an understanding and appreciation of them is a necessary part of toxicology.

The concept of toxicity is an important one: it involves a damaging, noxious or deleterious effect on the whole or part of a living system which may or may not be reversible. The toxic response may be a transient biochemical or pharmacological change or a permanent pathological lesion. The effect of a toxic substance on an organism may be immediate, as with a pharmacodynamic response such as a hypotensive effect, or delayed, as in the development of a tumour.

It has been said that there are no harmless substances, only harmless ways of using them, which underscores the concept of toxicity as a relative phenomenon. It depends on the dose and type of substance, the frequency of exposure and the organism in question. There is no absolute value for toxicity, although it is clear that botulinum toxin has a very much greater relative toxicity or potency than DDT on a weight-for-weight basis (table 1.1). The derivation and meaning of the LD_{50} will be discussed in detail in Chapter 2.

Table 1.1.   Acute LD_{50} values for a variety of chemical agents.

Agent	Species	LD_{50} (mg/kg body weight)
Ethanol	Mouse	10 000
Sodium chloride	Mouse	4000
Ferrous sulphate	Rat	1500
Morphine sulphate	Rat	900
Phenobarbital, sodium	Rat	150
DDT	Rat	100
Picrotoxin	Rat	5
Strychnine sulphate	Rat	2
Nicotine	Rat	1
d-Tubocurarine	Rat	0·5
Hemicholinium-3	Rat	0·2
Tetrodotoxin	Rat	0·1
Dioxin (TCDD)	Guinea-pig	0·001
Botulinum toxin	Rat	0·00001

Data from Loomis, T.A. (1974) *Essentials of Toxicology* (Philadelphia: Lea & Febiger).

There are many different types of toxic compound producing the various types of toxicity detailed in Chapter 6. One compound may cause several toxic responses. For instance, vinyl chloride (figure 4.6) is carcinogenic after low doses with a long latent period for the appearance of tumours, but it is narcotic and hepatotoxic after single large exposures.

Investigation of the sites and modes of action of toxic agents and the factors affecting their toxicity as briefly summarized here is fundamental for an understanding of toxicity and also for its prediction and treatment.

For example, the elucidation of the mechanism of action of the war gas Lewisite (figure 1.2), which involves interaction with cellular sulphydryl groups,

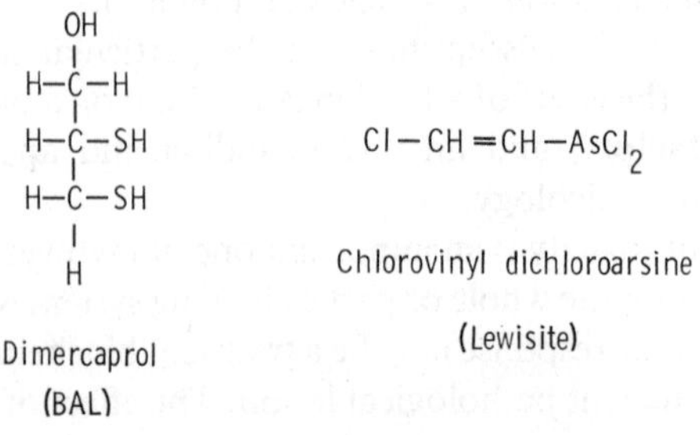

FIGURE 1.2. Structures of Lewisite and dimercaprol or British Anti-Lewisite.

allowed the **antidote**, British Anti-lewisite or **dimercaprol** (figure 1.2) to be devised. Without the basic studies performed by Sir Rudolph Peters and his colleagues, an antidote would almost certainly not have been available for the victims of chemical warfare.

Likewise, empirical studies with chemical carcinogens may have provided much interesting data but would have been unlikely to explain why such a diverse range of compounds cause cancer, until basic biochemical studies provided some of the answers.

Bibliography

ALBERT, A. (1979) *Selective Toxicity* (London: Chapman & Hall).

ALBERT, A. (1987) *Xenobiosis* (London: Chapman & Hall).

ANDERSON, D. and CONNING, D. M. (editors) (1988) *Experimental Toxicology. The Basic Principles* (Cambridge: Royal Society of Chemistry).

BRUIN, A. DE (1976) *Biochemical Toxicology of Environmental Agents* (Amsterdam: Elsevier).

CASARETT, L. J. and BRUCE, M. C. (1986) Origin and scope of toxicology. In *Casarett and Doull's Toxicology, The Basic Science of Poisons*, edited by C.D. Klaassen, M.O. Amdur and J. Doull (New York: Macmillan).

EFRON, E. (1984) *The Apocalyptics, Cancer and the Big Lie* (New York: Simon & Schuster).

GOLDBERG, L. (1979) Toxicology; Has a new era dawned? *Pharmac. Rev.*, **30**, 351.

HAYES, A. W. (editor) (1989) *Principles and Methods of Toxicology*, 2nd edition (New York: Raven Press).

HODGSON, E. and GUTHRIE, F. E. (1980) Biochemical toxicology; Definition and scope. In *Introduction to Biochemical Toxicology*, edited by E. Hodgson and F.E. Guthrie (New York: Elsevier-North Holland).

HODGSON, E. and LEVI, P. E. (1987) *A Textbook of Modern Toxicology* (New York: Elsevier).

LU, F. C. (1991) *Basic Toxicology*, 2nd edition (Washington, D.C.: Hemisphere).

MCCLELLAN, R. O. (editor) (1971) *Critical Reviews in Toxicology* (Boca Raton, Florida: CRC Press).

MORIARTY, F. (1988) *Ecotoxicology: The Study of Pollutants in Ecosystems*, 2nd edition (London: Academic Press).

PETERS, R. A. (1963) *Biochemical Lesions and Lethal Synthesis* (Oxford: Pergamon).

PRATT, W. B. and TAYLOR, P. (editors) (1990) *Principles of Drug Action, The Basis of Pharmacology*, 3rd edition (New York: Churchill Livingstone).

VOLANS, G. N., SIMS, J., SULLIVAN, F. M. and TURNER, P. (editors) (1990) *Basic Science in Toxicology, Proceedings of the Vth International Congress of Toxicology* (London: Taylor & Francis).

WEXLER, P. (1987) *Information Sources in Toxicology*, 2nd edition (New York: Elsevier).

ZBINDEN, G. (1979) Application of basic concepts to research in toxicology. *Pharmac. Rev.*, **30**, 605.

ZBINDEN, G. (1980) Predictive value of pre-clinical drug safety evaluation. In *Clinical Pharmacology and Therapeutics, Proceedings of the First World Conference on Clinical Pharmacology and Therapeutics* (1980) (Macmillan).

ZBINDEN, G. (1988) Biopharmaceutical studies, a key to better toxicology, *Xenobiotica*, **18**, Suppl. 1, 9.

Chapter 2

Dose–response relationships

Introduction

The relationship between the dose of a compound and its toxicity is central in toxicology. **Paracelsus** (1493–1541), who was the first to put toxicology on a scientific basis, clearly recognized this relationship. His well-known statement '*All substances are poisons; there is none that is not a poison. The right dose differentiates a poison and a remedy*', has immortalized the concept. Implicit in this statement is the premise that there is a dose of a compound which has no observable effect and another, higher dose, which causes the maximal response. The **dose–response relationship** involves quantifying the toxic effect and showing a correlation with exposure. The relationship underlies the whole of toxicology and an understanding of it is crucial. Parameters gained from it have various uses in both investigational and regulatory toxicology. It should be appreciated, however, that toxicity is a *relative* phenomenon and that the ways of measuring it are many and various.

Criteria of toxicity

The simplest measurement of toxicity is lethality, but this end-point is a relatively crude measure and gives little information about the underlying basis of the toxicity. The variability is usually considerable, as the end-point is often dependent on a number of physiological or biochemical processes. In either the traditional LD_{50} test or the replacement, however, there is scope for observation of toxic effects so that the information gained is more than a simple number. (For more details of the LD_{50} determination see below.) There are many other criteria of toxicity, not necessarily complex, which may be used in initial toxicity studies. An initial acute toxicity study may therefore simply observe the animals exposed to a range of doses without attempting to determine the lethal dose. Careful observation of the toxic signs may give valuable insight into possible mechanisms underlying the toxic effect. For example, an organophosphorus insecticide which affects the neuromuscular and central nervous systems in mammals may cause a variety of signs including salivation, diarrhoea, miosis, dyspnoea, tremors and convulsions, lack of reflexes and hind limb weakness (see Chapter 7). There are moves to *replace* the traditional lethality test in which the LD_{50} is determined as required by certain regulatory authorities, with one in which animals are exposed to particular doses simply in order to *classify* the compound and determine its *toxicity* rather than determine the lethal dose (van den Heuvel *et al.*, 1987.)

During such an initial study, observation of the animals may provide an indication of the toxic effect(s) produced by the compound, although it may or may not be the cause of death. It is therefore preferable to use a measure of toxicity which is as close as possible to the underlying mechanism. This may, of course, require prior knowledge of the *target site*, which may be a *receptor*, *enzyme* or *other macromolecule*, but an indication of the underlying cause of death or toxicity can sometimes be gained from an observation of the time-course of such effects after dosing. If an animal dies within minutes of the administration of a compound, it may well be that a major biochemical or physiological system has been affected.

For example, cyanide is rapidly lethal because the target is cytochrome aa$_3$ in the mitochondrial electron transport chain which is vital to all cells. Blockade of this enzyme system will therefore stop cellular respiration in many different tissues.

Clearly there are many different types of toxicity which can be assessed and form the basis of the study of a dose–response relationship (see also Chapter 6). Thus toxic effects may be direct or indirect, local or systemic, immediate or delayed and reversible or irreversible. For example a strong acid may have a direct, local toxic effect on skin which is immediate but reversible. Paracetamol (see Chapter 7) has a direct but systemic effect leading to liver damage, which is somewhat delayed. Alternatively penicillin (see Chapter 7) has an indirect and sometimes immediate systemic effect on several body systems which is reversible. Cancer caused by chemicals, such as dimethylnitrosamine or benzo(a)pyrene, is a response which may be local or systemic and is generally irreversible and often very delayed, but usually it is the result of a direct effect on a particular organ (see Chapter 7). An effect may be reversible if damage to a tissue can be repaired or an inhibited enzyme regenerated or resynthesized, but in the case of a vital tissue or enzyme this may not occur rapidly enough, and so the toxicity has a fatal outcome and is essentially irreversible. It is clear that these features of toxic effects will depend on many factors including the mechanism of the toxicity. Many of the later examples will exemplify these features and the factors governing them.

The selection of a measurable index of toxicity in the absence of an obvious pathological lesion may therefore be difficult, and the index may not relate to the overt toxicity or to the lethality. For instance, certain organophosphorus compounds may inhibit blood cholinesterase activity, but this change may not be directly related to the main toxic effect which is delayed neuropathy (see Chapter 7). Similarly, the accumulation of triglycerides in the liver caused by hydrazine (figure 2.1) is not directly related to the lethality, this being due to another and probably unrelated effect involving the central nervous system. However, fatty liver induced by hydrazine is an example of a toxic response which can be readily quantitated and which shows a clear dose–response relationship. In this case, there is a graded, rather than an 'all-or-none' response, between a normal level of

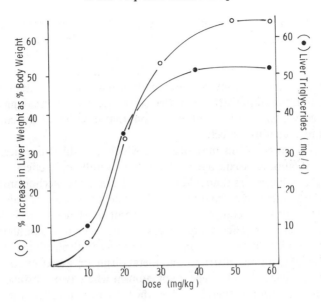

FIGURE 2.1. Increase in liver weight and liver triglycerides caused by hydrazine.
Data from Timbrell *et al.* (1982) *J. Toxicol. Environ. Health*, **10**, 955.

triglycerides in the liver and a maximum value. The **fatty liver** may be adequately assessed by measurement of the liver weight (expressed as % body weight), a relatively simple measure which shows a similar dose–response relationship to triglyceride levels and which can be carried out at the same time as an initial toxicity study.

Conversely, the lung damage and oedema (water accumulation) due to the compound ipomeanol, discussed in greater detail in Chapter 7, is directly related to the lethality. This can be seen from the dose–response curve (figure 7.25) and also when the time-course of death and lung oedema, measured as the wet weight: dry weight ratio, are compared (figure 7.14), strongly suggesting a causal relationship between them.

Changes in body weight and changes in organ weight are often sensitive indices of toxicity in animals which are readily determined in short-term toxicity tests.

It is appropriate at this point to mention in general terms interactions which may affect toxic responses. Many specific factors are discussed in detail later in this book and especially in Chapter 5.

When two toxic substances are given to an animal together the resulting toxic response may simply be the sum of the two individual responses. This situation where there is no interaction is known as an **additive** effect.

Conversely, if the overall toxic response following exposure to two toxic compounds is greater than the sum of the individual responses, this effect is known as **synergism**. For example carbon tetrachloride and ethanol together are more toxic to the liver than each is separately.

Potentiation is similar to synergism except that the two substances in question have different toxic effects or perhaps only one is toxic. For instance when the drug **disulphiram** is given to alcoholics, subsequent intake of ethanol causes toxic effects to occur due to the interference in the metabolism of ethanol by disulphiram. Disulphiram has no toxic effect at the doses administered however. There are many other examples of potentiation and a number are covered in this book (see paracetamol, bromobenzene, carbon tetrachloride, Chapter 7). It is conceivable that the administration of two substances to an animal may lead to a toxic response which is entirely different from that of either of the compounds. This would be a **coalitive** effect.

Alternatively, **antagonism** may occur in which one substance decreases the toxic effect of another toxic agent. Thus the overall toxic effect of the two compounds together is less than additive. There are many mechanisms whereby this occurs and these may involve the blockade of a receptor, alteration of metabolism or chemical complexation for example. Hence there is **functional antagonism** where the effects are opposite and therefore counterbalanced; **chemical antagonism** in which a complex is produced; **dispositional antagonism** in which the absorption, distribution, metabolism or excretion of the toxic compound is influenced and **receptor antagonism** where two substances interact with the same receptor and thereby reduce the toxic response. These mechanisms may be important in the action of antidotes. Several examples of such antagonism are discussed later in this book (see paracetamol, cyanide, carbon monoxide and lead, Chapter 7).

These interactive effects may be visualized graphically as **isoboles** (figure 2.2) or alternatively there are simple formulae which may be used for detecting them.

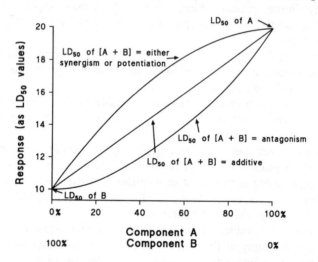

FIGURE 2.2. A set of isoboles for a mixture of two toxic compounds A and B.

$$V = \frac{\text{expected LD}_{50} \text{ of } (A + B)}{\text{observed LD}_{50} \text{ of } (A + B)}$$

If $V < 0.7$, there is antagonism; if $V = 0.7-1.3$, an additive effect is occurring; if

$V = 1 \cdot 3 - 1 \cdot 8$, the effect is more than additive; if $V > 1 \cdot 8$, there is synergism or potentiation. For further discussion see Brown (1988) and references therein. Note that these interactive effects may occur with single acute doses or repeat dosing, and may depend on the timing of the doses relative to each other.

The response of an organism to a toxic compound may become modified after repeated exposure. For example, **tolerance** or reduced responsiveness may develop when a compound is repeatedly administered. This may be the result of increasing or decreasing the concentration of a particular enzyme involved or by altering the number of receptors. For example, repeated dosing of animals with phenobarbital leads to tolerance to the pharmacological response as a result of enzyme induction (see Chapter 5). Conversely, tolerance to the hepatotoxic effect of a large dose of carbon tetrachloride results from the destruction of particular enzymes after small doses of the compound have been administered (for more details see Chapter 7).

Dose–response

It is clear from the preceding discussion that the measurable end-point of toxicity may be a pharmacological, biochemical or a pathological change which shows percentage or proportional change. Alternatively the end-point of toxicity may be an 'all-or-none' or quantal type of effect such as death or loss of consciousness. In either case, however, the dose–response relationship is graded between a dose at which no effect is measurable and one at which the maximal effect is demonstrated. The basic form of this relationship is shown in figure 2.3.

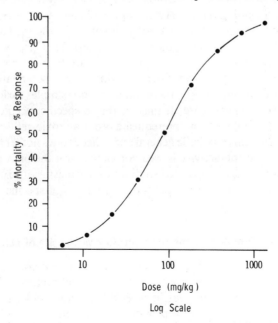

FIGURE 2.3. Dose-response curve.

The dose–response relationship is predicated on certain assumptions, however:

(a) that the toxic response is a function of the concentration of the compound at the site of action
(b) that the concentration at the site of action is related to the dose
(c) that the response is causally related to the compound given.

Examination of these assumptions indicates that there are various factors which may affect the relationship. Furthermore, it is also assumed that there is a method for measuring and quantifying the toxic effect in question. As already indicated there are many possible end-points or criteria of toxicity, but not all are appropriate.

Receptors

In some cases toxic effects are due to the interaction between the compound and a specific molecular receptor site. This receptor might be an enzyme which could be inhibited, or some other macromolecule, or a receptor with a normal physiological function, but in many cases its identity is unknown. Two examples of toxicity where the receptor is known are carbon monoxide which interacts specifically with **haemoglobin** and cyanide which interacts specifically with the enzyme **cytochrome a$_3$** of the electron transport chain (see Chapter 7). The toxic effects of these two compounds are a direct result of these interactions and, it is assumed, depend on the number of molecules of the toxic compound bound to the receptors. However, the final toxic effects involve cellular damage and death and also depend on other factors. Unlike in pharmacology the study of receptors has not yet featured prominently in toxicology, and there are few examples where specific receptors are known to be directly involved in the mediation of toxic effects (but see TCDD, Chapter 5, and various examples in Chapter 7). However, with some toxic effects such as the production of liver necrosis caused by paracetamol for instance, although a dose–response relation can be demonstrated (see Chapter 7), there may be no simple toxicant–receptor interaction in the classical sense. It may be that a specific receptor–xenobiotic interaction is not always a prerequisite for a toxic effect. Thus the pharmacological action of volatile general anaesthetics does not seem to involve a receptor, but instead the activity is well correlated with the oil–water partition coefficient. However, future detailed studies of mechanisms of toxicity will, it is hoped, reveal the existence of receptors or other types of specific targets where these are involved in toxic effects.

Toxic response is a function of the concentration at the site of action

The site of action may be an enzyme, a pharmacological receptor, another type of macromolecule, or a cell organelle or structure. The interaction of the toxic compounds at the site of action may be reversible or irreversible. The interaction is, however, assumed to initiate a proportional response. If the interaction is reversible, it may be described as follows:

$$R + T \underset{k_2}{\overset{k_1}{\rightleftharpoons}} RT \qquad (2.1)$$

where R = receptor; T = toxic compound; RT = receptor–compound complex; k_1 and k_2 = rate constants for formation and dissociation of the complex,

then:

$$\frac{[R] \, [T]}{[RT]} = \frac{k_2}{k_1} = K_T \qquad (2.2)$$

where K_T = dissociation constant of the complex. If $[R_t]$ is the total concentration of receptors and $[R_t] = [R] + [RT]$,

then:

$$\frac{[RT]}{[R_t]} = \frac{[T]}{K_T + [T]} \qquad (2.3)$$

If the response or effect (e) is proportional to the concentration of RT then:

$$e = k_3[RT]$$

and the maximum response ($E_{max.}$) occurs when all the receptors are occupied:

$$E_{max.} = k_3[R_t].$$

then:

$$e = E_{max.} \frac{[RT]}{[R_t]}$$

This may be transformed into:

$$e = \frac{E_{max.} \, [T]}{K_T + [T]} \qquad (2.4)$$

Thus, when $[T] = 0, e = 0$ and when $e = \frac{1}{2}E_{max.}$, $K_T = [T]$. Thus equation (2.4) is analogous to the **Michaelis–Menten equation** describing the interaction of enzyme and substrate.

Thus the more molecules of the receptor that are occupied by the toxic compound the greater the toxic effect. Theoretically there will be a concentration of the toxic compound at which all of the molecules of receptor (r), are occupied and hence there will be no further increase in the toxic effect.

$$\text{i.e. } \frac{[RT]}{[r]} = 1 \text{ or } 100\% \text{ occupancy.}$$

The relationship described above gives rise to the classical **dose–response curve** (figure 2.3). For more detail and the mathematical basis and treatment of the relationship between the receptor–ligand interaction and dose–response the reader is recommended to consult one of the texts indicated at the end of this chapter (Hathway, 1984; Pratt and Tayler, 1990).

However, the mathematics describes an idealized situation and the real situation *in vivo* may not be so straightforward. For example with carbon monoxide, as already indicated, the toxicity involves a reversible interaction with a receptor, the protein molecule haemoglobin (see Chapter 7 for further details of this example). This interaction will certainly be proportional to the concentration of carbon monoxide in the red blood cell. However, *in vivo* about 50% occupancy or 50% carboxyhaemoglobin may be sufficient for the final toxic effect which is cellular hypoxia and lethality. Duration of exposure is also a factor here because hypoxic cell death is not an instantaneous response. This time × exposure index is also very important in considerations of chemical carcinogenesis. Therefore *in vivo* toxic responses often involve several steps or sequelae, which may complicate an understanding of the dose–response relationship in terms of simple receptor interactions. Clearly it will depend on the nature of response measured. Thus, although an initial biochemical response may be easily measurable and explainable in terms of receptor theory, when the toxic response of interest and relevance is a pathological change which occurs over a period of time, this becomes more difficult.

The number of receptor sites and the position of the equilibrium (equation (2.1)) as reflected in K_T, will clearly influence the nature of the dose response, although the curve will always be of the familiar sigmoid type (figure 2.3). If the equilibrium lies far to the right (equation (2.1)) the initial part of the curve may be short and steep. Thus the *shape* of the dose–response curve depends on the type of toxic effect measured and the mechanism underlying it. For example, as already mentioned cyanide binds very strongly to cytochrome a_3 and curtails the function of the electron transport chain in the mitochondria, and hence stops cellular respiration. As this is a function vital to the life of the cell the dose–response curve for lethality is very steep for cyanide. The intensity of the response may also depend on the number of receptors available. In some cases, a proportion of receptors may have to be occupied before a response occurs. Thus there is a threshold for toxicity. With carbon monoxide for example, there are no toxic effects below a carboxyhaemoglobin concentration of about 20%, although there may be measurable physiological effects. A threshold might also occur when the receptor is fully occupied or saturated. For example, an enzyme involved in the biotransformation of the toxic compound may become saturated, allowing another metabolic pathway to occur which is responsible for toxicity. Alternatively a receptor involved in active excretion may become saturated, hence causing a disproportionate increase in the level of the toxic compound in the body when the dose is increased. Such saturable processes may determine the shape and slope of the dose–response curve.

However, when the interaction is irreversible, although the response may be proportional to the concentration at the site of action, other factors will also be important.

If the interaction is described as:

$$R + T \rightarrow RT$$
$$RT \rightarrow ?$$
(2.5)

the fate of the complex RT in equation (2.5) is clearly important. The repair or removal of the toxin–receptor complex RT may therefore be a determinant of the response and its duration.

From this discussion it is clear that the reversible and irreversible interactions may give rise to different types of response. With reversible interactions it is clear that at low concentrations occupancy of receptors may be negligible with no apparent response, and there may therefore be a threshold below which there is a **'no-effect level'**. The response may also be very short, as it depends on the concentration at the site of action which may only be transient. Also, repeated or continuous low-dose exposure will have no measurable effect.

With irreversible interactions, however, a single interaction will theoretically be sufficient. Furthermore, continuous or repeated exposure allows a cumulative effect dependent on the turnover of the toxin–receptor complex. An example of this is afforded by the organophosphorus compounds which inhibit cholinesterase enzymes (see Chapter 7). This inhibition involves reaction with the active site of the enzyme which is often irreversible. Resynthesis of the enzyme is therefore a major factor governing the toxicity. Toxicity only occurs after a certain level of inhibition is achieved (around 50%). The irreversibility of the inhibition allows cumulative toxicity to occur after repeated exposures over an appropriate period of time relative to the enzyme resynthesis rate.

With chemical carcinogens the interaction with DNA after a single exposure could be sufficient to initiate eventual tumour production with relatively few molecules of carcinogen involved, depending on the repair processes in the particular tissue. Consequently, chemical carcinogens may not show a measurable threshold, indicating that there may not be a 'no-effect level' as far as the concentration at the site of action is concerned. Although the DNA molecule may be the target site or receptor for a carcinogen, it now seems as though there are many subsequent events or necessary steps involved in the development of a tumour. There may therefore be more than one receptor–carcinogen interaction which will clearly complicate the dose–response relationship. For example DNA repair seems to be very important in some cases of chemical carcinogenesis and will contribute to the presence of a **dose threshold**.

The existence of 'no-effect doses' for toxic compounds is a controversial point, but it is clear that the ability to measure the exposure sufficiently accurately and to detect the response reliably are major problems (see below for further discussion). Suffice it to say that certain carcinogens are carcinogenic after exposure to concentrations measured in parts per million, and the dose–response curves for some nitrosamines and for ionizing radiation appear to pass through zero when the linear portion is extrapolated. At present, therefore, in some cases 'no-effect levels' cannot be demonstrated for certain types of toxic effect.

With chemical carcinogens time is also an important factor, both for the appearance of the effect, which may be measured in years, and for the length of exposure. It appears that some carcinogens do not induce tumours after single exposures or after low doses but others do. In some cases, there seems to be a relationship between exposure and dose, that is, low doses require longer exposure times to induce tumours than high doses, which is as would be expected for irreversible reactions with nucleic acids. For a further discussion of this topic the reader is referred to the Bibliography, particularly the articles by Williams and Weisburger (1986), Schmahl (1979), Zbinden (1979) and Lawley (Chapter 6).

Concentration at the site of action is related to the dose

Although the concentration in tissues is generally related to the dose of the foreign compound, there are various factors which affect this concentration. Thus, the absorption from the site of exposure, distribution in the tissues, metabolism and excretion all determine the concentration at the target site. However, the concentration of the compound may not be directly proportional to the dose, so the dose–response relationship may not be straightforward or marked thresholds may occur. For instance, if one or more of the processes mentioned is saturable or changed by dose, disproportionate changes in response may occur. For example, saturation of plasma-protein binding sites may lead to a marked increase in the plasma and tissue levels of the free compound in question. Similarly, saturation of the processes of metabolism and excretion, or accumulation of the compound, will have a disproportionate effect. This may occur with acute dose–response studies and also with chronic dosing, as for example with the drug chlorphentermine (figure 3.19), which accumulates in the adrenals but not in the liver after chronic dosing (figure 3.19). The result of this is accumulation of phospholipids, or phospolipidosis, in the tissues where accumulation of the drug occurs. Active uptake of a toxic compound into the target tissue may also occur. For example the herbicide paraquat is actively accumulated in the lung, reaches toxic concentrations in certain cells and then tissue damage occurs (see Chapter 7).

The relationship between the dose and the concentration of a compound at its site of action is also a factor in the consideration of the magnitude of the response and 'no-effect level'.

The processes of distribution, metabolism and excretion may determine that none of the compound in question reaches the site of action after a low dose, or only does so transiently. For both irreversible and reversible interactions, but particularly for the latter, this may be the major factor determining the threshold and the magnitude and duration of the response. For example, the dose required for a barbiturate to induce sleep in an experimental animal and the length of time that that animal remains unconscious can be drastically altered by altering the activity of the enzymes responsible for metabolizing the drug. Changes in the level of a toxic compound in the target tissue may occur due to changes in the pH of the blood or urine causing changes in distribution and excretion of the compound. This phenomenon is utilized in the treatment of poisoning to reduce the level of drug in the central nervous system after overdoses of barbiturates and salicylates (see Chapter 7). Both of these examples involve alteration of the concentration of drug at the site of action.

Response is causally related to the compound

Although this may seem straightforward, in some cases the response is only indirectly related and is therefore not a useful parameter of toxicity to use in a dose–response study. This may apply to situations where enzyme inhibition is a basic parameter but where it may not relate to the overall toxic effect. For example inhibition by lead of aminolaevulinic acid dehydrase, an enzyme which is involved in haem synthesis, can be readily demonstrated to be dose–related, but is clearly not an appropriate indicator of lead-induced renal toxicity *in vivo*. When more information has been gained about the toxicity or when the underlying mechanism of toxicity is understood then more precise indicators of toxicity can be measured. Similarly, this criterion must be rigorously applied to epidemiological studies where a causal relationship may not be apparent or indeed may not even exist.

Measurement of dose–response relationships

Dose–response curves may be derived by consideration of the population as a whole system or the consideration of the response in each individual.

The first type of treatment is obviously necessary where the end-point is an 'all-or-none' or quantal effect such as death. The second treatment may be applied to situations where there is a graded response to the toxic compound in the individual. Either treatment will give rise to the familiar dose–response curve (figure 2.3) when the percent response or percentage responding is plotted against dose. Thus, for the 'all-or-none' effect, the animals at each dose may be considered as parts of a single individual contributing towards the total percent response.

The shape of the dose–response curve depends on the number of factors, but it is basically derived from the familiar Gaussian curve (figure 2.4), which describes a normal distribution in biological systems. This bell-shaped distribution curve results from biological variation; in this case it represents the fact that a few animals respond at low doses, and others at high doses, but the majority respond at around the median dose. The perfect Gaussian distribution gives rise to a symmetrical sigmoid dose–response curve. The more animals used, the closer the curve is to a true sigmoid shape. The portion of the dose–response curve between 16% and 84% is the most linear, and may be used to determine the parameter **LD$_{50}$**. The LD$_{50}$ is that dose which, from the dose–response curve, is expected to be lethal to 50% of the animals. The linearity of the dose–response curve may often be improved by plotting the $\log_{10}$ of the dose, although this is an empirical

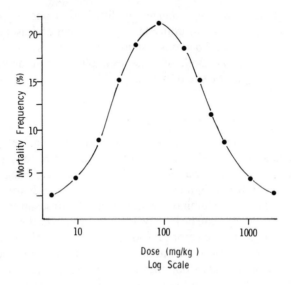

FIGURE 2.4. Dose–response relationship expressed as a frequency distribution.

transformation. In some cases dose–response curves may be linearized by applying other transformations. Thus for the conversion of the whole sigmoid dose–response curve into a linear relationship, probit analysis may be used, which depends upon the use of standard deviation units. The sigmoid dose–response curve may be divided into multiples of the standard deviation from the median dose, this being the point at which 50% of the animals being used respond. Within one standard deviation either side of the median, the curve is linear and includes 68% of the individuals; within two standard deviations fall 95.4% of the individuals.

Probit units define the median as probit five, and then each standard deviation unit is one probit unit above or below. The dose–response curve so produced is linear, when the logarithm of the dose is used (figure 2.5).

As well as mortality, other types of response can be plotted against dose. Similarly a median effective dose can be determined from these dose–response curves such as the **ED$_{50}$** where a pharmacological, biochemical or physiological response is measured, or the **TD$_{50}$** where a toxic response is measured. These parameters are analogous to the LD$_{50}$ (figure 2.6). The effective dose for 50% of the animals is used because the range of values encompassed is narrowest at this point compared with points at the extremities of the dose–response curve. A variation of the LD$_{50}$ is the **LC$_{50}$**, which is the concentration of a substance which is lethal to 50% of the organisms when exposed. This parameter is used in situations where an organism is exposed to a particular concentration of a substance in air or water, but the dose is unknown. Clearly the exposure time must be indicated in this case as well as the concentration.

The *slope* of the dose–response curve depends on many factors, such as the variability of measurement of the response and the variables contributing to the

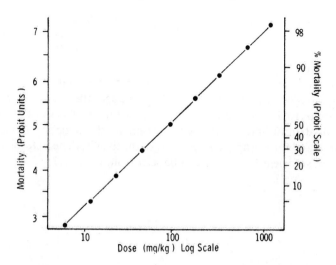

FIGURE 2.5. Dose–response relationship expressed in probit units.

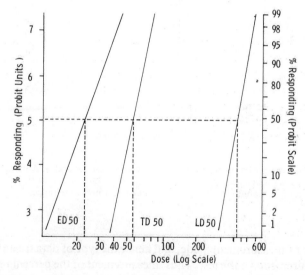

FIGURE 2.6. Dose–response curves for pharmacological effect, toxic effect and lethal effect, illustrating the ED_{50}, TD_{50} and LD_{50}. The proximity of the curves for efficacy and toxicity indicates the margin of safety for the compound and the likelihood of toxicity occuring in certain individuals after doses necessary for the desired effect.

response. The greater the number of animals or individual measurements and the more precise the measurement of the effect, the more accurate are the parameters determined from the dose–response curve. The slope of the curve also reflects the type of response. Thus when the response reflects a potent single effect, such as avid binding to an enzyme or interference with a vital metabolic function as is the case with cyanide or fluoroacetate for example, the dose–response curve will be steep and the value of the slope will be large. Conversely, a less specific toxic effect with more inherent variables results in a shallower curve with a greater standard deviation around the TD_{50} or LD_{50}. The slope therefore may give some indication of the mechanism underlying the toxic effect. Sometimes two dose–response curves may be parallel. Although they may have the same mechanism of toxicity this does not necessarily follow. The slope of the curve is also essential information for a comparison of the toxicity of two or more compounds and for a proper appreciation of the toxicity. The LD_{50} or TD_{50} value alone is not sufficient for this as can be seen from figure 2.7.

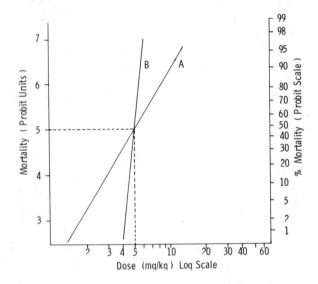

FIGURE 2.7. Comparison of the toxicity of two compounds, A and B. Although the LD_{50} is the same (5 mg/kg), toxicity occurs with A at a much lower dose than with B, but the minimum to maximum effect is achieved with B over a very much narrower dose range.

The type of measurement made, and hence the type of data treatment, depends on the requirements of the test. Thus, measurement of the percent response at the molecular level may be important mechanistically and more precisely measured. However, for the assessment of toxicity, measurement of the population response may be more appropriate.

Apart from possibly giving an indication of the underlying mechanism of toxicity, one particular value of quantitation of toxicity in the dose–response relationship is that it allows comparison. Thus comparisons may be made between different responses, between different substances and between different animal species for example.

Comparison of different responses underlies the useful parameter, **therapeutic index**, defined as follows:

$$\text{Therapeutic Index (TI)} = \frac{TD_{50}}{ED_{50}} \text{ or } \frac{LD_{50}}{ED_{50}}$$

which relates the pharmacologically effective dose to the toxic or lethal dose (figure 2.6). The therapeutic index gives some indication of the safety of the compound in use, as the larger the ratio, the greater is the relative safety. However, as already indicated, simple comparison of parameters derived from the dose–response curve such as the LD_{50} and TD_{50} may be misleading without some knowledge of the shape and slope of the curve. A more critical index is the margin of safety:

$$\text{Margin of safety} = \frac{TD_1}{ED_{99}} \text{ or } \frac{LD_1}{ED_{99}}$$

Similarly, comparison of two toxic compounds can be made using the LD_{50} (TD_{50}) (figure 2.7) and the dose–response curves, and this may also give information on possible mechanisms of toxicity. Thus, apart from the slope, which may be useful in a comparative sense, examination of ED_{50}, TD_{50} and LD_{50} may also provide useful information regarding mechanisms. Comparison of the LD_{50} or TD_{50} values of a compound after various modes of administration (table 2.1) may reveal differences in toxicity which might indicate what factors affect the toxicity of that particular compound. Thus with the anti-tubercular drug **isoniazid** there is little difference in toxicity after dosing by different routes of administration whereas with the local anaesthetic **procaine**, there is an 18-fold difference in the LD_{50} between intravenous and subcutaneous administration of the drug (table 2.1). Shifts in the dose–response curve or parameters derived from it caused by various factors may give valuable insight into the mechanisms underlying toxic effects (see table 2.2). The dose of the compound to which the animal is exposed is usually expressed as mg/kg body weight, or sometimes mg/m^2 of surface area. However, because of the variability of the absorption and distribution of compounds, it is preferable to relate the response to the plasma concentration or concentration at the target site. This may be particularly

Table 2.1.   Effect of route of administration on the toxicity of various compounds.

Route of administration	Pentobarbital[1] LD$_{50}$ (mg/kg)	Ratio to i.v.	Isoniazid[1] LD$_{50}$ (mg/kg)	Ratio to i.v.	Procaine[1] LD$_{50}$ (mg/kg)	Ratio to i.v.	DFP[2] LD$_{50}$ (mg/kg)	Ratio to i.v.
Oral	280	3.5	142	0.9	500	11	4.0	11.7
Subcutaneous	130	1.6	160	1.0	800	18	1.0	2.9
Intramuscular	124	1.5	140	0.9	630	14	0.85	2.5
Intraperitoneal	130	1.6	132	0.9	230	5	1.0	2.9
Intravenous	80	1.0	153	1.0	45	1	0.34	1.0

[1]Mouse toxicity data.
[2]Rabbit toxicity data on di-isopropylfluorophosphate.
Data from Loomis, T.A. (1968) *Essentials of Toxicology* (Philadelphia: Lea & Febiger).

Table 2.2.   Effect of bile duct ligation (BDL) on the toxicity of certain compounds.

Compound	LD$_{50}$:mg/kg		
	Sham operation	BDL	Sham:BDL ratio
Amitryptiline	100	100	1
Diethylstilboestrol	100	0.75	130
Digoxin	11	2.6	4.2
Indocyanine Green	700	130	5.4
Pentobarbital	110	130	0.8

Source: C.D. Klaassen (1974), *Toxicol. Appl. Pharmacol.* **24**, 27.

important with drugs used clinically which have a narrow therapeutic index or which show wide variation in absorption.

It will be clear from the discussion in the preceding pages, and should be noted, that the **LD$_{50}$** value is not an absolute biological constant as it depends on a large number of factors. Therefore, despite standardization of test species and conditions for measurement, the value for a particular compound may vary considerably between different determinations in different laboratories. Comparison of LD$_{50}$ values must therefore be undertaken with caution and regard for these limitations.

The value of the LD$_{50}$ test and the problems associated with it have recently been reviewed (see Bibliography, review article by Zbinden and Flury-Roversi).

Hazard and risk assessment

Another important role of the dose–response relationship is its use in the *extrapolation* of toxic effects seen at high doses to lower doses in order to undertake hazard and risk assessment.

As already indicated for some types of toxic effect there will clearly be a dose threshold, below which there is no detectable response. This 'No Observed Effect Level' (NOEL) may apply to either a quantal response such as death or a pathological lesion or to a response such as enzyme inhibition or binding to an endogenous receptor. The cause of the threshold may be the saturation of an enzyme or physiological repair system for example. However with some types of toxic effect such as chemical carcinogenesis for instance, the perceived mechanism supports the view that there may be no threshold for the biological effect (see also above). Thus a single molecule of carcinogen might, theoretically, be sufficient to interact with DNA and cause a permanent change in the genome of a single cell which could then lead to the development of a tumour (see Chapter 6). This one-hit type of model would give rise to a linear dose–response curve with no dose–threshold when extrapolated to the point of zero response (compound A in figure 2.8). There are various models which have been suggested for the dose–response relationships of chemical carcinogens as well as the one-hit model, which incorporate the concept of the **multistage** nature of chemical carcinogenesis, the possibility of repair and the requirement for metabolic activation.

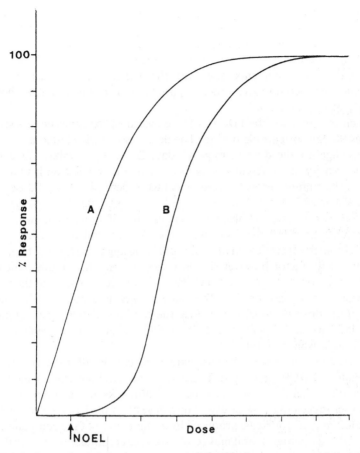

FIGURE 2.8. Comparison of the dose–response relationship for two compounds A and B. For compound A there is a response at all doses with no threshold. For compound B there is a dose or threshold below which there is a No Observed Effect Level (NOEL).

The concept of a threshold dose is an important one in toxicology, particularly in terms of the extrapolation of toxic doses derived from relatively small scale animal experiments and the subsequent assessment of risk to man.

The NOEL is used in setting exposure limits such as the **Acceptable Daily Intake (ADI)** for chemicals such as food additives or **Threshold Limit Values (TLV)** for industrial chemicals, usually with a 100-fold or sometimes greater safety factor to take account of species differences in response and human variability in response:

$$ADI = \frac{NOEL \text{ mg/kg/day}}{100}$$

Chronic toxicity

Chronic toxicity may be quantitated in a similar manner to acute toxicity, using the TD_{50} or LD_{50} concept. Measurement of chronic toxicity in comparison with acute-toxicity measurements may reveal that the compound is accumulating *in vivo*, and may therefore give a rough approximation of the probable whole-body half-life of the compound.

For chronic toxicity, the TD_{50} or LD_{50} is measured for a specific period of time, such as 90 days of chronic dosing. The dose–response is plotted as the percent response against the dose in (mg/kg)/day. If the TD_{50} values for acute and chronic toxicity are different it may indicate that accumulation is taking place. This may be quantitated as the **chronicity factor**, defined as LD_{50} 1 dose/LD_{50} 90 doses, where the LD_{50} 90 dose is expressed as (mg/kg)/day. If this value is 90, the compound in question is absolutely cumulative, if more than two, relatively cumulative, and if less than two, relatively non-cumulative.

The chronicity factor could of course utilize dosing periods other than 90 days. The chronicity factor however should be viewed only as a crude indication of accumulation of the response. It does not indicate accumulation of the substance and because it is based on the LD_{50} value it takes no account of the shape and slope of the dose–response curve. Also the conditions for determination of the acute and chronic LD_{50} may be different and this may introduce factors which make comparison uncertain.

An example of absolutely cumulative toxicity is afforded by tri-*o*-cresyl phosphate or **TOCP** (figure 2.9). This compound is a cholinesterase inhibitor and neurotoxin. In chickens, an acute dose of 30 mg/kg has a severe toxic effect, which is produced to the same extent by a dose of 1 (mg/kg)/d given for 30 days. This effect may of course be produced by accumulation of the compound *in vivo* to a threshold toxic level, or it may result from the accumulation of the effect, as it probably does in the case of TOCP.

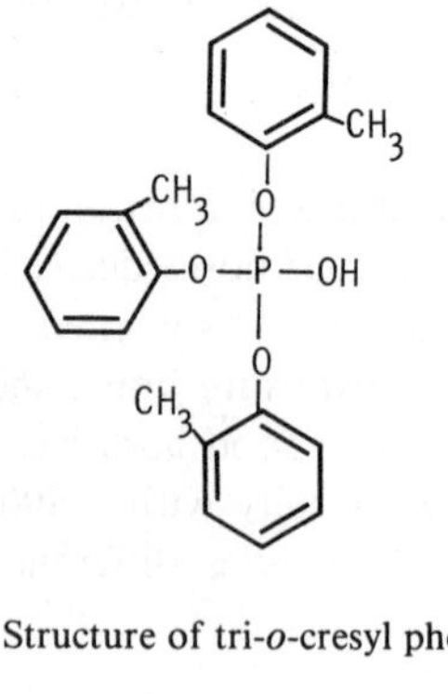

FIGURE 2.9. Structure of tri-*o*-cresyl phosphate (TOCP).

Thus, the inhibition of cholinesterase enzymes by organophosphorus compounds may last for several days or weeks, and repeated dosing at shorter intervals than the half-life of regeneration of the enzyme leads to accumulation of the inhibition until the toxic threshold of around 50% is reached.

Bibliography

BROWN, V. K. (1988) *Acute and Sub-acute Toxicology* (London: Edward Arnold).

DEICHMANN, W. B., HENSCHLER, D., HOLMSTEDT, B. and KEIL, G. (1986) What is there that is not a poison: a study of the Third Defense by Paracelsus. *Arch. Toxicol.*, **58**, 207.

GRIFFIN, J. P. (1985) Predictive value of animal toxicity studies. *ATLA*, **12**, 163.

HATHWAY, D. E. (1984) *Molecular Aspects of Toxicology* (London: The Royal Society of Chemistry).

HAYES, W. J. (1972) The 90 dose LD_{50} and a chronicity factor as measures of toxicity. *Toxic. Appl. Pharmac.*, **23**, 91.

HAYES, A. W. (editor) (1989) *Principles and Methods of Toxicology*, 2nd edition (New York: Raven Press).

KLAASSEN, C. D. (1986) Principles of toxicology. In *Toxicology: The Basic Science of Poisons*, edited by C.D. Klaassen, M.O. Amdur and J. Doull (New York: Macmillan).

NIEHS (1987) *Basic Research in Risk Assessment, Environmental Health Perspectives*, 76 (Dec.) (North Carolina: NIEHS).

PRATT, W. B. and Tayler, P. (editors) (1990) *Principles of Drug Action: The Basis of Pharmacology* (New York: Churchill Livingstone).

ROBERTS, C. N. (editor) (1989) *Risk Assessment–The Common Ground* (Eye, Suffolk: Life Science Research).

SCHMAHL, D. (1979) Problems of dose–response studies in chemical carcinogenesis with special reference to *N*-nitroso compounds. *CRC Crit. Rev. Toxicol.*, **6**, 257.

VAN DEN HEUVEL, M. J. DAYAN, A. D. and SHILLAKER, R. O. (1987) Evaluation of the BTS approach to the testing of substances and preparations for their acute toxicity, *Human Toxicol.*, **6**, 279.

WILKINSON, C. F. (1986) Risk assessment and regulatory policy, *Comments Toxicol.*, **1**, 1.

WILLIAMS, G. M. and WEISBURGER, J. H. (1986) Chemical carcinogens. In *Toxicology: The Basic Science of Poisons*, edited by C. D. Klaassen, M. O. Amdur and J. Doull (New York: Macmillan).

WORLD HEALTH ORGANISATION (1978) *Principles and Methods for Evaluating the Toxicity of Chemicals*, Part I, *Environmental Health Criteria 6* (Geneva: WHO).

ZBINDEN, G. (1979) The no-effect level, an old bone contention in toxicology. *Archs Toxicol.*, **43**, 3.

ZBINDEN, G. and FLURY-ROVERSI, M. (1981) Significance of the LD_{50} test for the toxicological evaluation of chemical substances. *Archs Toxicol.*, **47**, 77.

Chapter 3

Factors affecting toxic responses: disposition

The disposition of a toxic compound in a biological system may be conveniently divided into four interrelated phases:

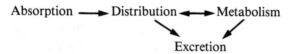

Each of these will be considered in turn.

Absorption

It is clear that to exert a toxic effect a compound must come into contact with the biological system under consideration. It may exert a local effect at the site of administration on initial exposure, but it must penetrate the organism in order to have a **systemic effect**. The most common means of entry for toxic compounds are via the gastrointestinal tract and the lungs, although in certain circumstances absorption through the skin may be an important route. Therapeutic agents may also enter the body after administration by other routes.

Transport across membranes

Although there are several sites of first contact between a foreign compound and a biological system, the absorption phase (and also distribution and excretion) necessarily involves the passage across cell membranes whichever site is involved. Therefore it is important first to consider membrane structure and transport in order to understand the absorption of toxic compounds.

Membranes are basically composed of phospholipids and proteins with the lipids arranged as a bilayer interspersed with proteins as shown simply in figure 3.1. A more detailed illustration is to be found in figure 3.2 which shows that the membrane, on average about 70 Å (7 nm) thick, is not symmetrical and that there are different types of phospholipid and proteins as indicated in the figure. Furthermore carbohydrates, attached to proteins (glycoproteins) and lipids

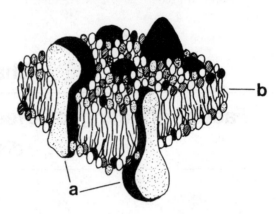

FIGURE 3.1. The three-dimensional structure of the animal cell membrane. Proteins (a) are interspersed in the phospholipid bilayer (b). From Timbrell, J. A., *Introduction to Toxicology*, Taylor and Francis, London, 1989.

(glycolipids), and cholesterol esters may also be constituents of the membrane. The presence of **cholesterol** seems to exert an influence on the fluidity and mechanical stability of the membrane, increasing the rigidity by intercalating between the phospholipid molecules. The particular proteins and phospholipids incorporated into the membrane, the proportions and their arrangement vary depending on the cell type in which the membrane is located and also the part of the membrane. For example the *ratio* of protein to lipid varies from $0 \cdot 25 : 1$ in the myelin membrane to $4 \cdot 6 : 1$ in the intestinal epithelial cell. Furthermore the particular proteins and phospholipids on the inside may be different from those on the outside of the membrane and in different parts of the cell reflecting differences in function of these molecules. This leads to differences in charge between outside and inside. For example, the liver cell membrane with a ratio of protein to lipid of about $1 : 1 \cdot 4$ has more phosphatidylcholine (neutral) in the exterior lipid layer than the interior layer, where there is more phosphatidylserine (negatively charged). Glycolipids are found only on the outer surface. Although the sinusoidal and canalicular surfaces of the liver cell are similar, having phosphatidyl ethanolamine and sphingomyelin in the exterior surface, on the contiguous surface of the liver cell, the exterior layer is almost entirely composed of phosphatidylcholine.

As well as the four basic types of phospholipid (figure 3.2), there are also variations in the fatty acid content which are very significant. The most common fatty acids in the phospholipids have 16–18 carbon atoms although C_{12}–C_{22} fatty acids may occur. However, not only does the chain length vary, but so also does the extent of saturation. Thus one or more double bonds may occur in the fatty acid chain and the greater the unsaturation the greater will be the fluidity of the membrane. The character of the membrane may change between different tissues and cells and even within the same cell such that fluidity and function will vary. The presence of double bonds in the membrane phospholipid is also significant from a toxicological point of view as these bonds are susceptible to peroxidation.

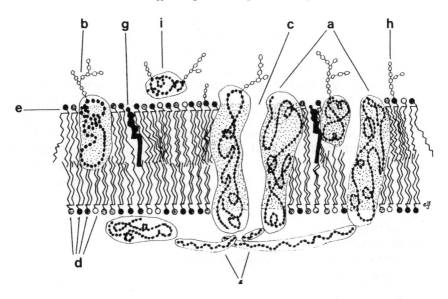

FIGURE 3.2. The molecular arrangement of the cell membrane, a: integral proteins: b: glycoprotein; c: pore formed from integral protein; d: various phospholipids with saturated fatty acid chains; e: phospholipid with unsaturated fatty acid chains; f: network proteins; g: cholesterol; h; glycolipid; i: peripheral protein. There are four different phosopholipids: phosphatidyl serine; phosphatidyl choline; phosphatidyl ethanolamine; sphingomyelin represented as O; Ø; ◎; ●. The stippled area of the protein represents the hydrophobic portion. From Timbrell, J.A., *Introduction to Toxicology*. Taylor and Francis, London, 1989.

Consequently peroxidation of membrane phospholipids may occur following exposure to toxic chemicals such as carbon tetrachloride (see Chapters 6 and 7). The membranes of cells found in the nervous system may contain a high proportion of lipid (**Type 1** membranes) and thereby allow the ready passage and accumulation of lipophilic substances. Membranes or regions of membranes which have a specific role in transport other than passive diffusion (**Type 2** and **Type 3** membranes) contain specific carriers. Membranes containing many pores such as those in the kidney glomerulus and liver parenchymal cells are known as **Type 4**.

The membrane proteins will have different characteristics and functions: structural, receptor or enzymatic. For example, some proteins which have a transport function may also have ATPase activity. The different surfaces of the membrane of a cell may contain proteins which reflect the function at that surface. For example, the *sinusoidal* surface of the liver cell will have proteins such as transferases which are involved in the transport of carbohydrates and amino acids and receptors for hormones such as insulin whereas the *bile canalicular membrane* surface will have specialized proteins for the transport of bile salts. Pores in the membrane will involve integral proteins, which span the entire membrane and have outer hydrophilic regions and inner hydrophobic regions. These pores will vary in both frequency and diameter ranging from 4 Å to perhaps 45 Å in the glomerulus of the kidney. The network proteins, such as

spectrin, are involved in the cytoskeleton which may also be a target for toxic substances (see Chapter 6). The proteins of the outer surface are often associated with carbohydrates as glycoproteins, which may be involved in cell–cell interactions and may help to maintain the orientation of the proteins in the membrane. For more details of plasma membrane structure see Bibliography.

Perhaps the most important feature of the plasma membrane is that it is *selectively* permeable. Therefore only certain substances are able to pass through the membrane, depending on particular physicochemical characteristics. It will be apparent throughout this book that the *physicochemical characteristics* of molecules are major determinants of their disposition and often of their toxicity.

Thus with regard to the passage of foreign, potentially toxic molecules through membranes the following physicochemical characteristics are important:

(a) size/shape
(b) lipid solubility/hydrophobicity
(c) structural similarity to endogenous molecules
(d) charge/polarity.

The role and importance of these characteristics will become apparent with the following discussion of the different ways in which foreign compounds may pass across membranes:

(1) filtration
(2) passive diffusion
(3) active transport
(4) facilitated diffusion
(5) phagocytosis/pinocytosis.

(1) Filtration

This process relies on diffusion through pores in the membrane down a concentration gradient. Only small, hydrophilic molecules with a molecular weight of 100 or less such as ethanol or urea will normally cross membranes by filtration. Ionized compounds and even small ions such as sodium will not pass through pores however; the latter will in fact be hydrated in aqueous environments and therefore too large for normal pores. Pore sizes vary between cells and tissues and in the kidney pores may be large enough (up to 45 Å) to allow passage of molecules with molecular weights of several thousand. The sinusoidal membrane in the liver is a specialized, discontinuous membrane which also has large pores, allowing the ready passage of materials into and out of the blood stream (see below Chapter 6).

(2) Passive diffusion

This is probably the most important mechanism of transport for foreign and toxic compounds. It does not show substrate specificity but relies on diffusion through the lipid bilayer. Passive diffusion requires certain conditions:

(a) there must be a concentration gradient across the membrane
(b) the foreign compound must be lipid soluble
(c) the compound must be non-ionized.

These conditions are embodied in the **pH-partition theory**: only non-ionized lipid soluble compounds will be absorbed by passive diffusion down a concentration gradient.

Let us examine the three conditions in turn.

THE CONCENTRATION GRADIENT

This is normally in the direction external to internal relative to the cell or organism. The rate of diffusion is affected by certain factors: it is proportional to the concentration gradient across the membrane, the area and thickness of the membrane and a diffusion constant which depends on the physicochemical characteristics of the compound in question. This relationship is known as **Fick's Law**:

$$\text{Rate of diffusion} = \frac{KA\,(C_2 - C_1)}{d} \tag{3.1}$$

where A is the surface area, C_2 is the concentration of compound outside and C_1 the concentration on the inside of the membrane, d is the thickness of the membrane and K is a constant, the diffusion coefficient. The concentration gradient is represented by $C_2 - C_1$ and for the above relationship it is assumed that the temperature is constant. In practice the diffusion coefficient, K, for a particular compound will incorporate physicochemical characteristics such as lipophilicity, size and shape. From this relationship it is clear that passive diffusion is a *first-order* rate process as the rate is directly proportional to the

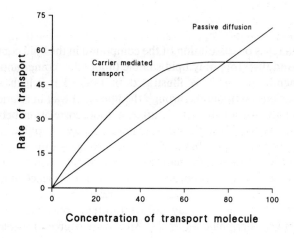

FIGURE 3.3. Comparison of the kinetics of carrier mediated transport and passive diffusion.

concentration of the compound at the membrane surface. This means that it is not a saturable process in contrast to active transport (see below and figure 3.3).

As biological systems are *dynamic*, the concentration gradient will normally be maintained and an equilibrium will not be reached. Thus the concentration on the inside of the membrane will be continuously *decreasing* as a result of ionization (see below), metabolism (see Chapter 4) and removal by distribution into other compartments such as via **blood flow** (figure 3.4). It is clear from this discussion that tissues such as the lungs, which have a large surface area, are served by an extensive vascular system and with few cell membranes to cross, will allow rapid passage of suitable foreign compounds.

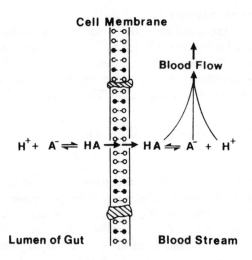

FIGURE 3.4. Role of blood flow and ionization in the absorption of foreign compounds. Both blood flow and ionization create a gradient across the membrane. From Timbrell, J.A., *Introduction to Toxicology*. Taylor and Francis, London, 1989.

LIPID SOLUBILITY

Passive diffusion relies on dissolution of the compound in the lipid component of the membrane and therefore only lipid soluble (lipophilic) compounds will pass through the membrane. This is illustrated in table 3.1, which shows the absorption of various compounds through the intestinal wall in relation to their partition coefficient. Although there is often a good correlation between lipid solubility and ability to diffuse through membranes very lipophilic compounds may become trapped in the membrane. Diffusion through the membrane will also depend on other factors such as the nature of the particular membrane or part of the membrane, especially the proportion of lipid and the presence of hydrophilic areas on integral proteins. Thus a lipid soluble foreign compound which also has a polar but non-ionized group may diffuse through some membranes more efficiently than a very lipophilic molecule. Also some degree of water solubility may assist the passage through membranes and in absorption from the gastro-

Table 3.1. Comparison between intestinal absorption and lipid: water partition of the non-ionized forms of organic acids and bases.

Drugs	Percentage absorbed	$K_{chloroform}$
Thiopental	67	100
Aniline	54	26.4
Acetanilide	43	7.6
Acetylsalicylic acid	21	2.0
Barbituric acid	5	0.008
Mannitol	<2	<0.002

$K_{chloroform}$ is the partition coefficient determined between chloroform and an aqueous phase, the pH of which was such that the drug was largely in the non-ionized form.
Data from Hogben *et al.* (1958) *J. Pharmac. Exp. Ther.*, **126**, 275.

intestinal tract especially may indeed be an important factor (see below). The lipid solubility of a compound is an intrinsic property of that compound, dependent on the structure and usually denoted by the **partition coefficient**, *P* (or log *P*). The larger the partition coefficient the greater is the lipophilicity of the compound. Compounds of similar structure and ionization may have very different partition coefficients. For example, **thiopental** and **pentobarbital** are very similar in structure and acidity but have very different lipophilicity (figure 3.5) and hence their disposition *in vivo* is different (see below).

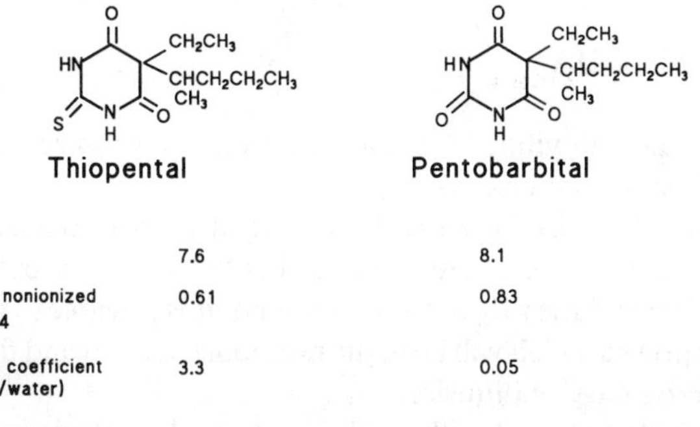

	Thiopental	Pentobarbital
pk$_a$	7.6	8.1
Fraction nonionized at pH 7.4	0.61	0.83
Partition coefficient (heptane/water)	3.3	0.05

FIGURE 3.5. Comparison of the structures and physicochemical characteristics of pentobarbital and thiopental.

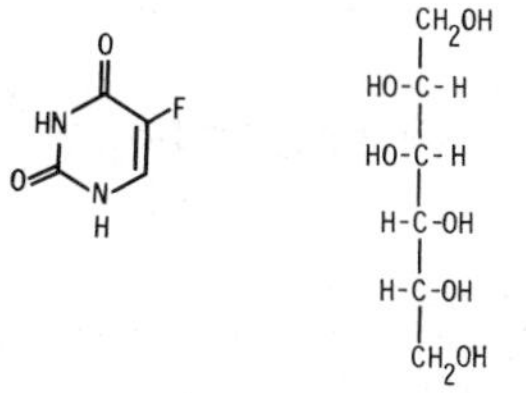

FIGURE 3.6. Structures of (A) 5-fluorouracil and (B) mannitol.

Solvents such as carbon tetrachloride, which are very lipid-soluble, are rapidly and completely absorbed from most sites of application, whereas more polar compounds such as the sugar, **mannitol** (figure 3.6), are very poorly absorbed as a consequence of limited lipid solubility (table 3.1).

THE DEGREE OF IONIZATION
This determines the extent of absorption, as only the non-ionized form will be able to pass through the lipid bilayer by passive diffusion. As already indicated, the lipid and water solubility of this non-ionized form is also a major factor.

The degree of ionization of a compound can be calculated from the **Henderson–Hasselbach equation:**

$$pH = pK_a + \text{Log} \frac{[A^-]}{[HA]} \qquad (3.2)$$

where pK_a, is the dissociation constant for an acid, HA and where:

$$HA \rightleftharpoons H^+ + A^-$$

For a base, A,

$$pH = pK_a + \text{Log} \frac{[A]}{[HA^+]}$$

where $A + H^+ \rightleftharpoons HA^+$

The pK_a of a compound, the pH at which it is 50% ionized, is a physicochemical characteristic of that compound.

Normally only HA in the case of an acid or A in the case of a base will be absorbed (figure 3.14). Therefore knowing the pH of the environment at the site of absorption and the pK_a of the compound, it is possible to calculate the amount of the compound which will be in the non-ionized form and therefore estimate the absorption by passive diffusion.

For example, an acid with a pK_a of 4 can be calculated to be mainly non-ionized in acidic conditions, at pH 1. Rearranging the Henderson–Hasselbach equation (3.2).

$$pH - pK_a = \log \frac{A^-}{HA}, \text{ and}$$

$$\text{anti-log } pH - pK_a = \frac{A^-}{HA}$$

For an acid with a pK_a 4 in an environment of pH 1,

$$\text{anti-log } 1-4 = \frac{A^-}{HA}$$

$$\text{i.e. anti-log } -3 = \frac{A^-}{HA} = 0.001$$

that is $A^-/HA = 1/1000$ or the acid is 99.9% non-ionized.

Conversely for a base, pK_a 5 at pH 1,

$$pH - pK_a = \log\frac{A}{HA^+}$$

$$1-5 = \log\frac{A}{HA^+}$$

$$\text{anti-log } -4 = \frac{A}{HA^+} = 0.0001$$

that is, $A/HA^+ = 1/10\,000$ or the base is 99·9% ionized.

The same calculations may be applied to calculate the degree of ionization of acids and bases under alkaline conditions. It can easily be seen that weak acids will be mainly non-ionized and will therefore, if lipid-soluble, be absorbed from an acidic environment, whereas bases will not, being mainly ionized under acidic conditions. Conversely, under alkaline conditions, acids will be mainly ionized, whereas bases will be mainly non-ionized and will therefore be absorbed.

Because the situation *in vivo* is normally dynamic, continual removal of the non-ionized form of the compound from the inside of the membrane causes continued ionization rather than the attainment of an equilibrium:

$$\text{membrane}$$
$$H^+ + A^- \rightleftharpoons HA - \| \rightarrow HA \rightarrow \text{removal}$$

If HA is continuously removed from the inside of the membrane, most of the compound will be absorbed from the site, provided its concentration at the site is not reduced by other factors.

(3) Active transport

This mechanism of membrane transport has several important features:

  (a) a specific membrane carrier system is required
  (b) metabolic energy is necessary to operate the system
  (c) transport occurs against a concentration gradient
  (c) the process may be inhibited by metabolic poisons
  (d) the process may be saturated at high substrate concentration
  (e) substrates may compete for uptake.

As active transport utilizes a carrier system, it is normally specific for a particular substance or group of substances. Thus the chemical structure of the compound and possibly even the spatial orientation are important. This type of transport is normally reserved for endogenous molecules such as amino acids, required nutrients, precursors or analogues. For example the anti-cancer drug **5-fluorouracil** (figure 3.6), an analogue of uracil, is carried by the pyrimidine transport system. The toxic metal **lead** is actively absorbed from the gut via the calcium transport system. Active uptake of the toxic herbicide **paraquat** into the lung is a crucial part of its toxicity to that organ (see Chapter 7). Polar and non-ionized molecules may be transported as well as lipophilic molecules. As active transport may be saturated, it is a *zero-order rate process* in contrast to passive diffusion (figure 3.3).

There are, however, various types of active transport systems, involving protein carriers and known as **uniports, symports** and **antiports** as indicated in figure 3.7. Thus symports and antiports involve the transport of two different molecules either in the same or a different direction. Uniports are carrier proteins which actively or passively (see facilitated diffusion below) transport one molecule through the membrane. Active transport requires a source of energy, usually ATP which is hydrolysed by the carrier protein, or the co-transport of ions such as Na^+ or H^+ down their electrochemical gradients. The transport proteins usually seem to traverse the lipid bilayer and appear to function like membrane-bound enzymes. Thus the protein carrier has a specific binding site for the solute or solutes to be transferred. For example, with the Na^+/K^+ ATPase antiport the solute (Na^+) binds to the carrier on one side of the membrane, an energy-mediated *conformational change* occurs involving phosphorylation of the protein via ATP and this allows the solute to be released on the other side of the membrane (figure 3.7). The second solute (K^+) then binds to the carrier protein and this then undergoes a conformational change following release of the phosphate moiety allowing the substance to be released to the other side of the membrane. **Glucose** is transported into intestinal cells via a symport along with Na^+, which enters the cell along its electrochemical gradient. The Na^+ is then transported out in the manner described above. Carrier-mediated transport can also involve **gated** channels, which may require the binding of a ligand to open the channel for instance.

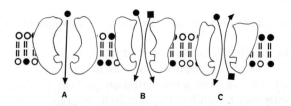

FIGURE 3.7. Mechanisms of active transport. A is a uniport, B a symport and C an antiport.

(4) Facilitated diffusion

This type of membrane transport has some similar features to active transport:

(a) it involves a specific carrier protein molecule
(b) the process may be saturated or competitively inhibited.

However:

(c) movement of the compound is only down a concentration gradient
(d) there is no requirement for metabolic energy.

Thus a specific carrier molecule is involved but the process relies on a concentration gradient as does passive diffusion. The transport of glucose out of intestinal cells into the bloodstream occurs via facilitated diffusion and utilizes a uniport.

(5) Phagocytosis/pinocytosis

These processes are both forms of **endocytosis** and involve the invagination of the membrane to enclose a particle or droplet respectively. The process requires metabolic energy and may be induced by the presence of certain molecules, such as ions, in the surrounding medium. The result is the production of a vesicle which may fuse with a primary lysosome to become a secondary lysosome in which the enzymes may digest the macromolecule. In some cases a particular part of the plasma membrane with specific receptors binds the macromolecule and then invaginates. Certain types of cells such as macrophages are especially important in the phagocytic process. Thus large molecules such as carrageenens with a molecular weight of about 40000 may be absorbed from the gut by this type of process. Insoluble particles such as those of uranium dioxide and asbestos are known to be absorbed by phagocytosis in the lungs (see below).

Sites of absorption

The following are the major routes of entry for foreign compounds:

(a) skin
(b) gastrointestinal tract
(c) lungs/gills
(d) intraperitoneal (i.p.)
(e) intramuscular (i.m.)
(f) subcutaneous (s.c.) and
(g) intravenous (i.v.)

Routes (d)–(g), known as **parenteral** routes, are normally confined to the administration of therapeutic agents or are used in experimental studies. The site of entry of a foreign, toxic compound may be important in the final toxic effect. Thus the acid conditions of the stomach may hydrolyse a foreign compound, or

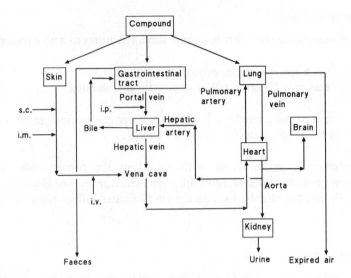

FIGURE 3.8. Blood flow and resulting distribution of a foreign compound from the three major sites of absorption or routes of injection. (i.v.: intravenous injection; s.c.: subcutaneous injection; i.m.: intramuscular injection; i.p.: intraperitoneal injection).

the gut bacteria may change the nature of the compound by metabolism and thereby affect the toxic effect. The site of entry may also be important to the final disposition of the compound. Thus absorption through the skin may be slow and will result in initial absorption into the peripheral circulation (figure 3.8). Absorption from the lungs, in contrast, is generally rapid and exposes major organs very quickly (figure 3.8). Compounds absorbed from the gastrointestinal tract first pass through the liver, which may mean that extensive metabolism takes place (figure 3.8). The toxicity of compounds after oral administration is therefore often less than after i.v. administration (table 2.1).

Skin

The skin is constantly exposed to foreign compounds such as gases, solvents and substances in solution and so absorption through the skin is potentially an important route. However, although the skin has a large surface area, some 18 000 cm2 in humans, fortunately it represents an almost continuous *barrier* to foreign compounds as it is not highly permeable. The outer layer of the non-vascularized **epidermis**, the stratum corneum, consists mainly of cells packed with **keratin** which limits the absorption of compounds, and a few hair follicles and sebaceous glands (figure 3.9).

The underlying dermis is more permeable and vascularized, but in order to reach the systemic circulation through the skin the toxic compound would have to traverse several layers of cells, in contrast to the situation in, for example, the gastrointestinal tract, where only two cells may separate the compound from the bloodstream.

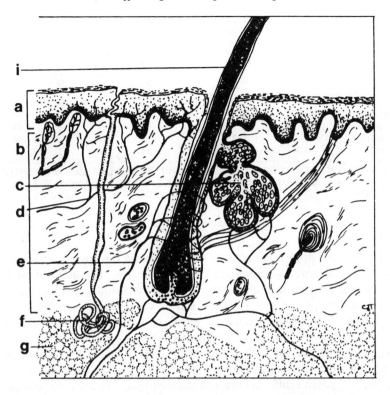

FIGURE 3.9. The structure of mammalian skin, a: epidermis; b: dermis; c: sebaceous gland; d: capillary; e: nerve fibre; f: sweat gland; g: adipose tissue; i: hair. From Timbrell, J. A., *Introduction to Toxicology*. Taylor and Francis, London, 1989.

Absorption through the skin is by passive diffusion mainly through the epidermis. Consequently compounds which are well absorbed percutaneously are generally lipophilic, such as solvents like carbon tetrachloride which may cause systemic toxicity (liver damage) following absorption by this route. Indeed insecticides such as **parathion** have been known to cause death in agricultural workers following skin contact and absorption. However lipophilicity as indicated by a large partition coefficient is not always a prerequisite for extensive absorption and there is not necessarily a good correlation (table 3.2). Thus polar compounds, such as the small, water soluble compound hydrazine, may also be absorbed through the skin sufficiently to cause a systemic toxic effect as well as a local reaction. The absorption of this compound may reflect its small molecular size (figure 5.28). Damage to the outer, horny layer of the epidermis increases absorption and a toxic compound might facilitate its own absorption in this way. Absorption through the skin will, however, vary depending on the site, and hence nature, of the skin and thickness of the stratum corneum. Thus penetration through skin of the foot is at least an order of magnitude less than that through the skin of the scalp. It should also be noted that the epidermis has significant metabolic activity and so may metabolize substances as they are absorbed.

Table 3.2. Physicochemical properties of various pesticides and their oral absorption and skin penetration in mice.

Compound	Partition coefficient	Water solubility (ppm)	Penetration Half-life		% Penetrated	
			Skin	Oral	Skin	Oral
DDT	1775	0.001	105	62	34	55
Parathion	1738	24	66	33	32	57
Chlorpyrifos	1044	2	20	78	69	47
Permethrin	360	0.07	6	178	80	39
Nicotine	0.02	Miscible	18	23	71	83

Data of Shah *et al.* (1981) *Toxicol. Appl. Pharmacol.*, **59**, 414 and Ahdaya *et al.* (1981) *Pestic. Biochem. Physiol.*, **16**, 38 from Hodgson and Levi (1987) *A Textbook of Modern Toxicology* (New York: Elsevier).

Lungs

Exposure to and absorption of toxic compounds via the lungs is toxicologically important and more significant than skin absorption. The ambient air in the environment whether it is industrial, urban or household, may contain many foreign substances such as toxic gases, solvent vapours and particles. The lungs have a large surface area, around 50–100 m² in humans; they have an excellent blood supply and the barrier between the air in the alveolus and the blood stream may be as little as two cell membranes thick (figure 3.10). Consequently, absorption from the lungs is usually *rapid* and *efficient*. The main process of absorption is passive diffusion through the membrane for lipophilic compounds (solvents such as chloroform), small molecules (gases such as carbon monoxide), and also solutions dispersed as aerosols. The substance will generally dissolve in the blood and may also react with plasma proteins or some other constituent. Therefore as the blood flow is *rapid* there will be a continuous removal of the substance and consequently a constant concentration gradient. However the solubility in the blood is a major factor in determining the rate of absorption. For compounds with low solubility the rate of transfer from alveolus to blood will be mainly dependent on blood flow (*perfusion limited*), whereas if there is high solubility in the blood the rate of transfer will be mainly dependent on respiration rate (*ventilation limited*).

As well as gases, vapours and aerosols, particles of toxic compounds may also be taken into the lungs. However the fate of these particles will depend on a number of factors, but especially the size (figure 3.11). Thus the larger particles will be retained in the respiratory tract initially to a greater extent than smaller particles because of rapid sedimentation under the influence of gravity whereas small particles will be exhaled more easily. Overall, approximately 25% of particles will be exhaled, 50% retained in the upper respiratory tract and 25% deposited in the lower respiratory tract. It can be seen from figure 3.11 that the larger particles (20 μm) tend to be retained in the upper parts of the respiratory system, whereas the smaller particles (< 6 μm) are confined to the alveolar ducts and terminal bronchioles. The optimum size for retention in the alveolar sacs is around 6 μm. Particles trapped by the mucus on the walls of the bronchi will be

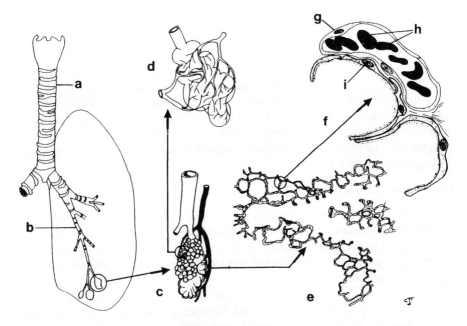

FIGURE 3.10. The structure of the mammalian respiratory system. a: trachea; b: bronchiole; c: alveolar sac with blood supply; d: arrangement of blood vessels around alveoli; e: arrangement of cells and airspaces in alveoli showing the large surface area available for absorption; f: cellular structure of alveolus showing the close association between the endothelial cell of the capillary, g, with erythrocytes, h, and the epithelial cell of the alveolar sac, i. The luminal side of the epithelial cell is bathed in fluid which also facilitates absorption and gaseous exchange. From Timbrell, J. A., *Introduction to Toxicology*. Taylor and Francis, London, 1989.

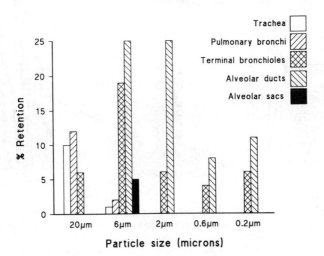

FIGURE 3.11. The retention of inhaled particles in various regions of the human respiratory tract in relation to size. Data from Hatch and Gross, *Pulmonary Deposition and Retention of Inhaled Aerosols*. New York: Academic Press, 1964.

removed by the ciliary escalator, whereas those of around $1\,\mu$m or less, which penetrate the alveolus, may be absorbed by phagocytosis, and hence may remain in the respiratory system for a long time. For example, **asbestos** fibres are phagocytosed in the lungs, remain there and eventually cause fibrosis and possibly lung tumours. It is known that **uranium dioxide** particles of less than $3\,\mu$m diameter can enter the blood stream and cause kidney damage after inhalation. Similarly particles of **lead** of about $0\cdot25\,\mu$m diameter enter the blood stream and cause biochemical effects following absorption via the lungs (see Chapter 7). The lymphatic system also seems to be involved in the movement of phagocytosed particles.

Gastrointestinal tract

Many foreign substances are ingested orally, either in the diet, or as drugs and poisonous substances taken either accidentally or intentionally. Most suicidal poisonings involve oral intake of the toxic agent. Consequently, the gastrointestinal tract is a very important site and perhaps the *major* route of absorption for foreign compounds.

The internal environment of the gastrointestinal tract varies throughout its length, particularly with regard to the pH. Substances taken orally first come into contact with the lining of the mouth (buccal cavity), where the pH is normally

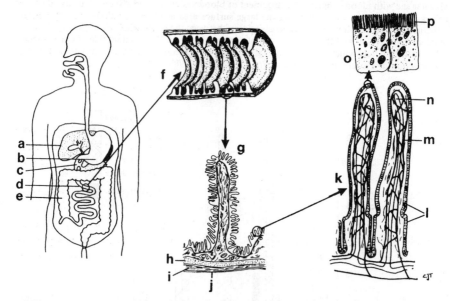

FIGURE 3.12. The mammalian gastrointestinal tract showing important features of the small intestine, the major site of absorption for orally administered compounds. a: liver; b: stomach; c: duodenum; d: ileum; e: colon; f: longitudinal section of the ileum showing folding which increases surface area; g: detail of fold showing villi with circular and longitudinal muscles, h and i respectively, bounded by the serosal membrane, j; k: detail of villi showing network of capillaries, m, lacteals, n, and epithelial cells, l; o: detail of epithelial cells showing brush border or microvilli, p. The folding, vascularization and microvilli all facilitate absorption of substances from the lumen. From Timbrell, J. A., *Introduction to Toxicology*. Taylor and Francis, London, 1989.

around 7 in man, but more alkaline in some other species such as the rat. The next region of importance is the stomach where the pH is around 2 in man and certain other mammals. The substance may remain in the stomach for some time, particularly if it is taken in with food. In the small intestine where the pH is around 6, there is a good blood supply and a large surface area due to the folding of the lining and the presence of villi (figure 3.12). The lining of the gastrointestinal tract essentially presents a continuous lipoidal barrier, passage through which is governed by the principles discussed above. Because of the change in pH in the gastrointestinal tract, different substances may be absorbed in different areas depending on their physicochemical characteristics, although absorption may occur along the whole length of the tract. Lipid soluble, non-ionized compounds may be absorbed anywhere in the tract, but ionizable substances will generally only be absorbed by passive diffusion if they are non-ionized at the pH of the particular site and are also lipid soluble. However, despite the fact that the gastrointestinal tract is well adapted for the absorption of compounds and lipophilic substances should be readily and rapidly absorbed by passive diffusion, this is perhaps simplistic and not always the case as illustrated by comparative data from mice (table 3.2).

Thus, in the acidic areas of the tract such as the stomach (pH 1–3), compounds which are lipid soluble in the non-ionized form, such as weak acids, are absorbed, whereas in the more alkaline (pH 6) small intestine, weak bases are more likely to be absorbed. The fact that in practice weak acids are also absorbed in the small intestine despite being ionized (figure 3.15) is due to the following:

(a) the large surface area of the intestine (figure 3.12)
(b) removal of compound by blood flow (figure 3.4)
(c) ionization of the compound in blood at pH 7.4 (figure 3.15).

These factors ensure that the concentration gradient is maintained and so weak acids are often absorbed to a significant extent in the small intestine if they have not been fully absorbed in the stomach. Using the Henderson–Hasselbach equation, the degree of ionization can be calculated and the site and likelihood of absorption may be indicated.

Let us consider the situation in the gastrointestinal tract, using benzoic acid and aniline as examples.

The pH of gastric juice is 1–3. The pK_a of benzoic acid, a weak acid, is 4. Taking the pH in the stomach as 2, and using the Henderson–Hasselbach equation as described above, it can be calculated that benzoic acid is almost completely non-ionized at this pH (figure 3.13):

$$\text{anti-log pH} - pK_a = \frac{A^-}{HA} = \text{anti-log } 2 - 4$$

$$\frac{A^-}{HA} = \frac{1}{100}$$

or 99% non-ionized.

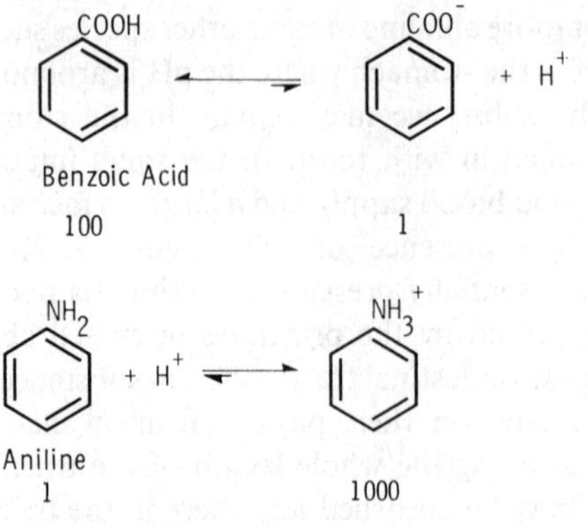

FIGURE 3.13. Ionization of benzoic acid and aniline at pH 2.

Benzoic acid should therefore be absorbed under these conditions and pass across the cell membranes into the plasma. Here, the pH is 7·4, which favours more ionization of benzoic acid.

Using the same calculation, at pH 7·4:

$$\frac{A^-}{HA} = \frac{1000}{1}$$

or 99·9% ionized.

Therefore, the overall situation is as shown in figure 3.14.

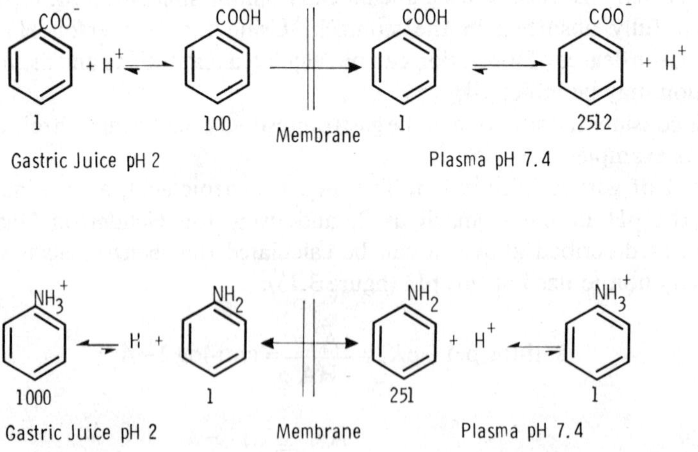

FIGURE 3.14. Disposition of benzoic acid and aniline in gastric juice and plasma. Figures represent proportions of ionized and non-ionized forms.

The non-ionized form of the benzoic acid crosses the membrane, but the continual removal by ionization in the plasma ensures that no equilibrium is reached. Therefore, the ionization in the plasma facilitates the absorption by removing the transported form.

Considering the situation for aniline in the same way (figure 3.13), for a base in the gastric juice of pH 2:

$$\text{anti-log } 2 - 5 = \frac{A}{HA^+} = 0 \cdot 001$$

$$\frac{A}{HA^+} = \frac{1}{1000}$$

or 99·9% ionized.

Aniline is therefore not absorbed under these conditions (figure 3.14). Furthermore, the ionization in the plasma does not facilitate diffusion across the membrane, and with some bases secretion from the plasma back into the stomach may take place. The situation in the small intestine, where the pH is around 6, is the reverse, as shown in figure 3.15.

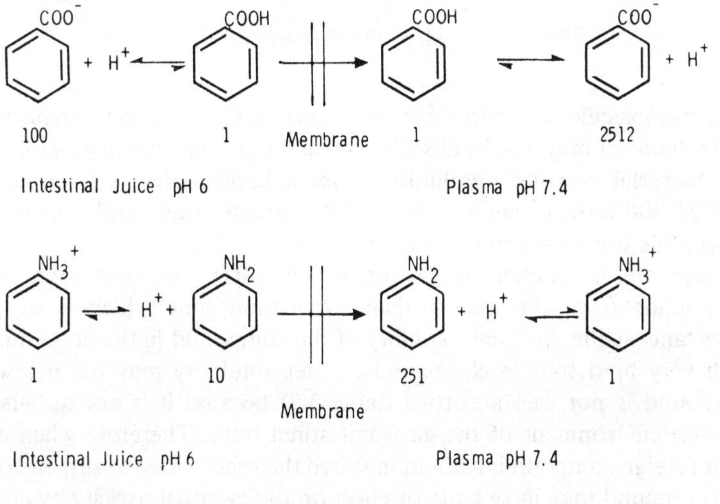

FIGURE 3.15. Disposition of benzoic acid and aniline in the small intestine and plasma. Figures represent proportions of ionized and non-ionized forms.

Therefore, it is clear that a weak base will be absorbed from the small intestine (figure 3.15) and, although the ionization in the plasma does not favour removal of the non-ionized form, other means of redistribution ensure removal from the plasma side of the membrane.

With the weak acid, however, it can be appreciated that although most is in the ionized form in the small intestine, ionization in the plasma facilitates removal of the transported form, maintaining the concentration gradient across the gastrointestinal membrane (figure 3.15). Consequently, weak acids are generally fairly well absorbed from the small intestine.

In contrast, strong acids and bases are not usually appreciably absorbed from the gastrointestinal tract by passive diffusion. However, some highly ionized compounds are absorbed from the gastrointestinal tract such as the quaternary ammonium compounds pralidoxime, an antidote (figure 3.16) which is almost entirely absorbed from the gut, and paraquat, a highly toxic herbicide (figure 3.16). Sufficient paraquat is absorbed from the gastrointestinal tract after oral ingestion for fatal poisoning to occur, but the nature of the transport systems for both of these compounds are currently unknown although carrier-mediated transport is perhaps the most likely (see Chapter 7). Carrier-mediated transport systems important for foreign, toxic compounds are known to operate in the gastrointestinal tract. For example cobalt is absorbed via the system that transports iron and lead by the calcium uptake system.

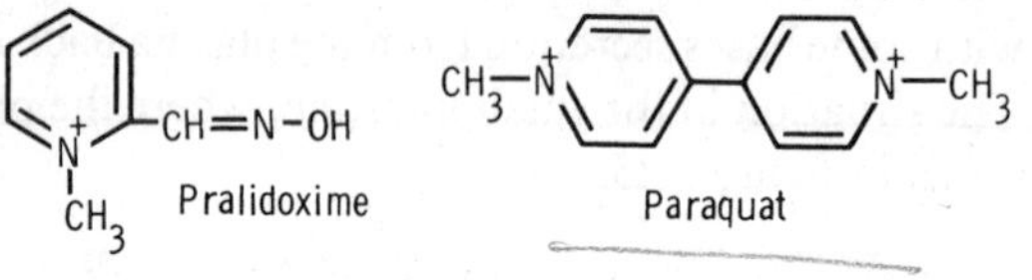

FIGURE 3.16. Structures of paraquat and pralidoxime.

Large molecules and particles such as carrageenen and polystyrene particles of $22\,\mu m$ diameter may also be absorbed from the gut, presumably by phagocytosis. The bacterial product **botulinum toxin**, a large molecule (molecular weight 200000–400000) is sufficiently well absorbed after oral ingestion to be responsible for toxic and often fatal effects.

There are a number of factors which affect the absorption of foreign compounds from the gut or their disposition, one which is of particular importance is the aqueous solubility of the compound in the non-ionized form. With very lipid soluble compounds, water solubility may be so low that the compound is not well absorbed (table 3.2) because it is not dispersed in the aqueous environment of the gastrointestinal tract. Therefore when drugs and other foreign compounds are administered the vehicle used to suspend or dissolve the compound may have a major effect on the eventual toxicity by affecting the rate of absorption. Also the physical form of the substance may be important, for example large particle size may decrease absorption. Similarly, when large masses of tablets are suicidally ingested, even those with reasonable water solubility, such as **aspirin** (acetylsalicylic acid), the bolus of tablets may remain in the gut for many hours after ingestion. Another factor which may affect absorption from the gastrointestinal tract is the presence of food. This may *facilitate* absorption if the substance in question dissolves in any fat present in the foodstuff. Alternatively, food may *delay* absorption if the compound binds to food or constituents, or if it is only absorbed in the small intestine, as food prolongs gastric emptying time.

Allied to this is gut motility which may be altered by disease, infection or other chemical substances present, and hence change the absorption of a compound from the gut.

Apart from influencing the absorption of foreign compounds, the environment of the gastrointestinal tract may also affect the compound itself, making it more or less toxic. For example, **gut bacteria** may enzymically alter the compound, and the pH of the tract may affect its chemical structure.

The natural occurring carcinogen cycasin, which is a glycoside of methylazoxy-methanol (figure 1.1) is hydrolysed by the gut bacteria after oral administration. The product of the hydrolysis is methylazoxymethanol, which is absorbed from the gut and which is the compound responsible for the carcinogenicity. Given by other routes, cycasin is not carcinogenic as it is not hydrolysed.

The gut bacteria may also reduce nitrates to nitrites, which can cause methaemoglobinaemia or may react with secondary amines in the acidic environment of the gut, giving rise to carcinogenic nitrosamines.

Conversely, the acidic conditions of the gut may inactivate some toxins, such as snake-venom, which is hydrolysed by the acidic conditions.

The absorption from the gastrointestinal tract is of particular importance because compounds so absorbed are transported directly to the liver via the hepatic-portal vascular system (figure 3.8). Extensive metabolism in the liver may alter the structure of the compound, making it more or less toxic. Little of the parent compound reaches the systemic circulation in these circumstances. This **'first-pass' effect** is very important if hepatic metabolism can be saturated; it may lead to markedly different toxicity after administration by different routes. Highly cytotoxic compounds given orally may consequently selectively damage the liver by exposing it to high concentrations, whereas other organs are not exposed to such high concentrations, as the compound is distributed throughout body tissues after leaving the liver.

The gastrointestinal tract itself has significant metabolic activity and can metabolize foreign compounds *en route* to the liver and systemic circulation giving rise to a 'first-pass effect'. For example the drug **isoprenaline** undergoes significant metabolism in the gut after oral exposure which effectively inactivates the drug. Therefore administration by aerosol into the lungs, the target site, is the preferred route.

Distribution

Following absorption by one of the routes described, foreign compounds will enter the bloodstream. The part of the vascular system into which the compound is absorbed will depend on the site of absorption (figure 3.8). Absorption through the skin leads to the peripheral blood supply whereas the major pulmonary circulation will be involved if the compound is absorbed from the air via the lungs. For most compounds oral absorption will be followed by entry of the compound into the portal vein supplying the liver with blood from the gastrointestinal tract. Once in the bloodstream the substance will distribute around the body and be diluted by the blood. Although only a small proportion of a compound in the body may be in contact with the *receptor* or *target site*, it is the distribution of the bulk of the compound which governs the concentration and disposition of that critical proportion. The plasma concentration of the compound is therefore very important, because it often directly relates to the concentration at the site of action. Blood circulates through virtually all tissues and some equilibration between blood and tissues is therefore expected. The distribution of foreign toxic compounds throughout the body is affected by the factors already discussed in connection with absorption. This distribution involves the passage of foreign compounds across cell membranes. The passive diffusion of foreign compounds across membranes is restricted to the non-ionized form, and the proportion of a compound in this form is determined by its pK_a and the pH of the particular tissue.

The passage of compounds out of the plasma through capillary membranes into the extravascular water occurs fairly readily, the major barrier being molecular size. Even charged molecules may therefore pass out of capillaries by movement through pores or epithelial cell junctions and driven by a concentration gradient. The passage of substances through pores in arterial capillaries may also be assisted by hydrostatic pressure. For lipid soluble

Table 3.3.   Major proteins in human plasma.

Protein and electrophoresis region	g/100g plasma protein (% total)	Function
Prealbumin	0.1–0.5	Binds thyroxine
Albumin	50–65	Colloid osmotic pressure; binds hormones, fatty acids, bilirubin, drugs
α_1 Region		
α_1-Acid glycoprotein	0.5–1.5	Tissue breakdown product
α_1-Antitrypsin	1.9–4.0	Trypsin inhibitor
α_1-Lipoprotein	4.5–8	Lipid transport
α_2 Region		
α_2-Macroglobulin	1.5–4.5	Proteinase inhibitor
α_2-Ceruloplasmin	0.3–0.5	Copper transport
α_2-Haptoglobulin	0.3–1.9	Binds haemoglobin
α_2-Lipoprotein	0.5–1.5	Lipid transport
β_1 Region		
Transferrin	3–6.5	Iron transport
β_1-Lipoprotein	4–14	Lipid transport
β_2 Region		
Fibrinogen	2.5–5	Blood clotting
γRegion (γ-globulins; immunoglobulins)		
IgA	0.8–2.8	Antibodies involved in immune reactions
IgM	0.6–1.7	
IgG	13–22	
IgD	<0.5	
IgE	<0.002	

Data from *Documenta Geigy, Scientific Tables* (1970), edited by K. Diem and C. Lenter (Basle: CIBA-GEIGY).

compounds a major determinant of the rate of movement across capillaries will be the *partition coefficient*. Therefore most small molecules, whether ionized or non-ionized, pass readily out of the plasma, either through pores in the capillary membranes or by dissolving in the lipid of the membrane.

Large molecules pass out very slowly, possibly by pinocytosis. The pores in the capillary membranes vary considerably in size, and therefore some capillaries are more permeable than others. Pore sizes of about 30 Å correspond to a molecular weight of 60 000. The exceptions to this are the capillaries of the brain, which are relatively impermeable. Passage across cell membranes from extravascular or interstitial water into cells is, however, much more restrictive. Again, the *physicochemical characteristics* of the compound will be a crucial determinant of its disposition in conjunction with the particular environment.

A particularly important interaction in the bloodstream, which the foreign compound may undergo, is reaction with **plasma proteins**. In some cases, such as with compounds of low water solubility, this interaction may be essential for the transport of the compound in the blood and may facilitate transport to the tissues although usually it will restrict distribution. There are many different types of proteins but the most abundant and important with regard to binding of foreign compounds is **albumin** (table 3.3). However, other plasma proteins may be important in binding foreign compounds; the **lipoproteins**, for example, which bind lipophilic compounds such as **DDT**. In general the interactions are non-covalent although some drugs, such as **captopril** are known to bind covalently to plasma proteins and even to cause immune responses as a result of this interaction (see Chapter 6). There may be a specific interaction in the plasma between particular foreign molecules and antibodies.

The non-covalent binding to plasma proteins may involve four types of interaction (figure 3.17):

(a) ionic binding, in which there is bonding between charged groups or atoms, such as metal ions and the opposite charge on the protein

(b) hydrophobic interactions, occur when two nonpolar, hydrophilic groups associate and mutually repel water.

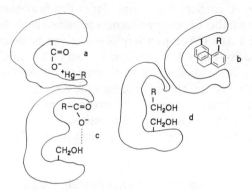

FIGURE 3.17. Types of bonding to plasma proteins which foreign compounds can undergo. a: ionic bonding; b: hydrophobic bonding; c: hydrogen bonding; d: van der Waals forces.

(c) hydrogen bonding, where a hydrogen atom attached to an electronegative atom (e.g. O) is shared with another electronegative atom (e.g. N).

(d) Van der Waals forces, are weak, acting between the nucleus of one atom and the electrons of another

The nature and strength of the binding will depend on the physicochemical characteristics of the foreign compound. For example lipophilic substances such as DDT will bind to proteins which have hydrophobic regions such as lipoproteins and albumin. Ionized compounds may bind to a protein with available charged groups such as albumin by forming ionic bonds. The albumin molecule has approximately 100 positive and a similar number of negative charges at its isoelectric point (pH 5) and has a net negative charge at the pH of normal plasma (7·4).

The non-covalent binding to plasma proteins is a reversible reaction which may be simply represented as:

$$T + P \underset{k_2}{\overset{k_1}{\rightleftharpoons}} TP$$

where T is the foreign compound and P is the protein and k_1 and k_2 are the rate constants for association and dissociation.

The overall **dissociation constant**, K_d is derived from:

$$\frac{1}{K_d} = \frac{[TP]}{[T][P]} \text{ or } K_d = k_2/k_1$$

When K_d is small then binding is tight.

Plotting $1/[TP]$ *vs.* $1/[T]$ may give some indication of the specificity of binding. Thus if the plot passes through the origin as is the case with DDT, then 'infinite', non-saturable binding is implied. If the concentration of bound/free compound is plotted against the concentration of bound compound a straight line with a negative gradient may result. From this **Scatchard plot** the **affinity constant** and the number of binding sites may be gained from the slope $(1/k_a)$ and intercept on the x axis (N) respectively. Binding can be described either as (i) specific, low capacity, high affinity or (ii) non-specific, low affinity, high capacity. When there is a specific binding site, mathematical treatments derived from the above relationship may be applied to determine the nature of the site. Usually however binding to plasma proteins will involve several different binding sites on one protein and maybe several different proteins.

The binding of foreign compounds to plasma proteins has several important implications.

(a) The *concentration* of the free compound in the plasma will be reduced. Indeed this removal of a portion of the compound from free solution may contribute to a concentration gradient.

(b) *Distribution* to the tissues may be restricted. Although an equilibrium exists between the non-bound, free portion of the compound in the plasma and the bound portion, only the free compound will distribute into tissues.

(c) Similarly, *excretion* by filtration and passive diffusion will be restricted to the free portion and hence the *half-life* may be extended by protein binding. However when a compound is excreted by an active process, then protein binding may have no significant effect. For example *p*-**aminohippuric acid** (figure 3.18) is more than 90% bound to plasma proteins yet it is cleared from the blood by a single pass through the kidney, being excreted by the organic acid transport system.

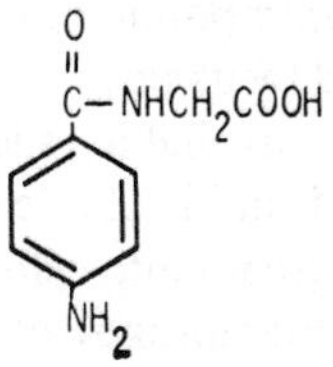

$$\underset{\substack{\\ \\ NH_2}}{\overset{\substack{O \\ \| \\ C-NHCH_2COOH}}{\bigcirc}}$$

FIGURE 3.18. Structure of *p*-aminohippuric acid.

(d) *Saturation* may occur. When a specific binding site is involved, there will be a limited number of sites. As the dose or exposure to the compound increases and the plasma level rises, these may become fully saturated. When this occurs the concentration of the non-bound, free, portion of the compound will rise. This may be the cause of a toxic *dose threshold*. The importance of this will depend partly on the extent of binding. Thus with highly bound compounds saturation will lead to a dramatic increase in the free concentration. For example if a compound is 99% bound to plasma proteins, at a total concentration of say 100 mg/l, the free concentration of compound in the bloodstream will be only 1 mg/l. If all the plasma protein binding sites are saturated at this concentration then any increase in dosage can dramatically increase the free plasma concentration. Thus doubling the dosage could increase the free concentration to 101 mg/l if all other factors remained the same. Clearly such a massive increase in the free concentration could result in the appearance of toxic effects.

(e) *Displacement* of one compound by another may occur. This may apply between two foreign compounds and between a foreign and an endogenous compound. For example when some **sulphonamide** drugs bind to plasma proteins they displace others such as **tolbutamide**, a hypoglycaemic drug. The resultant increased plasma concentration of free drug can give rise to an excessive reduction in blood sugar. Some metals may compete for the binding sites on the protein **metallothionein**. Displacement of **bilirubin** in premature infants by **sulphisoxazole** competing for the same plasma protein binding sites may lead to toxic levels of bilirubin entering the brain. The binding affinity will influence the consequences of this. Thus if a toxic compound has relatively low binding (e.g. 30%) and 10% is displaced by another, the increase in free compound is only 3% (from 70% to 73%). If, however, the compound is 98% bound, then a similar displacement would cause a more dramatic increase in the free concentration (from 2% to 12%).

It is clear that the physicochemical characteristics of the compound and hence the extent and avidity of binding will determine the importance of plasma protein binding in the disposition of the compound.

Tissue localization

The distribution of a foreign compound out of the bloodstream into tissues is, as already discussed, determined by its physicochemical characteristics. Thus unless specific transport systems are available, such as for analogues of endogenous compounds, only non-ionized, lipid soluble compounds or small molecules, will readily pass out of the bloodstream into cells throughout body tissues. The pores and spaces in the capillary membranes of certain tissues can be quite large, however, allowing much larger molecules to pass out of the bloodstream. For example, the **sinusoids** of the liver have a discontinuous basement membrane giving rise to large pores or fenestrations about $2 \mu m$ in diameter. The **Kupffer cells** lining these sinusoids may also be involved in removal of foreign substances from the bloodstream by phagocytosis. Compounds which have some water solubility will distribute throughout body water whereas very lipophilic substances may become preferentially localized in fat tissue. For example, the drug **chlorphentermine** (figure 3.19) is found to accumulate in fatty tissues. Thus, although the blood : tissue ratio is 1 : 1 for the liver after both single and multiple

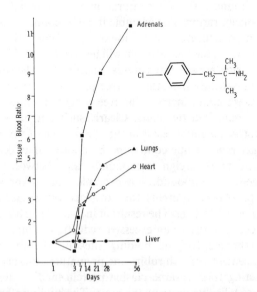

FIGURE 3.19. Tissue distribution of chlorphentermine on chronic dosing. Data from Lullman *et al.* (1975) *CRC Crit. Rev. Toxicol.*, **4**, 185.

doses, in the heart, lungs and especially the adrenal gland, the ratio becomes much greater than 1:1 after several doses. The accumulation of chlorphentermine results in a disturbance of lipid metabolism and a dramatic accumulation of phospholipids, especially in the adrenals. This toxic response, **phospholipidosis**, is due to the amphiphilic nature of the molecule. This is the possession of both lipophilic and hydrophilic groups in the same molecule. This characteristic is responsible for the disruption of lipid metabolism. Thus it is believed that the hydrophobic portion of the chlorphentermine molecule associates with lipid droplets in target cells. The hydrophilic portion of the chlorphentermine molecule alters the surface charge of lipid droplets and so decreases their breakdown by lipases giving rise to accumulation.

Another, more dramatic example of this is the case of the **polybrominated biphenyls (PBBs)** (figure 3.20). These are industrial chemicals which accidentally became added to animal feed in Michigan in 1973. Because of high lipophilicity the PBBs became localized in the body fat of all the livestock primarily exposed and the humans subsequently exposed. There was little, if any, elimination of these compounds from the body and so the humans unfortunately exposed will be so for long, perhaps indefinite, periods of time. Sequestration of foreign compounds in particular tissues may in some cases be protective if the tissue is not the target site for toxicity. For example, highly lipophilic compounds such as DDT become localized in adipose tissue where they seem to exert little effect. Although this sequestration into adipose tissue reduces exposure of other tissues, mobilization of the fat in the adipose tissue may cause a sudden release of compound into the bloodstream with a dramatic rise in concentration and toxic consequences.

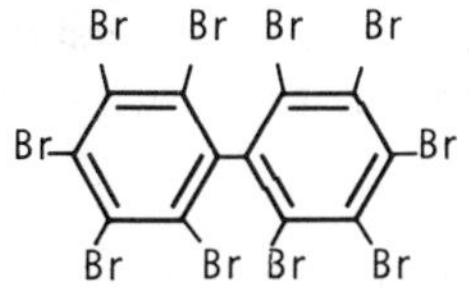

FIGURE 3.20. Structure of decabromobiphenyl.

Lead is another example of a toxic compound which is localized in a particular tissue, bone, which is not the major target tissue (see Chapter 7).

Some compounds accumulate in a specific tissue because of their affinity for a particular macromolecule. For instance, carbon monoxide specifically binds to haemoglobin in red cells, this being the target site (see Chapter 7), whereas when cadmium binds to metallothionein in liver and kidney this is initially a detoxication process (see Chapter 7). Sequestration into tissues may occur due to the action of an endogenous substance such as the protein **ligandin** (now identified as **glutathione transferase B**; see Chapter 4) which binds and transports both endogenous and exogenous compounds, such as carcinogenic azo dyes, into the liver.

Passage of foreign compounds into particular tissues can also occur as a result of a specialized active transport system. For example, the herbicide paraquat (figure 3.16) is taken up by the polyamine transport system into the lungs and thereby reaches a toxic concentration (see Chapter 7).

Therefore, foreign compounds may be distributed throughout all the tissues of the body or they may be restricted to certain tissues. Two areas for special consideration are the foetus and the brain. Because of the organization of the placenta, the blood in the **embryo** and **foetus** is in intimate contact with the maternal bloodstream, especially at the later stages of pregnancy where the tissue layers between the two blood supplies may be only $2\,\mu$m thick (see Chapter 6 and figure 3.21). Consequently, movement of non-ionized lipophilic substances from the maternal bloodstream into the embryonic circulation by passive diffusion or filtration of small molecules through pores is facilitated. Specialized transport systems also exist for endogenous compounds and ions. Consequently many foreign compounds achieve the same concentration in foetal as in maternal plasma. However, if metabolism *in utero* converts the compound into a more polar metabolite, accumulation may occur in the foetus. Despite extensive blood flow (16% cardiac output; $0\cdot5\,$ml/min/g of tissue) entry of foreign compounds into the **brain** takes place much less readily than passage into other tissues. Hence the term '**blood-brain barrier**'. Ionized compounds will not penetrate the brain in appreciable quantities unless they are carried by active transport systems. The reasons for this restricted entry are due to reduced permeability of the capillaries of the central nervous system because of:

(a) coverage of the basement membrane of the capillary endothelium by the processes of **glial cells (astrocytes)**
(b) **tight junctions** between capillaries leaving few, if any pores.

The low concentration of protein in the interstitial fluid has been suggested as another factor which may reduce the distribution of some substances in the

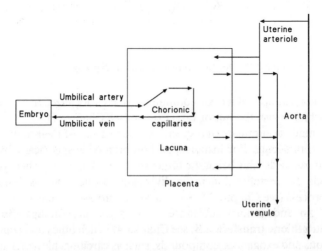

FIGURE 3.21. The blood supply to the mammalian embryo and placenta.

central nervous system. Lipid soluble compounds, such as **methyl mercury** which is toxic to the central nervous system (see Chapter 7), can enter the brain readily, the facility being reflected by the partition coefficient. Another example which illustrates the importance of the lipophilicity in the tissue distribution and duration of action of a foreign compound is afforded by a comparison of the drugs **thiopental** and **pentobarbital** (figure 3.5). These drugs are very similar in structure, only differing by one atom. Their pK_a values are similar and consequently the proportion ionized in plasma will also be similar (figure 3.5). The partition coefficients, however, are very different and this accounts for the different rates of anaesthesia due to each compound. Thiopental being much more lipophilic enters the brain very rapidly and thereby causes its pharmacological effect, anaesthesia. The duration of this, however, is short as redistribution into other tissues, including body fat, causes a loss from the plasma and hence by equilibration, the central nervous system also. Pentobarbital, conversely, distributes more slowly and so, although the concentration in the brain does not reach the same level as thiopental, it is maintained for a longer period.

Changes in plasma pH may also affect the distribution of toxic compounds by altering the proportion of the substance in the non-ionized form, which will cause movement of the compound into, or out of tissues. This may be of particular importance in the treatment of salicylate poisoning (see Chapter 7) and barbiturate poisoning for instance. Thus the distribution of **phenobarbital**, a weak acid (pK_a 7.2), shifts between the brain and other tissues and the plasma, with changes in plasma pH (figure 3.22). Consequently, the depth of anaesthesia varies depending on the amount of phenobarbital in the brain. Alkalosis, which increases plasma pH, causes plasma phenobarbital to become more ionized, alters the equilibrium between plasma and brain and causes phenobarbital to diffuse back into the plasma (figure 3.22). Acidosis will cause the opposite shift in distribution. Administration of bicarbonate is therefore used to treat overdoses of phenobarbital. This treatment will also cause alkaline diuresis and therefore facilitate excretion of phenobarbital into the urine (see below).

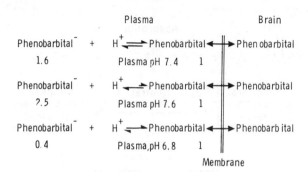

FIGURE 3.22. Disposition of phenobarbital in plasma at different pH values. Figures represent proportions of ionized and non-ionized forms.

Plasma level

The plasma level of a toxic compound is a particularly important parameter as:

(a) it reflects and is affected by the absorption, distribution, metabolism and excretion of the compound

(b) it often reflects the concentration of compound at the target site more closely than the dose; it should be noted that this may not always be the case such as when sequestration in a particular tissue occurs which may or may not be the target tissue, for example, chlorphentermine (see figure 3.19), lead and polybrominated biphenyls (figure 3.20)

(c) it may be used to derive other parameters

(d) it may give some indication of tissue exposure and so expected toxicity in situations of intentional or accidental overdosage

(e) it may indicate accumulation is occurring on chronic exposure

(f) it is central to any kinetic study of the disposition of a foreign compound.

The plasma level *profile* for a drug or other foreign compound is therefore a composite picture of the disposition of the compound, being the result of various dynamic processes. The processes of disposition can be considered in terms of *'compartments'*. Thus, absorption of the foreign compound into the central compartment will be followed by distribution, possibly into one or more peripheral compartments, and removal from the central compartment by excretion and possibly metabolism (figure 3.23). A very simple situation might only consist of one, central compartment. Alternatively there may be many compartments. For such multicompartmental analysis and more details of pharmacokinetics and toxicokinetics see references in the Bibliography. The central compartment may be, but is not *necessarily*, identical with the blood. It is really the compartment with which the compound is in rapid equilibrium. The distribution to peripheral compartments is reversible, whereas the removal from the central compartment by metabolism and excretion is irreversible.

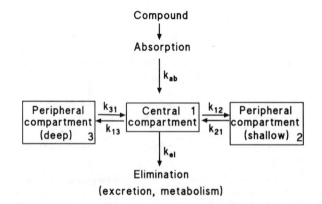

FIGURE 3.23. Distribution of a foreign compound into body compartments.

The rates of movement of foreign compound into and out of the central compartment are characterized by **rate constants**, k_{ab} and k_{el} (figure 3.23). When a compound is administered intravenously, the absorption is effectively instantaneous and is not a factor. The situation after a single, intravenous dose, with distribution into one compartment, is the most simple to analyse kinetically as only distribution and elimination are involved. With a rapidly distributed compound then this may be simplified further to a consideration of just elimination. When the plasma (blood) concentration is plotted against time the profile normally encountered is an exponential decline (figure 3.24). This is because the rate of removal is proportional to the concentration remaining, it is a *first-order process* and so a constant fraction of the compound is excreted at any given time. When the plasma concentration is plotted on a $\log_{10}$ scale, the profile will be a straight line for this simple, one compartment model (figure 3.25). The equation for this line is:

$$\log_{10} C = \log_{10} C_0 - \frac{k_{el} \times t}{2 \cdot 303}$$

where C is plasma concentration, t is time, C_0 is the intercept on the y axis and the gradient or slope is $-k_{el}/2 \cdot 303$. The units of k_{el} are h^{-1} or min^{-1}.

The elimination process is represented by the **elimination rate constant** k_{el} which may be determined from the gradient of the plasma profile (figure 3.25). The reasons for the overall process of elimination being first order are that the processes governing it (excretion by various routes, and metabolism) are *irreversible processes* and are also first order. When the latter is not the case then this model does not apply.

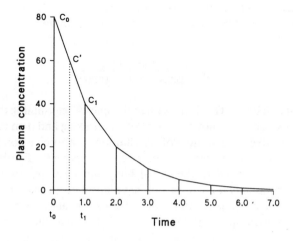

FIGURE 3.24. Profile of the decline in the plasma concentration of a foreign compound with time after intravenous administration. The Area Under the Curve (AUC) may be determined by dividing the curve into a series of trapezoids as shown and calculating the area of each. Thus the $AUC_{C0 \to C1}$ is $(C' \times T_1 - T_0)$. The $AUC_{C0 \to C7}$ is then the sum of all the individual areas.

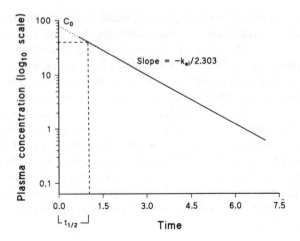

FIGURE 3.25. Log_{10} plasma concentration time profile for a foreign compound after intravenous administration. The plasma half-life ($t_{\frac{1}{2}}$) and the elimination rate constant (K_{el}) of the compound can be determined from the graph as shown.

Volume of distribution

The total water in the body of an animal can be conveniently divided into three compartments: the plasma water, interstitial water and the intracellular water. The way a foreign compound distributes into these compartments will profoundly affect the plasma concentration. If a compound is only distributed in the plasma water (which is approximately three litres in man) the plasma concentration will obviously be much higher than if it distributed in all extracellular water (approximately 14 l) or the total body water (approximately 40 l). This may be quantified as a parameter known as the **volume of distribution** (V_D), which can be calculated as follows:

$$V_D = \frac{\text{dose (mg)}}{\text{plasma conc. (mg/l)}}$$

and is expressed in litres. This is the same principle as determining the volume of water in a jar by adding a known amount of dye, mixing and then measuring the concentration of dye. The volume of distribution of a foreign compound is the volume of body fluids into which the compound is apparently distributed. It does not necessarily correspond to the actual body water compartments, as it is a mathematical parameter and does not have absolute physiological meaning. However, the V_D may yield important information about a compound. For example, a very high apparent V_D, perhaps higher than the total body water, indicates that the compound is localized or sequestered in a storage site such as fat or bone. If the value is low and similar to plasma water it indicates that the compound is retained in the plasma. A compound with a high V_D will tend to be excreted more slowly than a compound with a low V_D. From the V_D and plasma concentration, the total amount of foreign compound in the body or the **total body burden** may be estimated:

Total body burden (mg) = Plasma concentration (mg/l) $\times V_D$ (l)

The determination of V_D will depend on the type of model which describes the distribution of the particular compound. Thus, if the compound is given intravenously and is distributed into a one-compartment system, the V_D can be determined from the starting plasma concentration, C_0. This may be determined from the graph of plasma concentration against time by extrapolation (figure 3.25). Thus:

$$V_D = \frac{\text{Dose}_{iv}}{C_0}$$

For compounds whose distribution fits more complex models, a more rigorous method utilizes the area under the plasma concentration versus time curve (**AUC**). Thus:

$$V_D = \frac{\text{Dose}_{iv}}{\text{AUC}_0 - \infty \times k_{el}}$$

The AUC is determined by plotting the plasma concentration versus time on normal rectilinear graph paper, dividing the area up into trapezoids and calculating the area of each (figure 3.24). The total area is then the sum of the individual areas. Although the curve theoretically will never meet the x axis, the area from the last plasma level point to infinity may be determined from C_t/k_{el}.

When a compound is administered by a route other than intravenously, the plasma level profile will be different as there will be an *absorption phase*, and so the profile will be a composite picture of absorption in addition to distribution and elimination (figure 3.26). Just as first-order elimination is defined by a rate constant, so also is absorption, k_{ab}. This can be determined from the profile by the method of *residuals*. Thus, the straight portion of the semilog plot of plasma level against time is extrapolated to the y axis. Then each of the actual plasma level points which deviate from this during the absorptive phase are subtracted from the equivalent time point on the extrapolated line. The differences are then plotted and a straight line should result. The slope of this line can be used to calculate the absorption rate constant, k_{ab} (figure 3.26). The volume of distribution should not really be determined from the plasma level after oral administration (or other routes except intravenous) as the administered dose may not be the same as the absorbed dose. This may be because of **first-pass metabolism** (see above), or incomplete absorption and will be apparent from a comparison of the plasma level profile of the compound after oral and intravenous administration (figure 3.27). First-pass metabolism may occur at the site of absorption such as in the gastrointestinal tract after oral administration, or it may occur in the liver during the passage of the substance through the portal

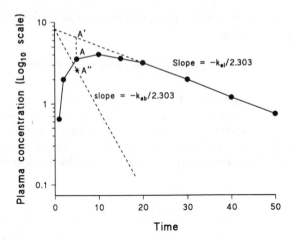

FIGURE 3.26. The $\log_{10}$ plasma concentration against time profile for a compound after oral administration. The absorption rate constant k_{ab} can be determined from the slope of the line (dotted) plotted using the method of residuals as described in the text. (Thus $A'' = A' - A$, etc.)

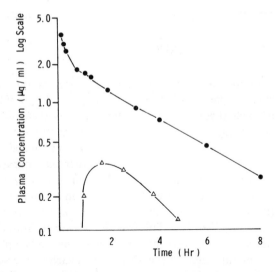

FIGURE 3.27. Effect of first-pass metabolism. Plasma concentration of a compound after oral (△) and intravenous (●) administration to the dog. Adapted from Smyth, R. D. & Hottendorf, G. H. (1980) *Toxicol. Appl. Pharmac.*, **53**, 179.

circulation and liver before it reaches the systemic circulation. The extent of this first-pass metabolism may be as high as 70% as with the drug **propranolol**, when given orally. This means that the systemic circulation is only exposed to 30% of the original dose as parent drug. The dose actually absorbed may be quantified as **bioavailability**:

$$\text{Bioavailability} = \frac{AUC_{oral}}{AUC_{iv}} \times 100$$

Plasma half-life

The plasma half-life is another important parameter of a foreign compound which can be determined from the plasma level. It is defined as *the time taken for the concentration of the compound in the plasma to decrease by half from a given point.* It reflects the rates at which the various dynamic processes, distribution, excretion and metabolism, are taking place *in vivo*. The value of the half-life can be determined from the semilog plot of plasma level against time (figure 3.25), or it may be calculated:

$$\text{Half-life } (t_{\frac{1}{2}}) = \frac{0 \cdot 693}{k_{\text{el}}}$$

The half-life will be independent of the dose provided that the elimination is first order and therefore should remain constant. Changes in the half-life, therefore, may indicate alteration of elimination processes due to toxic effects because the half-life of a compound reflects the ability of the animal to metabolize and excrete that compound. When this ability is impaired, for example by saturation of enzymic or active transport processes, or if liver or kidneys are damaged, the half-life may be prolonged. For example, after overdoses of paracetamol, the plasma half-life increases several-fold as the liver damage reduces the metabolic capacity, and in some cases kidney damage may reduce excretion (see Chapter 7).

Another indication of the ability of an animal to metabolize and excrete, and therefore of the elimination of, a foreign compound which can be gained from the plasma level data is **total body clearance**. This may be calculated from the parameters already described:

$$\text{Total body clearance} = V_{\text{D}} \times k_{\text{e}1}$$

or a better calculation is:

$$\text{Total body clearance} = \frac{\text{Dose}}{\text{AUC}}$$

as it may be applied to multi as well as single compartments models.

The units are $1/\text{h}$ if, for example, the plasma concentration is measured in mg/l and the time scale is in hours. Again this applies to compounds given intravenously.

If the semilog plot of the plasma level against time after an intravenous dose is not a straight line, then the compound may be distributing in accordance with a *two-compartment* or *multicompartment model* (figures 3.28 and 3.23). If a two-compartment model is appropriate, then the semilog plot can be resolved into two straight lines using the method of residuals ('feathering') already described. The first part of the plasma level profile is the *α-phase* and the second is the *β-phase*. The straight line which describes the α-phase is the difference between the observed points and the back extrapolated β-phase line. The α-phase represents

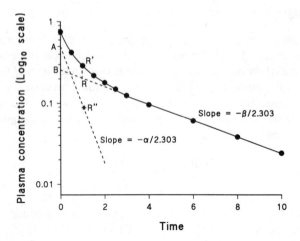

FIGURE 3.28. Log_{10} plasma concentration against time profile for a foreign compound after· intravenous administration. The distribution of this compound fits the two-compartment model. The dotted line is determined by the method of residuals as described in the text. (Thus $R'' = R'$, etc.)

the initial, rapid distribution into the peripheral compartment, and the second, slower β-phase represents elimination from the central compartment.

The individual rate constants α and β can be determined from the graph. The rate constants for movement into the peripheral compartment, k_{12} and k_{21} (figure 3.23) and the overall k_{e1} can be determined as follows:

$$k_{21} = \frac{\alpha B + \beta A}{A + B}$$

$$k_{e1} = \frac{\alpha B}{K_{21}}$$

$$K_{12} = \alpha + \beta - k_{21} - k_{e1}$$

A knowledge of these rate constants allows an assessment of the contribution of distribution and elimination.

For a further discussion of multicompartmental analysis, which is beyond the scope of this book, the reader is referred to the Bibliography.

The kinetics described so far have been based on first-order processes, yet often in toxicology, the situation after large doses are administered has to be considered when such processes do not apply. This situation may arise when excretion or metabolism is saturated and hence the rate of elimination decreases. This is known as **Michaelis–Menten** or **saturation kinetics**. Excretion by active transport (see below) and enzyme-mediated metabolism are saturable processes. In some cases cofactors are required and their concentration may be limiting (see Chapter 7, salicylate poisoning). When the concentration of foreign compound in the relevant tissue is lower than the k_m then linear, first-order kinetics will apply, but when the concentration is greater, non-linear, *zero-order kinetics* are observed. There are a number of consequences including an increase in half-life, the AUC is not proportional to the dose, a threshold for toxic effects may become apparent, a constant amount of compound is excreted (independent of the dose) and proportions of metabolites may change. When an arithmetic, as opposed to a semilog, plot of plasma concentration against time is drawn, this is linear. As the elimination changes with dose, a true half-life and k_{e1} cannot be calculated.

Saturation kinetics are important in toxicology because they may herald the point at which toxicity occurs.

The plasma level and half-life are also important parameters if a compound is to be administered chronically or there is repeated exposure. Thus if the dosing interval is *shorter* than the half-life, the compound *accumulates*, whereas if the half-life is very short compared with the dosing interval, the compound does not accumulate. The effects on the plasma concentration of the compound of these two situations are shown in figure 3.29.

In order rapidly to achieve a steady-state level of the compound in the plasma

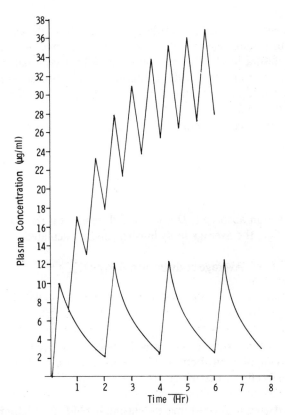

FIGURE 3.29. Effect of chronic dosing with a foreign compound on its plasma level. Half-life of compound is 1 h. Upper curve shows the effect of dosing every 30 min, lower curve the effect of a dosage interval of 2 h.

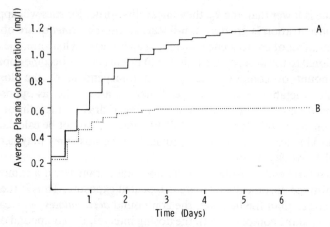

FIGURE 3.30. Average plasma concentrations of two foreign compounds after multiple dosing. Compound A, half-life 24 h; compound B, half-life 12 h; dosage interval in each case is 8 h. The accumulation plateau is directly proportional to the half-life, while the rate of accumulation is inversely proportional to the half-life. Adapted from van Rossum, J. M. (1968) *J. Pharm. Sci.*, **57**, 2162.

so that the organism is exposed to a fairly constant level, the dosage interval and half-life should be similar (figure 3.30). For *steady-state conditions*, the half-life *determines* the plasma level. That is, substrates with a long half-life attain a higher steady-state plasma level than compounds with shorter half-lives. It is obviously important to measure this plasma concentration for an assessment of chronic toxicity (figure 3.30).

The average *plateau concentration* after repeated oral administration can be calculated:

$$C_{av} = \frac{f \times D}{V_D \times k_{e1} \times t}$$

where f is the fraction absorbed, D is the oral dose and t is the dosing interval. From this value C_{av}, the **average body burden** can be calculated:

$$\text{Average body burden} = V_D \times C_{av}$$

Excretion

The elimination of toxic substances from the body is clearly a very important determinant of their biological effects. Rapid elimination will reduce both the likelihood of toxic effects occurring and their duration. Removal of a toxic compound may help to reduce the extent of damage. The elimination of foreign compounds is reflected in the parameters **plasma half-life** ($t_{1/2}$), **elimination rate constant** (k_{e1}) and **total body clearance**. The plasma half-life also reflects other processes as well as excretion; the **whole body half-life** is the time required for half of the compound to be eliminated from the body and therefore reflects the excretion of the compound. It can be readily measured by administering a radiolabelled compound and determining amount excreted over time.

The most important route of excretion for most non-gaseous or non-volatile compounds is through the kidneys into the urine. Other routes are secretion into the bile, expiration via the lungs for volatile and gaseous substances, and secretion into the gastrointestinal tract, or into fluids such as milk, saliva, sweat, tears and semen.

Urinary excretion

Many toxic substances and other foreign compounds are removed from the blood as it passes through the kidneys. The kidneys receive around 25% of the cardiac output of blood and so they are exposed to and filter out a significant proportion of foreign compounds. However excretion into the urine from the bloodstream applies to relatively small, water soluble molecules; large molecules such as proteins do not normally pass out through the intact glomerulus and lipid soluble molecules such as bilirubin are reabsorbed from the kidney tubules (figure 3.31).

Excretion into the urine involves one of three mechanisms: **filtration** from the blood through the pores in the glomerulus, **diffusion** from the bloodstream into the tubules and **active transport** into the tubular fluid. The principles governing these processes are essentially the same as already described and depend on the

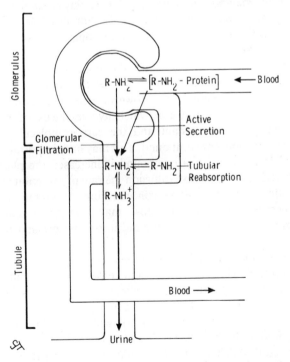

FIGURE 3.31. Schematic representation of the disposition of foreign compounds in the kidney.

physicochemical properties of the compound in question. The structure of the kidney facilitates the elimination of compounds from the bloodstream (figure 3.31). The basic unit of the kidney, the nephron, allows most small molecules to pass out of the blood in the glomerulus into the tubular ultrafiltrate aided by large pores in the capillaries and the hydrostatic pressure of the blood. The **pores** in the glomerulus are relatively large (40 Å) and will allow molecules with a molecular weight of less than 60 000 to pass through. This filtration therefore only applies to the non-protein bound form of the compound, and so the concentration of the compound in the glomerular filtrate will approximate to the free concentration in the plasma. Lipid soluble molecules will passively diffuse out of the blood provided there is a concentration gradient. However, such compounds, if they are not ionized at the pH of the tubular fluid, may be reabsorbed from the tubule by passive diffusion back into the blood which flows through the vessels surrounding the tubule because there will be a concentration gradient in the direction tubule→blood. Water soluble molecules which are ionized at the pH of the tubular fluid will not be reabsorbed by passive diffusion and will pass out into the urine.

Certain molecules, such as **p-aminohippuric acid** (figure 3.18), a metabolite of p-aminobenzoic acid are actively transported from the bloodstream into the tubules by a specific **anion transport system**. Organic anions and cations appear to be transported by separate transport systems located on the proximal convoluted tubule. Active transport is an energy requiring process and therefore may be inhibited by metabolic inhibitors, and there may be competitive inhibition between endogenous and foreign compounds. For example, the competitive inhibition of the active excretion of uric acid by compounds such as **probenecid** may precipitate gout.

Passive diffusion of compounds into the tubules is proportional to the concentration in the bloodstream, so the greater the amount in the blood, the greater will be the rate of elimination. However, when excretion is mediated via active transport or facilitated diffusion, which involves the use of specific carriers, the rate of elimination is constant and the carrier molecules may become saturated by large amounts of compound. This may have important toxicological consequences. As the dose of a compound is increased the plasma level will increase. If excretion is via passive diffusion the rate of excretion will increase as this is proportional to the plasma concentration. If excretion is via active transport, however, increasing the dose may lead to saturation of renal elimination and a toxic level of compound in the plasma and tissues may be reached.

Another factor which may affect excretion is binding to plasma proteins. This may reduce excretion via passive diffusion, especially if binding is tight and extensive, as only the free portion will be filtered or will passively diffuse into the tubule. Protein binding does not affect active transport however and a compound such as p-aminohippuric acid (figure 3.18), which is 90% bound to plasma proteins, is cleared in the first pass of blood through the kidney.

One of the factors which affects excretion in the **urinary pH**. If the compound which is filtered or diffuses into the tubular fluid is ionized at the pH of that fluid, it will not be reabsorbed into the bloodstream by passive diffusion. For example, an acidic drug such as phenobarbital is ionized at alkaline urinary pH and a basic drug such as amphetamine is ionized at an acidic urinary pH. This factor is utilized in the treatment of poisoning with barbiturates and salicylic acid for example (see Chapter 7). Thus by giving sodium bicarbonate to the patient, the urine becomes more alkaline and excretion of acidic metabolites is increased. The pH of urine may be affected by diet; high protein diet for instance causes urine to become more acid. The rate of urine flow from the kidney into the bladder is also a factor in the excretion of foreign compounds; high fluid intake, and therefore production of copious urine, will tend to facilitate excretion. Factors which affect kidney function such as age and disease will influence urinary excretion and may therefore increase toxicity by reducing elimination from the body. For example, sale of the antiarthritic drug **benoxaprofen** was stopped as it caused liver damage and other adverse effects. This was probably a result of accumulation following repeated and inappropriate doses of the drug given to elderly patients whose kidney function was reduced.

Biliary excretion

Excretion into the bile is an important route for certain foreign compounds, especially large polar and amphipathic substances and may indeed be the predominant route of elimination for such compounds.

Bile production occurs in the liver where it is secreted by the hepatocytes into the canaliculi, flows into the bile duct and eventually into the intestine after storage in the gall bladder (figure 3.32). Consequently, compounds which are excreted into the bile are usually eliminated in the faeces. The factors which affect biliary excretion are *molecular weight*, *charge* and the *species of animal* (tables 3.4 and 5.8). Consequently for polar compounds with a molecular weight of 300

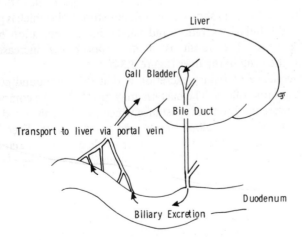

FIGURE 3.32. Enterohepatic circulation. The circulation of the compound is indicated by the arrows.

Table 3.4.   Effect of molecular weight on the route of excretion of biphenyls by the rat.

Compound	Molecular weight	Percent of total excretion	
		Urine	Faeces
Biphenyl	154	80	20
4-Monochlorobiphenyl	188	50	50
4,4'-Dichlorobiphenyl	223	34	66
2,4,5,2',5'-Pentachlorobiphenyl	326	11	89
2,3,6,2',3',6'-Hexachlorobiphenyl	361	1	99

Data from Matthews, H. B. (1980) In *Introduction to Biochemical Toxicology*, edited by E. Hodgson and F. E. Guthrie (New York: Elsevier-North Holland).

or so, such as glutathione conjugates (see Chapter 4), secretion into the bile can be a major route of excretion. Excretion into the bile is usually, although not exclusively, an active process and there are three specific transport systems, one for neutral compounds, one for anions and one for cations. Quaternary ammonium compounds may be actively secreted into the bile by a separate process. Compounds which undergo biliary excretion have been divided into *classes A, B and C*. *Class A* compounds are excreted by diffusion and have a bile to plasma ratio of around 1; *class B* compounds are actively secreted into bile and have a bile to plasma ratio of greater than 1; *class C* compounds are not excreted into bile and have a bile to plasma ratio of less than 1. The latter type of compound are usually macromolecules such as proteins or phospholipids.

Clearly, large ionic molecules would be poorly absorbed from the lumen of the gastrointestinal tract if ionized at the prevailing pH. If such compounds are extensively secreted into the bile during the first pass through the liver little may reach the systemic circulation after an oral dose. As with renal excretion via active transport, biliary excretion may be saturated and this may lead to an increasing concentration of compound in the liver. For example, the diuretic drug **furosemide** (figure 3.33) was found to cause hepatic damage in mice after doses of about 400 mg/kg. The toxicity shows a marked dose threshold which is partly due to saturation of the biliary excretion and partly due to saturation of plasma protein binding sites. The result is a *disproportionate* increase in the concentration of the drug in the liver (figure 3.34).

Another consequence of biliary excretion is that the compound comes into contact with the gut microflora. The bacteria may metabolize the compound and convert it into a more lipid soluble substance which can be reabsorbed from the intestine into the portal venous blood supply, and so return to the liver. This may lead to a cycling of the compound known as **enterohepatic recirculation** which may increase the toxicity (figure 3.32). If this situation occurs the plasma level

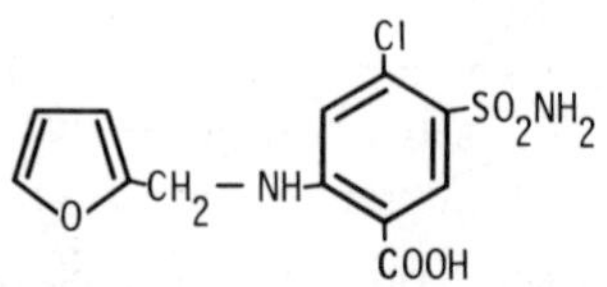

FIGURE 3.33. Structure of furosemide.

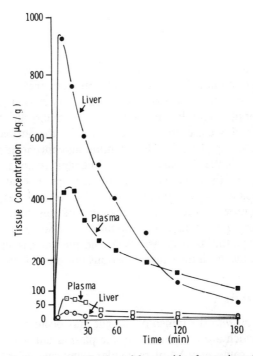

FIGURE 3.34. Changes in the tissue distribution of furosemide after toxic and non-toxic doses were administered to mice. Closed symbols represent a toxic dose (400 mg/kg), open symbols a non-toxic dose (80 mg/kg). Squares are the plasma concentration, circles the liver concentration. Adapted from Mitchell *et al.* (1975). In *Handbook of Experimental Pharmacology*, Vol. 28, Part 3, Concepts of Biochemical Pharmacology, edited by J.R. Gillette and J.R. Mitchell (Berlin: Springer).

profile may show peaks at various times corresponding to reabsorption rather than the smooth decline expected. If the compound is taken orally, and therefore is transported directly to the liver and is extensively excreted into the bile, it may be that none of the parent compound ever reaches the systemic circulation. Alternatively the gut microflora may metabolize the compound to a more toxic metabolite which could be reabsorbed and cause a systemic toxic effect. An example of this is afforded by the hepatocarcinogen **2,4-dinitrotoluene** discussed in more detail in Chapter 5. Compounds taken orally may also come directly into contact with the gut bacteria. For example, the naturally occurring glycoside cycasin is hydrolysed to the potent carcinogen methylazoxymethanol by the gut bacteria when it is ingested orally. Biliary excretion may therefore: (i) increase the *half-life* of the compound when followed by reabsorption from the gut; (ii) lead to the production of *toxic metabolites* in the gastrointestinal tract; (iii) increase hepatic exposure via the *enterohepatic recirculation*; (iv) be *saturated* and lead to hepatic damage.

The importance of biliary excretion in the toxicity of compounds can be seen from table 2.2 which shows that ligation of the bile duct increases the toxicity of certain chemicals many times.

Excretion via the lungs

The lungs are an important route of excretion for volatile compounds and gaseous and volatile metabolites of foreign compounds. For example about 50–60% of a dose of the aromatic hydrocarbon **benzene** is eliminated in the expired air. Excretion is by passive diffusion from the blood into the alveolus assisted by the concentration gradient. This is a very efficient, usually rapid, route of excretion for lipid soluble compounds as the capillary and alveolar membranes are thin and in very close proximity to allow for the normal gaseous exchange involved in breathing. There will be a continuous concentration gradient between the blood and air in the alveolus because of the rapid removal of the gas or vapour from the lungs and the rapid blood flow to the lungs. The rate of elimination also depends on the blood : gas solubility ratio. The effect of these factors on the rate of elimination may be crucial in the treatment of poisoning by such gases as the highly toxic carbon monoxide. Compounds may also be metabolized to gaseous toxic metabolites, for example methylene chloride which is metabolized to carbon monoxide and has been the cause of serious poisoning (see Chapter 7).

Gastrointestinal tract

Passive diffusion *into* the lumen of the gut may occur for compounds, such as weak bases, which are non-ionized in the plasma but which are ionized in the stomach. This route may be of particular significance for highly lipid soluble compounds.

Milk

Excretion into breast milk can be a very important route for certain types of foreign compounds, especially lipid soluble substances, because of the high lipid content in milk. Clearly new-born animals will be specifically at risk from toxic compounds excreted into milk. For example, nursing mothers exposed to **DDT** secrete it into their milk and the infant may receive a greater dose, on a body weight basis, than the mother. Also because the pH of milk $(6 \cdot 5)$ is lower than the plasma, basic compounds may be concentrated in the fluid.

Other routes

Foreign compounds may be excreted into other body fluids such as sweat, tears, semen or saliva by passive diffusion, depending on the lipophilicity of the compound. Although these routes are generally of minor importance quantitatively, they may have a toxicological significance such as the production of dermatitis by compounds secreted into the sweat for example.

Bibliography

Absorption and distribution

ADAMSON, R. H. and DAVIES, D. S. (1973) Comparative aspects of absorption, distribution, metabolism and excretion of drugs. In *International Encyclopaedia of Pharmacology and Therapeutics*, Section 85 (Comparative Pharmacology), p. 851 (Oxford: Pergamon Press).

ALBERTS, B., BRAY, D., LEWIS, J., RAFF, M., ROBERTS, K. and WATSON, J. D. (1989) *Molecular Biology of the Cell*, 2nd edition (New York: Garland).

BRODIE, B. B., GILLETTE, J. R. and ACKERMAN, H. S. (editors) (1971) *Handbook of Experimental Pharmacology*, Vol. 28, Part 1, *Concepts in Biochemical Pharmacology* (Berlin: Springer-Verlag).

CHASSEAUD, L. F. (1970) Processes of absorption, distribution and excretion. In *Foreign Compound Metabolism in Mammals*, Vol. 1, edited by D. E. Hathway. S. S. Brown, L. F. Chasseaud and D. H. Hutson (London: The Chemical Society).

FINDLAY, J. W. A. (1983) The distribution of some commonly used drugs in human breast milk. *Drug. Metab. Rev.,* **14**, 653.

GINSBURG, J. (1971) Placental drug transfer. *Annu. Rev. Pharmacol.,* **11**, 387.

GUTHRIE, F. E. (1980) Absorption and distribution. In *Introduction to Biochemical Toxicology*, edited by E. Hodgson and F. E. Guthrie (New York: Elsevier-North Holland).

HODGSON, E. and LEVI, P. E. (1987) *A Textbook of Modern Toxicology* (New York: Elsevier).

JOLLOW, D. J. (1973) Mechanisms of drug absorption and drug solution. *Rev. Can. Biol.,* **32**, 7.

KLAASSEN, C. D. (1986) Distribution, excretion and absorption of toxicants. In *Toxicology: The Basic Science of Poisons*, edited by C. D. Klaassen, M. O. Amdur and J. Doull (New York: Macmillan).

LA DU, B. N., MANDEL, H. G. and WAY, E. L. (editors) (1971) *Fundamentals of Drug Metabolism and Drug Disposition*, Chapters 1–7 (Baltimore: Williams & Wilkins).

PRATT, W. B. (1990) The entry, distribution and elimination of drugs. In *Principles of Drug Action*, 3rd edition, edited by W. B. Pratt and P. Taylor (New York: Churchill Livingstone).

SCHANKER, L. S. (1978) Drug absorption from the lung. *Biochem. Pharmac.,* **27**, 381.

SHEELER, P. and BIANCHI, D. E. (1983) The plasma membrane and the endoplasmic reticulum. In *Cell Biology. Structure, Biochemistry and Function*, 2nd edition (New York: John Wiley).

SMYTH, R. D. and HOTTENDORF, G. H. (1980) Application of pharmacokinetics and biopharmaceutics in the design of toxicological studies. *Toxic. Appl. Pharmac.,* **53**, 179.

WILKINSON, G. R. (1983) Plasma and tissue binding considerations in drug disposition. *Drug. Metab. Rev.,* **14**, 427.

Excretion

CAFRUNY, E. J. (1971) Renal excretion of drugs. In *Fundamentals of Drug Metabolism and Disposition*, edited by B. N. La Du, H. G. Mandel and E. L. Way (Baltimore: Williams & Wilkins).

KLAASSEN, C. D. (1984) Mechanisms of bile formation, hepatic uptake and biliary excretion. *Pharmacol. Rev.,* **36**, 1.

LEVINE, W. G. (1983) Excretion mechanisms. In *Biological Basis of Detoxication*, edited by J. Caldwell and W. B. Jakoby (New York: Academic Press).

MATTHEWS, H. B. (1980) Elimination of toxicants and their metabolites. In *Introduction to Biochemical Toxicology*, edited by E. Hodgson and F. E. Guthrie (New York: Elsevier-North Holland).

PRITCHARD, J. B. and JAMES, M. O. (1982) Metabolism and urinary excretion. In *Metabolic Basis of Detoxication*, edited by W. B. Jakoby, J. R. Bend and J. Caldwell (New York: Academic Press).

SMITH, R. L. (1973) *The Excretory Function of Bile* (London: Chapman & Hall).

STOWE, C. M. and PLAA, G. L. (1968) Extrarenal excretion of drugs and chemicals. *Annu. Rev. Pharmacol.*, **8**, 337.

Pharmacokinetics

CLARK, B. and SMITH, D. A. (1986) *An Introduction to Pharmacokinetics*, 2nd edition (Oxford: Blackwell).

GIBALDI, M. and PERRIER, D. (1982) *Pharmacokinetics*, 2nd edition (New York: Marcel Dekker).

HOUSTON, J. B. (1986) Role of pharmacokinetics in rationalizing tissue distribution. In *Target Organ Toxicity*, Vol. I, edited by G. M. Cohen (Boca Raton, Florida: CRC Press).

LEVY, G. and GIBALDI, M. (1975) Pharmacokinetics. In *Handbook of Experimental Pharmacology*, Vol. 28, Part 3, *Concepts in Biochemical Pharmacology*, edited by J. R. Gillette and J. R. Mitchell (Berlin: Springer).

NEUBIG, R. R. (1990) The time course of drug action. In *Principles of Drug Action*, 3rd edition, edited by W. B. Pratt and Palmer Taylor (New York: Churchill Livingstone).

ROWLAND, M. and TOZER, T. N. (1988) *Clinical Pharmacokinetics. Concepts and Applications*, 2nd edition (Lea and Febiger: Philadelphia).

TUEY, D. B. (1980) Toxicokinetics. In *Introduction to Biochemical Toxicology*, edited by E. Hodgson and F. E. Guthrie (New York: Elsevier-North Holland).

ZBINDEN, G. (1988) Biopharmaceutical studies, a key to better toxicology. *Xenobiotica*, **18**, Suppl. 1, 9.

Chapter 4

Factors affecting toxic responses: metabolism

Introduction

As discussed in the preceding chapter, foreign and potentially toxic compounds absorbed into biological systems are generally lipophilic substances. They are therefore not ideally suited to excretion as they will be reabsorbed in the kidney or from the gastrointestinal tract after biliary excretion. For example, highly lipophilic substances such as polybrominated biphenyls and DDT are poorly excreted and therefore may remain in the animal's body for years.

The **biotransformation** of foreign compounds, however, attempts to convert such lipophilic substances into more polar, and consequently more readily excreted metabolites. The exposure of the body to the compound is hence reduced and potential toxicity decreased. This process of biotransformation is therefore a crucial aspect of the disposition of a toxic compound *in vivo*. Furthermore, as will become apparent later in the book, biotransformation may also underlie the toxicity of a compound.

The metabolic fate of a compound can therefore have an important bearing on its toxic potential, disposition in the body and eventual excretion.

The primary results of biotransformation are therefore:

(1) the parent molecule is transformed into a more polar metabolite, often by the addition of ionizable groups
(2) molecular weight and size are often increased
(3) the excretion is facilitated, and hence elimination of the compound from the tissues and the body is increased.

The consequences of metabolism are:

(a) the biological half-life is decreased
(b) the duration of exposure is reduced
(c) accumulation of the compound in the body is avoided
(d) the biological activity may be changed
(e) the duration of the biological activity may be affected.

For example, the analgesic drug *paracetamol* (see Chapter 7) has a renal clearance value of 12 ml/min whereas one of its major metabolites, the sulphate conjugate, is cleared at the rate of 170 ml/min. However, biotransformation may not always increase water solubility. For example, many **sulphonamide drugs** are acetylated *in vivo*, but the acetylated metabolites can be less water soluble (table 4.1), precipitate in the kidney tubules and cause toxicity. As the chemical structure is changed from that of the parent compound, there may be consequential changes to the pharmacological and toxicological activity of the compound. For some drugs the pharmacological activity resides in the metabolite rather than the parent compound. A classic example of this is the antibacterial drug sulphanilamide which is released from the parent compound, prontosil, by bacterial metabolism in the gut (figure 4.38). For other drugs it is the parent compound that is active, as is the case with the muscle relaxant drug succinylcholine (suxamethonium). The action of this drug normally only lasts a few minutes because metabolism rapidly cleaves the molecule to yield inactive products (figure 7.36). The duration of action is therefore determined by metabolism in this case. Although biotransformation is usually regarded as a detoxication process, this is not always so. Thus the pharmacological or toxicological activity of the metabolite may be greater or different from that of the parent compound. Later in the chapter, and especially in the final chapter, we will examine ways and consider examples in which toxicity is increased by metabolism.

Table 4.1.   Solubility data for two sulphonamides and their acetylated metabolites.

Drug	Solubility in Urine (mg/ml at 37°C)	Urinary pH
Sulphisomidine	254	5·0
	282	6·8
Acetylsulphisomidine	9	5·0
	10	7·5
Sulphisoxazole	150	5·5
	1200	6·5
Acetylsulphisoxazole	55	5·5
	450	6·5

Data from Weinstein (1970). In *The Pharmacological Basis of Therapeutics*, edited by L.S. Goodman and A. Gilman (New York: Macmillan).

Metabolism is therefore an important determinant of the *activity* of a compound, the *duration* of that activity and the *half-life* of the compound in the body.

The metabolism of foreign compounds is catalysed by enzymes, some of which are specific for the metabolism of xenobiotics. The metabolic pathways involved may be many and various but the major determinants of which transformations take place are:

   (i) the structure and physicochemical properties of the compound in question
   (ii) the enzymes available in the exposed tissue.

Thus the *partition coefficient*, the *stereochemistry* and the *functional groups* present on a molecule may all influence the particular metabolic transformation which takes place and these factors are discussed in more detail later.

The enzymes specifically involved in the metabolism of foreign compounds are necessarily often flexible and non-specific. However, it follows from the above two conditions that if the structure of a foreign compound is similar to a normal endogenous molecule, then the foreign compound may be a suitable substrate for an enzyme primarily involved in intermediary metabolic pathways if the enzyme is present in the exposed tissue. Thus foreign compounds are not exclusively metabolized by specific enzymes.

The organ most commonly involved in the biotransformation of foreign compounds is the liver because of its *position*, *blood supply* and *function* (figure 6.2). Most foreign compounds are taken into the organism via the gastrointestinal tract, and the blood that drains the tract flows through the portal vein directly to the liver. Therefore the liver represents a portal to the tissues of the body and is exposed to foreign compounds at higher concentrations than most other tissues. Detoxication in this organ and possible removal by excretion into the bile are therefore protective measures. The role of the liver in endogenous metabolism and its structure (see Chapter 6) make it an ideal site for the biotransformation of xenobiotics. As already mentioned in the previous chapter, metabolism in the liver may be so extensive during the 'first pass' of the compound through the organ that little or none of the parent compound reaches the systemic circulation. However most other organs and tissues possess some metabolic activity with regard to foreign compounds and in some cases may be quantitatively more important than the liver.

The enzymes involved in biotransformation also have a particular subcellular localization: many are found in the **smooth endoplasmic reticulum (SER)**. Some are located in the cytosol and a few are found in other organelles such as the mitochrondria. These subcellular localizations may have important implications for the mechanism of toxicity of compounds in some cases.

Types of metabolic change

Metabolism can be simply and conveniently divided into two phases: **phase 1** and **phase 2**. Phase 1 is the alteration of the original foreign molecule so as to add on a functional group which can then be conjugated in phase 2. This can best be understood by examining the example in figure 4.1. The foreign molecule is **benzene**, a highly lipophilic molecule which is not readily excreted from the animal except in the expired air as it is volatile. Phase 1 metabolism converts benzene into a variety of metabolites, but the major one is phenol. The insertion of a hydroxyl group allows a phase 2 conjugation reaction to take place with the polar sulphate group being added. Phenyl sulphate, the final metabolite is very water soluble and is readily excreted in the urine.

Most biotransformations can be divided into phase 1 and phase 2 reactions, although the products of phase 2 biotransformations may be further metabolized in what is sometimes termed **phase 3** reactions.

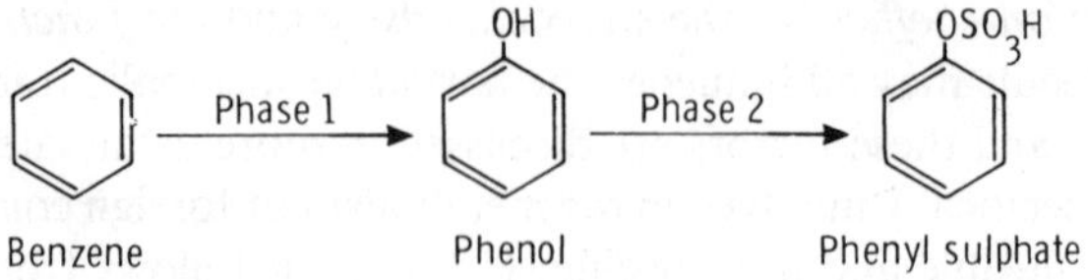

Principles of Biochemical Toxicology

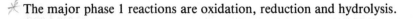

Benzene   Phenol   Phenyl sulphate

FIGURE 4.1. Metabolism of benzene to phenyl sulphate.

If the foreign molecule already possesses a functional group suitable for a phase 2 reaction, a phase 1 reaction will be unnecessary. Thus, if phenol is administered to an animal then it may immediately undergo a phase 2 reaction, such as conjugation with sulphate. Alternatively it may undergo another phase 1 type of reaction. The major types of reaction are shown in table 4.2.

Table 4.2. The major biotransformation reactions.

Phase 1	Phase 2	Phase 3
Oxidation	Sulphation	
Reduction	Glucuronidation	
Hydrolysis	Glutathione conjugation	Further metabolism of
Hydration	Acetylation	glutathione conjugates
Dehalogenation	Amino acid conjugation	
	Methylation	

Generally, therefore, the function of phase 1 reactions is to *modify* the *structure* of a xenobiotic so as to introduce a *functional group* suitable for conjugation with glucuronic acid, sulphate or some other highly-polar moiety, so making the entire molecule water-soluble.

Phase 1 reactions

The major phase 1 reactions are oxidation, reduction and hydrolysis.

Oxidation

For foreign compounds the majority of oxidation reactions are catalysed by **mono-oxygenase enzymes** found in the SER and known as **microsomal enzymes**. Other enzymes involved in the oxidation of xenobiotics are found in other organelles such as the mitochondria and the cytosol. Thus amine oxidases located in the mitochondria, xanthine oxidase, alcohol dehydrogenase in the cytosol, the prostaglandin synthetase system and various other peroxidases may all be involved in the oxidation of foreign compounds.

Microsomal oxidations may be subdivided into: aromatic hydroxylation; aliphatic hydroxylation; alicyclic hydroxylation; heterocyclic hydroxylation; N-,

S- and O-dealkylation; N-oxidation; N-hydroxylation; S-oxidation; desulphuration; and deamination; and dehalogenation.

Non-microsomal oxidations may be subdivided into: amine oxidation; alcohol and aldehyde oxidation; dehalogenation; purine oxidation; and aromatization.

Microsomal oxidations

CYTOCHROMES P-450 MONO-OXYGENASE SYSTEM

The majority of these reactions are catalysed by one enzyme system, the cytochromes P-450 mono-oxygenase system, which is located particularly in the SER of the cell. The enzyme system is isolated in the so-called **microsomal** fraction which is formed from the endoplasmic reticulum when the cell is homogenized and fractionated by differential ultracentrifugation. Microsomal vesicles are thus fragments of the endoplasmic reticulum in which most of the enzyme activity is retained. The endoplasmic reticulum is composed of a convoluted network of channels and so has a large surface area. Apart from cytochromes P-450, the endoplasmic reticulum has many enzymes and functions besides the metabolism of foreign compounds. These include the synthesis of proteins and triglycerides, and other aspects of lipid metabolism and fatty acid metabolism. Specific enzymes present on the endoplasmic reticulum include cholesterol esterase, azo reductase, glucuronosyl transferase, NADPH cytochromes P-450 reductase and NADH cytochrome b_5 reductase and cytochrome b_5. A FAD-containing mono-oxygenase is also found in the endoplasmic reticulum and this is discussed later in this chapter.

The cytochromes P-450 mono-oxygenase system is actually a collection of *isoenzymes* all of which possess an iron protoporphyrin IX as the prosthetic group. The monomer of the enzyme has a molecular weight of 45 000–55 000. The enzyme is membrane bound within the endoplasmic reticulum. Cytochromes P-450 is closely associated with another vital component of the system, **NADPH cytochrome P-450 reductase**. This is a **flavoprotein** which has 1 mol of FAD and 1 mol of FMN per mol of apoprotein. The monomeric molecular weight of the enzyme is 78 000. The enzyme transfers two electrons to cytochromes P-450, but one at a time. There only seems to be one reductase which serves a group of isoenzymes of cytochromes P-450, and consequently its concentration is 1/10 to 1/30 that of cytochromes P-450.

Phospholipid is also required in the enzyme complex, seemingly this is important for the integrity of the overall complex and the interrelationship between the cytochromes P-450 and the reductase. The individual components can be separated and reconstitution of these components results in a functional enzyme.

The overall reaction is:

$$SH + O_2 + NADPH + H^+ \rightarrow SOH + H_2O + NADP^+$$

where S is the substrate. The reaction therefore also requires NADPH and molecular oxygen.

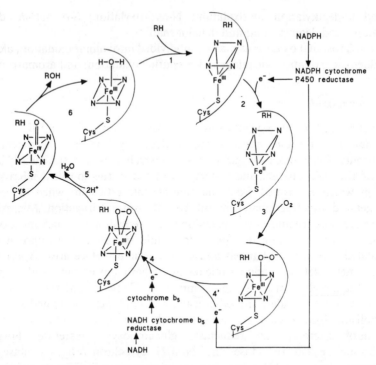

FIGURE 4.2. The catalytic cycle of the cytochrome(s) P450 mono-oxygenase (mixed function oxidase) system. Adapted from Tukey and Johnson (1990) in *Principles of Drug Action*, edited by W.B. Pratt and P. Taylor (New York: Churchill Livingstone).

The sequence of metabolic reactions is shown in figure 4.2 and involves at least four distinct steps:

(i)   addition of substrate to the enzyme
(ii) donation of an electron
(iii) addition of oxygen and rearrangement
(iv) donation of a second electron and loss of water.

Now let us look at these steps in more detail:

Step 1. *Binding* of the substrate to the cytochromes P-450. This takes place when the iron is in the oxidized, ferric state. There are three types of binding: **Type I, II and reverse Type I (or modified Type II)** which give rise to particular spectra. These are now known to be the result of perturbations in the spin equilibrium of cytochromes P-450. Thus, ethylmorphine gives rise to a Type I spectrum, aniline a type II and phenacetin a modified Type II. There are two spin states: low spin, hexa coordinated or high spin, penta coordinated (figure 4.2). Most P-450s are in the low spin state. When certain substrates bind (Type I) they bind to a hydrophobic site on the protein close enough to the haem to interact with oxygen and cause a perturbation and a conformational change. The change results in a high spin configuration. This change from low to high spin gives rise to spectral change with an increase in absorbance at 390 nm and a decrease at 420 nm. The difference spectrum therefore has a peak at 390 nm and a trough around 420 nm.

Type II substrates are compounds such as nitrogenous bases, with sp^2 or sp^3 non-bonded electrons. These bind to iron and give rise to a 6-coordinated, low spin haemoprotein. Such compounds may also be inhibitors of cytochromes P-450. The spectrum shows a peak at 420–435 nm and a trough at 390–410 nm in the difference spectrum.

Step 2. First electron *reduction* of the substrate–enzyme complex. The iron atom is reduced from the ferric to the ferrous state. The reducing equivalents are transferred from NADPH via the reductase. The equilibrium between the high spin and low spin states may be important in this step, but it is not yet clear.

Thus the reaction is:

$$\text{NADPH} + \text{H}^+ \rightarrow \underline{(\text{FAD NADPH cytochrome FMN})} \rightarrow \text{Cytochrome P-450}$$
$$\text{P-450 reductase}$$

In the reduced state cytochromes P-450 may also bind certain ligands to give particular difference spectra. The most well known is that which occurs when **carbon monoxide** binds giving an absorption maximum at 450 nm. A **Type III** spectrum gives two peaks at 430 and 455 nm after binding of certain compounds such as ethyl isocyanide or methylenedioxyphenyl compounds to the reduced enzyme. The latter form stable complexes with the enzyme and are also inhibitors.

Step 3. Addition of molecular oxygen and rearrangement of the ternary ferrous oxygenated cytochromes P-450–substrate complex. The reduced cytochromes P-450–substrate complex binds oxygen and undergoes a rearrangement. The oxidation state of the oxygen and the iron in the ternary complex is not entirely clear, but the oxygen may exist as **hydroperoxide ($O_2{}^{2-}$)** or **superoxide ($O_2{}^-$)**. It has been shown by experiments using $^{18}O_2$ that the oxygen bound is molecular oxygen and is not derived from water.

Step 4. Addition of the second electron from NADPH via P-450 reductase.
Step 4'. Alternatively the electron may be donated by NADH via cytochrome b_5 reductase and cytochrome b_5. This alternative source of the second electron step is still controversial.

The complex then rearranges with insertion of one atom of oxygen into the substrate to yield the product. The mechanism is obscure but seems to involve oxygen in an activated form. The other atom of oxygen is reduced to water, the other product. It has been suggested that the hydroxylation of hydrocarbons by cytochromes P-450 may be a two-step mechanism involving a radical intermediate formed by removal of a hydrogen atom from the substrate and then transference to the oxygen bound to the iron. The hydroxyl may then react with the substrate radical to produce hydroxylated substrate. Under certain circumstances cytochromes P-450 produces hydrogen peroxide. This seems to be when the cycle becomes uncoupled and the oxygenated P-450 complex breaks down differently to give the oxidized cytochrome–substrate complex and hydrogen peroxide.

Cytochromes P-450 may be found in other organelles as well as the SER including the rough endoplasmic reticulum and nuclear membrane. In the adrenal gland it is also found in the mitochondria, although here **adrenodoxin** and **adrenodoxin reductase** are additional requirements in the overall system. Although the liver has the highest concentration of the enzyme, cytochromes P-450 are found in most, if not all, tissues.

Cytochromes P-450 is now known to exist in a *multiplicity* of forms with overlapping substrate specificities. This multiplicity accounts for the diversity of the reactions catalysed and the substrates accommodated. The forms or isoenzymes can be separated using chromatographic and electrophoretic techniques and the DNA sequences determined using sophisticated molecular biological techniques. These isoenzymes may vary in distribution both within the cell and in the tissues of the whole organism. The *proportions* of the isoenzymes in any given tissue may change as a result of treatment with various compounds as described in the next chapter.

There are now a considerable number of isoenzymes which have been identified and some indication of the particular substrates and their characteristics is emerging. Those involved in the metabolism of xenobiotics can be seen in table 4.3. The different isoenzymes are coded for by distinct genes and the nomenclature used in this table is the current internationally accepted standard.

Table 4.3.   Characteristics of cytochromes P-450 families 1–4.

Isozyme	Substrate examples	Reactions
CYP 1A1	benzo(a)pyrene	Hydroxylation
	7-ethoxyresorufin	*O*-de-ethylation
1A2	acetylaminofluorene	*N*-hydroxylation
	phenacetin	*O*-de-ethylation
CYP 2A1	testosterone	7α-hydroxylation
2A2	testosterone	15α-hydroxylation
2A3		
CYP 2B1	hexobarbital	hydroxylation
	7-pentoxyresorufin	*O*-de-ethylation
2B2	7-pentoxyresorufin	*O*-de-ethylation
	7,12-dimethylbenzanthracene	12-methyl-hydroxylation
CYP 2C	S-mephenytoin	hydroxylation
CYP 2D	Debrisoquine	alicyclic hydroxylation
CYP 2E1	p-nitrophenol	hydroxylation
	aniline	hydroxylation
CYP 3A	ethylmorphine	*N*-demethylation
	aminopyrine	*N*-demethylation
CYP 4A1	lauric acid	ω-hydroxylation
	lauric acid	ω-1-hydroxylation

It should be noted that this is not an exhaustive list and that in different species different numbers of isozymes exist. It serves solely to illustate the multiplicity of the forms of cytochrome P-450 generally involved with xenobiotic metabolism and the differences and similarities between them.

For a more detailed discussion see Nebert, D. W. and Gonzales, F. (1990) The P450 gene superfamily. In *Frontiers of Biotransformation*, Vol. 2, *Principles, Mechanisms and Biological Consequences of Induction*, edited by K. Ruckpaul and H. Rein (London: Taylor & Francis), or other references in the Bibliography.

This groups enzymes into **gene families** based on primary amino acid sequence resulting from sequencing of cytochromes P-450 proteins, cDNAs and genes (Nebert *et al.*, 1991). Prior to the introduction of this nomenclature a confusing array of names was in use based on substrates or inducers and this may still be encountered.

Thus the **cytochromes P-450 gene superfamily** currently consists of 27 gene families. The various proteins and the genes coding for them are designated by CYP with the families indicated by Arabic numerals. *Subfamilies* are designated by capital letters and proteins and genes within these by Arabic numerals. For example the CYP 2 gene family of the rat is further divided into five subfamilies, A–E, each of which is divided into one or more genes coding for one or more proteins. Thus CYP 2E is currently known to have one gene, designated CYP 2E1, coding for P-450 2E1 (previously known as P-450j). The first four gene families, P-450s 1–4, are primarily hepatic, microsomal enzymes involved in the catabolism of foreign compounds. The four families CYP17, CYP19, CYP20 and CYP22 code for P-450s involved in steroid biosynthesis and found mainly in extrahepatic tissues. The mitochondrial P-450 is CYP11. There are also other forms of P-450 to be found in insects, yeast and bacteria. For further details on this topic the reader is referred to the Bibliography.

As already indicated, however, there is overlap between some of the P-450s in terms of substrates and types of reaction catalysed. These different isoenzymes may be separated on the basis of certain criteria. Thus they may have different monomeric molecular weights, the carbon monoxide difference spectra may show different maxima, the amino acid composition and terminal sequences may be different, substrate specificities may be different and they may be distinguished by specific antibodies. The importance of these different isoenzymes is that they may catalyse different biotransformations and this may be crucial to the toxicity of the compound in question. Variations in the proportions and presence of particular isoenzymes may underly differences in metabolism due to species, sex, age, nutritional status and interindividual variability. The comparison and identification of particular forms (orthologues) of cytochrome P-450 between species has proved to be difficult in some cases, however. The presence or absence of a particular isoenzyme may be the cause of toxicity in one organ or tissue. The change in the proportion of isoenzymes caused by exposure to substances in the environment, or drugs may explain changes in toxicity or other biological activity attributed to the compound of interest. These questions will be considered in greater detail in the following chapter.

Cytochromes P-450 has, unlike most enzymes, a low degree of substrate specificity. Thus there are not only many different substrates which may be metabolized, but one substrate may be metabolized to several products. Despite this lack of substrate specificity however, the enzyme may display significant *stereoselectivity* with chiral substrates (see below). There is an enormous variety of substrates for cytochromes P-450, and the only seemingly common factor is a degree of lipophilicity. Indeed there is a correlation between the metabolism of xenobiotics by microsomes and the *lipophilicity* of the compound. This is not surprising in that if the purpose of metabolism is to increase the water solubility of a foreign compound and hence its excretion, the compounds most needing this biotransformation are the lipophilic compounds. Furthermore, the lack of substrate specificity requires some control if many vital endogenous molecules are not to be wastefully metabolized. This control is exercised by the lipoidal character of the enzyme complex which effectively excludes many endogenous molecules. Cytochromes P-450 may also be involved in the metabolism of endogenous compounds, particularly in some tissues where the appropriate isoenzyme is located, but these substrates again tend to be lipophilic. For example in the kidney fatty acids are substrates, undergoing ω-1 hydroxylation, and prostaglandins also undergo this type of hydroxylation. In the adrenal cortex, steroids are hydroxylated by a mitochondrial cytochromes P-450.

Although in the majority of cases, cytochromes P-450 catalyse oxidation reactions, under certain circumstances the enzyme may catalyse other types of reaction such as reduction.

MICROSOMAL FLAVIN-CONTAINING MONO-OXYGENASES

As well as the cytochromes P-450 mono-oxygenase system there is also a system which involves FAD. This enzyme system is found particularly in the microsomal fraction of the liver and the monomer has a molecular weight of around 65 000. Each monomer has one molecule of FAD associated with it. The enzyme may accept electrons from either NADPH or NADH although the former is the preferred cofactor. It also requires molecular oxygen and the overall reaction is as written for cytochromes P-450:

$$NADPH + H^+ + O_2 + S \rightarrow NADP^+ + H_2O + SO$$

which includes the following steps:

$$Enzyme - FAD + NADPH\ H^+ \rightarrow NADP^+ - Enzyme - FADH_2 \qquad (1)$$

$$NADP^+ - Enzyme\text{-}FADH_2 + O_2 \rightarrow NADP^+ - Enzyme - FADH_2 - O_2 \quad (2)$$

$$NADP^+ - Enzyme\text{-}FADH_2 - O_2 \rightarrow NADP^+ - Enzyme - FAD - OOH \quad (3)$$

$$NADP^+ - Enzyme\text{-}FAD\text{-}OOH + S \rightarrow NADP^+ - Enzyme - FAD - OOH - S \quad (4)$$

$$NADP^+ - Enzyme\text{-}FAD\text{-}OOH\text{-}S \rightarrow SO + NADP^+ - Enzyme - FAD + H_2O \quad (5)$$

$$NADP^+ - Enzyme\text{-}FAD \rightarrow Enzyme - FAD \qquad (6)$$

Thus step 1 involves addition of NADPH and reduction of the flavin, step 2 the addition of oxygen. At step 3 an internal rearrangement results in the formation of a peroxy complex which then binds the substrate at step 4. The substrate is oxygenated and released at step 5. Step 6 regenerates the enzyme.

In the absence of a suitable substrate the complex at step 3 may degrade to produce hydrogen peroxide.

This enzyme system catalyses the oxidation of various nitrogen, sulphur and phosphorus containing compounds, which tend to be nucleophilic, although compounds with an anionic group are not substrates. For example the *N*-oxidation of **trimethylamine** (see figure 4.19) is catalysed by this enzyme, but also the hydroxylation of secondary amines, imines and arylamines and the oxidation of hydroxylamines and hydrazines. Various sulphur containing compounds including thioamides, thioureas, thiols, thioethers and disulphides are oxidized by this enzyme system. However, unlike cytochromes P-450, it cannot catalyse hydroxylation reactions at carbon atoms. It is clear that this enzyme system has an important role in the metabolism of xenobiotics and examples will appear in the following pages. Just as with the cytochromes P-450 system, there appear to be a number of isoenzymes which exist in different tissues which have overlapping substrate specificities.

Let us look at the major types of oxidation reaction catalysed by the cytochromes P-450 system.

AROMATIC HYDROXYLATION
Aromatic hydroxylation such as that depicted in figure 4.3 for the simplest aromatic system, benzene, is an extremely important biotransformation. The major products of aromatic hydroxylation are phenols, but catechols and quinols may also be formed, arising by further metabolism. One of the toxic effects of **benzene** is to cause **aplastic anaemia**, which is believed to be due to an intermediate metabolite, possibly hydroquinone. As a result of further metabolism of epoxide intermediates (see below), other metabolites such as diols and glutathione conjugates can also be produced. Because epoxides exist in one of two enantiomeric forms, various isomeric metabolites can result. These may have significance in the toxicity of compounds such as the pulmonary damage

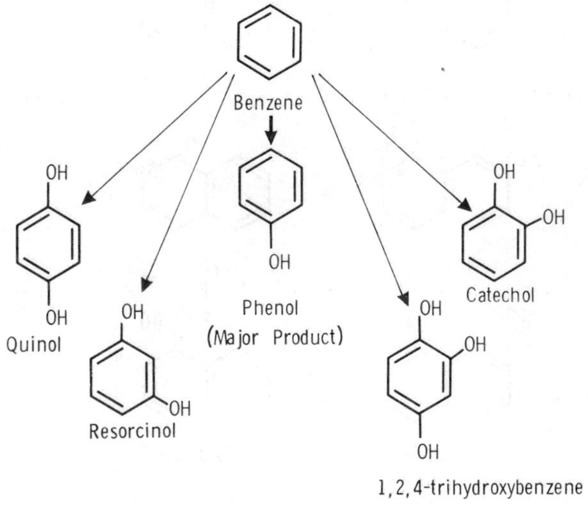

FIGURE 4.3. Aromatic products of the enzymatic oxidation of benzene. Phenol is the major metabolite.

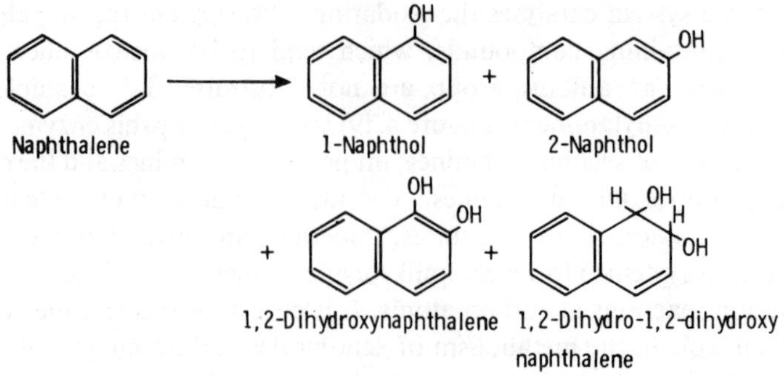

FIGURE 4.4. Hydroxylated products of naphthalene metabolism.

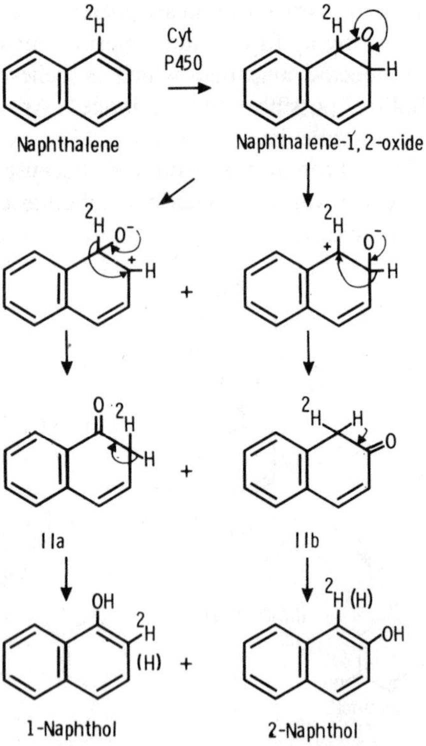

FIGURE 4.5. Metabolism of deuterium-labelled naphthalene via the 1,2-oxide (epoxide) intermediate, illustrating the NIH shift.

caused by naphthalene and the carcinogenicity of benzo[*a*]pyrene (see below and Chapter 7 for more details). Consequently, a number of hydroxylated metabolites may be produced from the aromatic hydroxylation of a single compound (figures 4.4 and 4.5).

Aromatic hydroxylation generally proceeds via the formation of an epoxide intermediate. This is illustrated by the metabolism of **naphthalene**, labelled in the 1 position with deuterium (^{2}H), via the 1,2-oxide as shown in figure 4.5. The shift in the deuterium atom that occurs during metabolism is the so-called **NIH shift**. This indicates that formation of an epoxide intermediate has occurred, and is one method of determining whether such an epoxide intermediate is involved. The phenolic products, 1- and 2-naphthols, retain various proportions of deuterium, however. The proposed mechanism involves the formation of an epoxide intermediate, which may break open chemically in two ways, leading to phenolic products (figure 4.5).

Each naphthol product may have deuterium or hydrogen in the adjacent position. The hydrogen and deuterium atoms (in 11a and 11b; figure 4.5) are equivalent, as the carbon atom to which they are attached is tetrahedral. Consequently, either hydrogen or deuterium may be lost, theoretically resulting in 50% retention of deuterium. However, in practice this may not be the case, as an isotope effect may occur. This effect results from the strength of the carbon–deuterium bond being greater than that of a carbon–hydrogen bond. Therefore, more energy is required to break the C-^{2}H bond than C-^{1}H bond, with a consequent effect on the rate-limiting chemical reactions involving bond breakage. Also, the direction of the opening of the epoxide ring is affected by the substituents, and the proportions of products therefore reflect this. The production of phenols occurs via a chemical rearrangement, and depends on the stability of the particular epoxide.

The further metabolism of suitably *stable* epoxides may occur, with the formation of dihydrodiols as discussed later. Dihydrodiols may also be further metabolized to catechols. Other products of aromatic hydroxylation via epoxidation are glutathione conjugates. These may be formed by enzymic or non-enzymic means or both, depending on the reactivity of the epoxide in question.

The products of epoxidation *in vivo* depend on the reactivity of the particular epoxide. Stabilized epoxides react with nucleophiles and undergo further enzymic reactions, whereas destabilized ones undergo spontaneous isomerization to phenols. Epoxides are generally reactive intermediates, however, and in a number of cases are known to be responsible for toxicity by reaction with cellular constituents.

With carcinogenic polycyclic hydrocarbons, dihydrodiols are further metabolized to epoxides as shown in figure 7.2. This epoxide-diol may then react with weak nucleophiles such as nucleic acids.

Unsaturated aliphatic compounds and heterocyclic compounds may also be metabolized via epoxide intermediates as shown in figures 4.6 and 5.26. Note that when the epoxide ring opens the chlorine atom shifts to the adjacent carbon atom (figure 4.6). In the case of the furan ipomeanol and vinyl chloride, the epoxide intermediate is thought to be responsible for the toxicity (see below, Chapter 7).

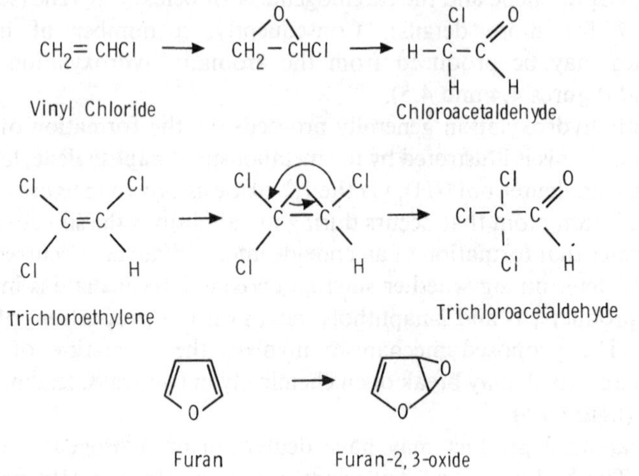

FIGURE 4.6. Metabolism of unsaturated aliphatic and heterocyclic compounds via epoxides.

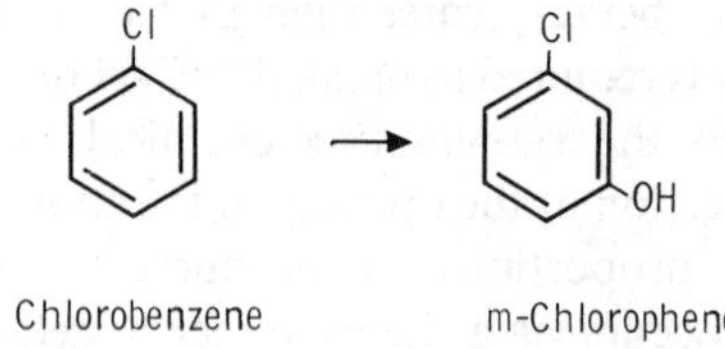

Chlorobenzene → m-Chlorophenol

FIGURE 4.7. *m*-Hydroxylation of chlorobenzene.

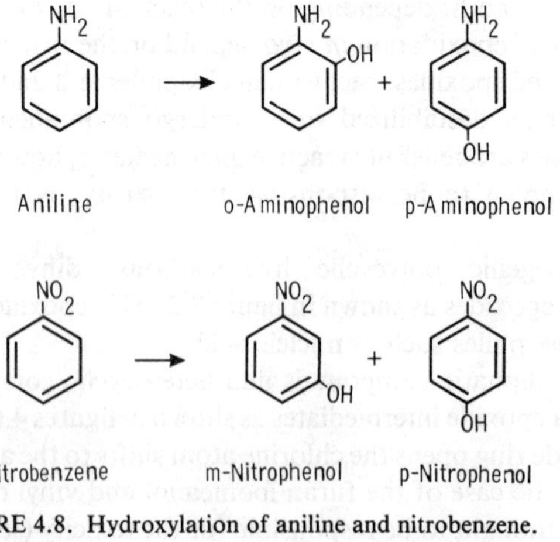

Aniline → o-Aminophenol + p-Aminophenol

Nitrobenzene → m-Nitrophenol + p-Nitrophenol

FIGURE 4.8. Hydroxylation of aniline and nitrobenzene.

Other examples of unsaturated aliphatic compounds which may be toxic and are metabolized via epoxides are diethylstilboestrol, **allylisopropyl acetamide**, which destroys cytochrome P-450, sedormid and secobarbital.

Aromatic hydroxylation may also take place by a mechanism other than epoxidation. Thus, the *m*-hydroxylation of chlorobenzene is thought to proceed via a direct insertion mechanism (figure 4.7).

The nature of the substituent in a substituted aromatic compound influences the position of hydroxylation. Thus, *o-p*-directing substituents, such as amino groups, result in *o*- and *p*-hydroxylated metabolites such as the *o*- and *p*-aminophenols from aniline (figure 4.8). Meta-directing substituents such as nitro groups lead to *m*- and *p*-hydroxylated products, for example **nitrobenzene** is hydroxylated to *m*- and *p*-nitrophenols (figure 4.8).

ALIPHATIC HYDROXYLATION

As well as unsaturated aliphatic compounds such as vinyl chloride mentioned above which are metabolized by epoxidation, saturated aliphatic compounds also undergo oxidation. The initial products will be primary and secondary alcohols. For example the solvent **n-hexane** is known to be metabolized to the secondary alcohol hexan-2-ol and then further to hexane-2,5-dione (figure 4.9) in occupationally exposed humans. The latter metabolite is believed to be responsible for the neuropathy caused by the solvent. Other toxicologically important examples are the nephrotoxic petrol constituents, 2,2,4- and 2,3,4-trimethylpentane, which are hydroxylated to yield primary and tertiary alcohols. However, aliphatic hydrocarbon chains are more readily metabolized if they are side-chains on aromatic structures. Thus, **n-propylbenzene** may be hydroxylated in three positions, giving the primary alcohol 3-phenylpropan-1-ol,

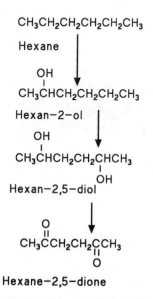

FIGURE 4.9. The aliphatic oxidation of hexane.

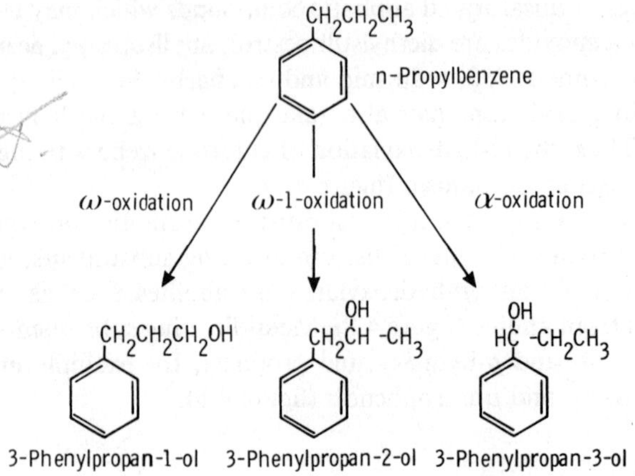

FIGURE 4.10. Aliphatic oxidation of *n*-propylbenzene.

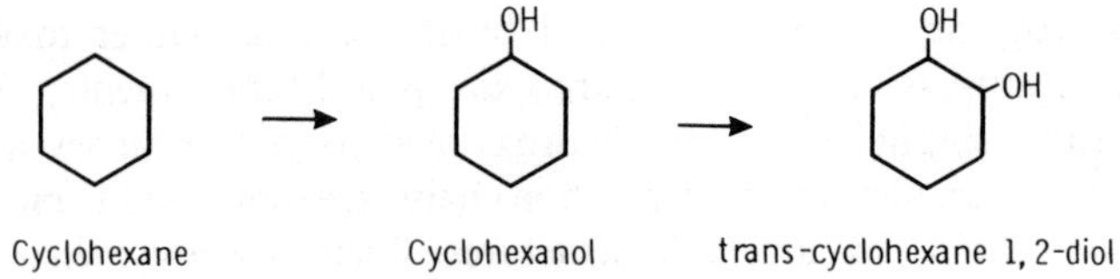

FIGURE 4.11. Hydroxylation of cyclohexane.

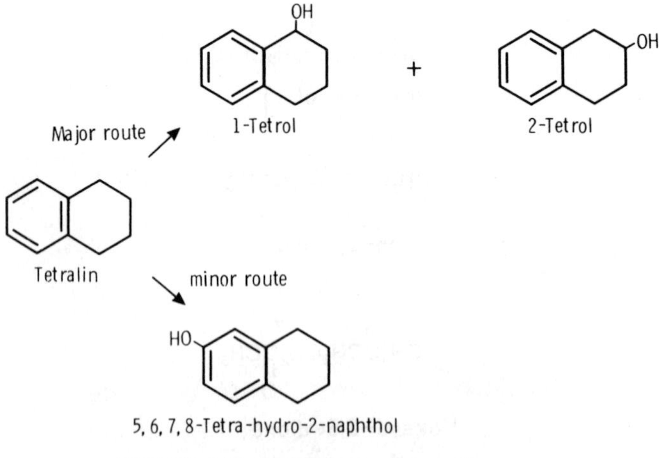

FIGURE 4.12. Hydroxylation of tetralin.

and two secondary alcohols (figure 4.10). Further oxidation of the primary alcohol may take place to give the corresponding acid phenyl-propionic acid, which may be further metabolized to benzoic acid, probably by oxidation of the carbon β to the carboxylic acid.

ALICYCLIC HYDROXYLATION

Hydroxylation of saturated rings yields monohydric and dihydric alcohols. For instance, **cyclohexane** is metabolized to cyclohexanol, which is further hydroxylated to *trans*-cyclohexane-1,2-diol (figure 4.11). With mixed alicyclic/aromatic, saturated and unsaturated systems, alicyclic hydroxylation appears to predominate, as shown for the compound **tetralin** (figure 4.12).

HETEROCYCLIC HYDROXYLATION

Nitrogen heterocycles such as **pyridine** and **quinoline** (figure 4.13) undergo microsomal hydroxylation at the 3 position. In quinoline, the aromatic ring is also hydroxylated in positions o- and p- to the nitrogen atom. Aldehyde oxidase, a non-microsomal enzyme discussed in more detail below, may also be involved in the oxidation of **quinoline** to give 2-hydroxyquinoline (figure 4.14). The heterocyclic **phthalazine** ring in the drug hydralazine is oxidized by the microsomal enzymes to phthalazinone. The mechanism, which may involve nitrogen oxidation, is possibly involved in the toxicity of this drug (see Chapter 7). Again other enzymes may also be involved (figure 4.35).

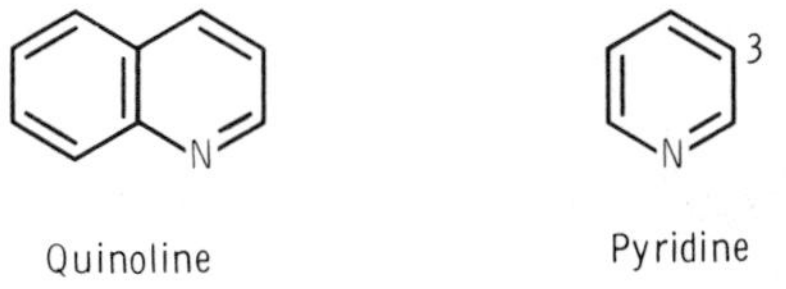

Quinoline   Pyridine

FIGURE 4.13. Structures of pyridine and quinoline.

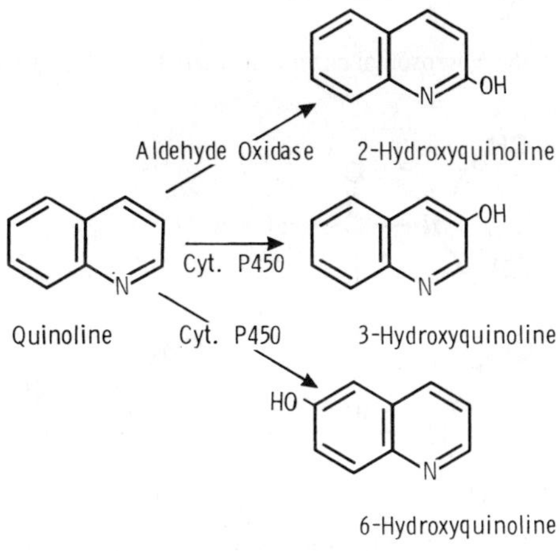

FIGURE 4.14. Hydroxylation of quinoline.

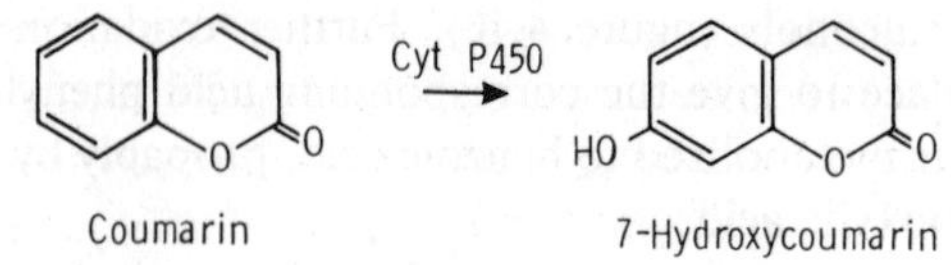

FIGURE 4.15. Microsomal enzyme-mediated hydroxylation of coumarin.

Another example of heterocyclic oxidation is the microsomal oxidation of **coumarin** to 7-hydroxycoumarin (figure 4.15).

N-DEALKYLATION

Dealkylation is the removal of alkyl groups from nitrogen, sulphur and oxygen atoms, and is catalysed by the microsomal enzymes. Alkyl groups attached to ring nitrogen atoms or those in amines, carbamates or amides are removed oxidatively by conversion to the corresponding aldehyde as indicated in figure 4.16. The reaction proceeds via a hydroxyalkyl intermediate which is usually *unstable* and spontaneously *rearranges* with loss of the corresponding aldehyde. However in some cases the hydroxyalkyl intermediate is more stable and may be

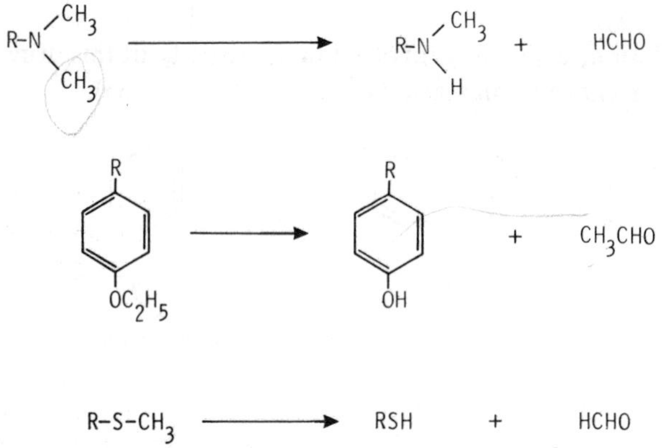

FIGURE 4.16. Microsomal enzyme-mediated *N*-, *O*- and *S*-dealkylation.

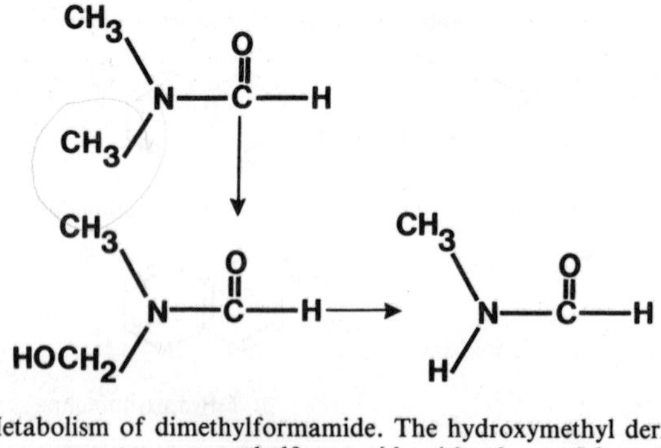

FIGURE 4.17. Metabolism of dimethylformamide. The hydroxymethyl derivative is stable but may subsequently rearrange to monomethylformamide with release of formaldehyde.

isolated, as for example with the solvent **dimethylformamide** (figure 4.17). *N*-Dealkylation is a commonly encountered metabolic reaction for foreign compounds which may have important toxicological consequences, as in the metabolism of the carcinogen dimethylnitrosamine (figure 7.3).

S-DEALKYLATION

A microsomal enzyme system catalyses *S*-dealkylation with oxidative removal of the alkyl group to yield the corresponding aldehyde, as with *N*-dealkylation (figure 4.16). However, as there are certain differences from the *N*-dealkylation reaction it has been suggested that another enzyme system, such as the **microsomal FAD-containing mono-oxygenase** system may be involved.

Figure 4.18 shows the *S*-demethylation of **6-methylthiopurine** to 6-mercapto-purine.

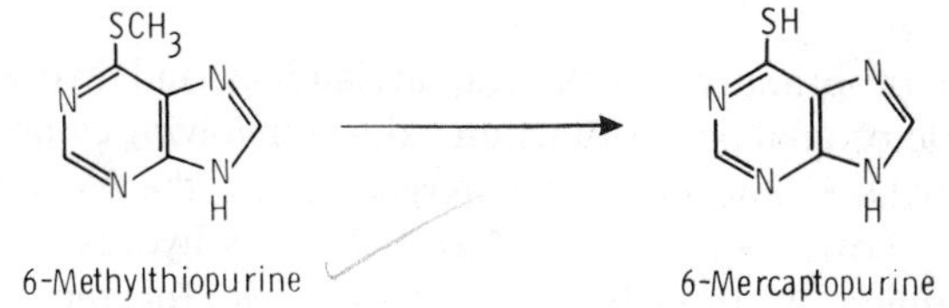

6-Methylthiopurine   6-Mercaptopurine

FIGURE 4.18. Oxidative *S*-demethylation of 6-methylthiopurine.

O-DEALKYLATION

Aromatic methyl and ethyl ethers may be metabolized to give the phenol and corresponding aldehyde (figure 4.16), as illustrated by the de-ethylation of **phenacetin** (figure 5.20). Ethers with longer alkyl chains are less readily *O*-dealkylated, the preferred route being ω-1-hydroxylation.

N-OXIDATION

The oxidation of nitrogen in tertiary amines, amides, imines, hydrazines and heterocyclic rings may be catalysed by microsomal enzymes or by other enzymes (see below). Thus the oxidation of **trimethylamine** to an *N*-oxide (figure 4.19) is catalysed by the microsomal FAD-containing mono-oxygenase. The *N*-oxide so formed may undergo enzyme-catalysed decomposition to a secondary amine and aldehyde. This *N* to *C transoxygenation* is mediated by cytochromes P-450. The *N*-oxidation of 3-methylpyridine, however, is catalysed by cytochromes P-450.

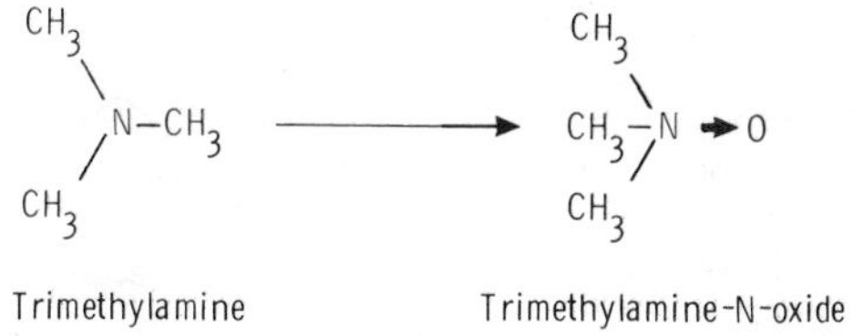

Trimethylamine   Trimethylamine-N-oxide

FIGURE 4.19. *N*-Oxidation of trimethylamine.

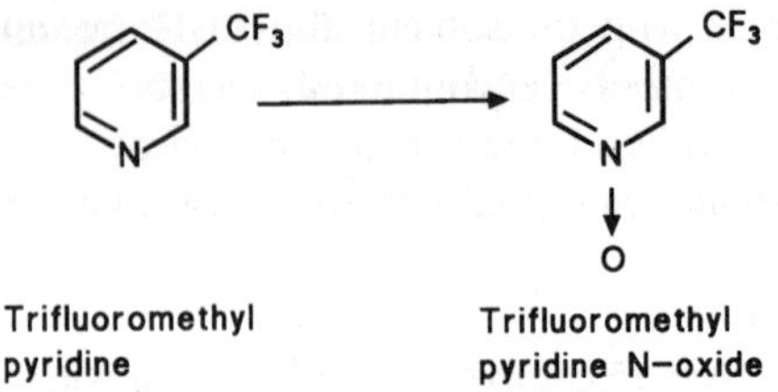

FIGURE 4.20. *N*-Oxidation of trifluoromethylpyridine.

This reaction may be involved in the toxicity of the analogue, **trifluoromethyl pyridine**, which is toxic to the nasal tissues. It is believed to be metabolized to the *N*-oxide by cytochromes P-450 present in the olfactory epithelium (figure 4.20).

N-HYDROXYLATION

N-hydroxylation of primary arylamines, arylamides and hydrazines is also catalysed by a microsomal mixed-function oxidase involving cytochromes P-450 and requiring NADPH and molecular oxygen. Thus, the *N*-hydroxylation of **aniline** is as shown in figure 4.21. The *N*-hydroxylated product, phenylhydroxylamine, is thought to be responsible for the production of **methaemoglobinaemia** after aniline administration to experimental animals. This may occur by further oxidation of phenylhydroxylamine to nitrosobenzene which may then be reduced back to phenylhydroxylamine. This reaction lowers the reduced glutathione concentration in the red blood cell, removing the protection of haemoglobin against oxidative damage.

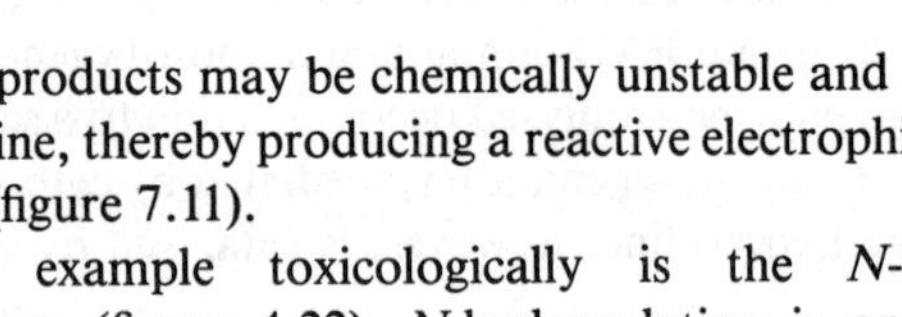

FIGURE 4.21. *N*-Hydroxylation of aniline.

N-hydroxylated products may be chemically unstable and dehydrate, as does phenylhydroxylamine, thereby producing a reactive electrophile such as an imine or imino-quinone (figure 7.11).

An important example toxicologically is the *N*-hydroxylation of **2-acetylaminofluorene** (figure 4.22). *N*-hydroxylation is one of the reactions

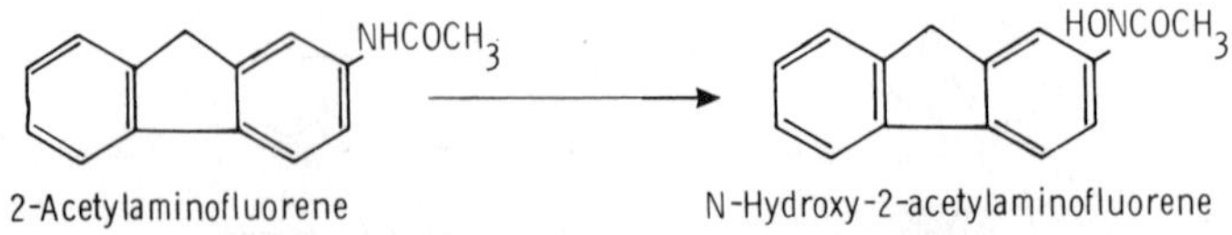

FIGURE 4.22. *N*-Hydroxylation of 2-acetylaminofluorene.

responsible for converting the compound into a potent carcinogen. A second example is the *N*-hydroxylation of **isopropylhydrazine**, thought to be involved in the production of a hepatotoxic intermediate (figure 7.16).

S-OXIDATION

Aromatic and aliphatic sulphides, thioethers, thiols, thioamides and thiocarbamates may undergo oxidation to form sulphoxides and then, after further oxidation, sulphones (figure 4.23). This is catalysed by a microsomal mono-oxygenase requiring NADPH and cytochromes P-450. The FAD-containing mono-oxygenases will also catalyse *S*-oxidation reactions.

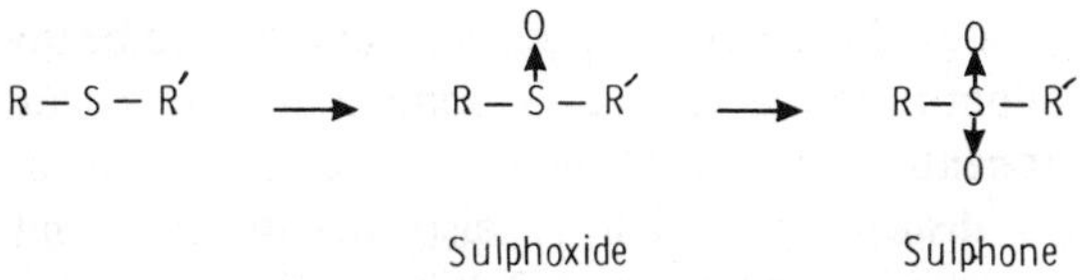

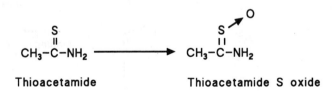

$$R - S - R' \longrightarrow R - \overset{O}{\overset{\uparrow}{S}} - R' \longrightarrow R - \overset{O}{\underset{O}{\overset{\uparrow}{\underset{\downarrow}{S}}}} - R'$$

Sulphoxide   Sulphone

FIGURE 4.23. *S*-Oxidation to form a sulphoxide and sulphone.

A number of foreign compounds, for example drugs like chlorpromazine and various pesticides such as temik undergo this reaction. An important toxicological example is the oxidation of the hepatotoxin **thioacetamide** (figure 4.24).

$$\overset{S}{\overset{\parallel}{CH_3-C-NH_2}} \longrightarrow \overset{S \nearrow^O}{\overset{\parallel}{CH_3-C-NH_2}}$$

Thioacetamide   Thioacetamide S oxide

FIGURE 4.24. *S*-Oxidation of thioacetamide.

P-OXIDATION

In an analogous manner to nitrogen and sulphur, phosphorus may also be oxidized to an oxide, as in the compound **diphenylmethylphosphine**. This is catalysed by microsomal mono-oxygenases, both cytochromes P-450 and the FAD-requiring enzyme.

DESULPHURATION

Replacement of sulphur by oxygen is known to occur in a number of cases, and the oxygenation of the insecticide parathion to give the more toxic paraoxon is a good example of this (figure 4.25). This reaction is also important for other phosphorothionate insecticides. The toxicity depends upon inhibition of cholinesterases and the oxidized product is much more potent in this respect. The reaction appears to be catalysed by either cytochromes P-450 or the FAD-

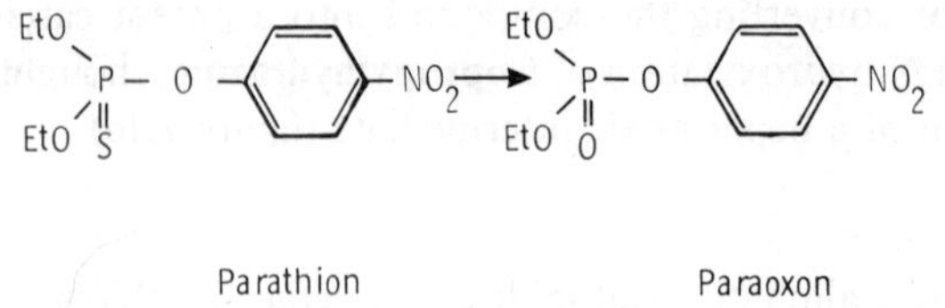

EtO — P — O —⟨benzene⟩— NO₂ ⟶ EtO — P — O —⟨benzene⟩— NO₂

Parathion   Paraoxon

FIGURE 4.25. Oxidative desulphuration of parathion.

containing mono-oxygenases and therefore requires NADPH and oxygen. The mechanism of desulphuration seems to involve formation of a phospho-oxithirane ring which rearranges with loss of 'active atomic sulphur'. This is highly reactive and is believed to bind to the enzyme and also to be involved in *toxicity*. Oxidative desulphuration at the C–S bond may also occur, such as in the barbiturate **thiopental** (figure 3.5), which is metabolized to pentobarbital. The solvent **carbon disulphide** which is also hepatotoxic undergoes oxidative metabolism catalysed by cytochromes P-450 and giving rise to carbonyl sulphide, thiocarbonate and again atomic sulphur which is thought to be involved in the toxicity (figure 4.26).

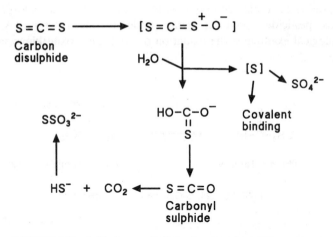

FIGURE 4.26. Oxidative metabolism of carbon disulphide.

DEAMINATION

Amine groups can be removed oxidatively via a deamination reaction which may be catalysed by cytochromes P-450. Other enzymes, such as monoamine oxidases, may also be involved in deamination reactions (see below). The product of deamination of a primary amine is the corresponding ketone. For example **amphetamine** is metabolized in the rabbit to phenylacetone (figure 4.27). The mechanism probably involves oxidation of the carbon atom to yield a carbinolamine which can rearrange to the ketone with loss of ammonia. Alternatively, the reaction may proceed via phenylacetoneoxime which has been isolated as a metabolite and for which there are several possible routes of

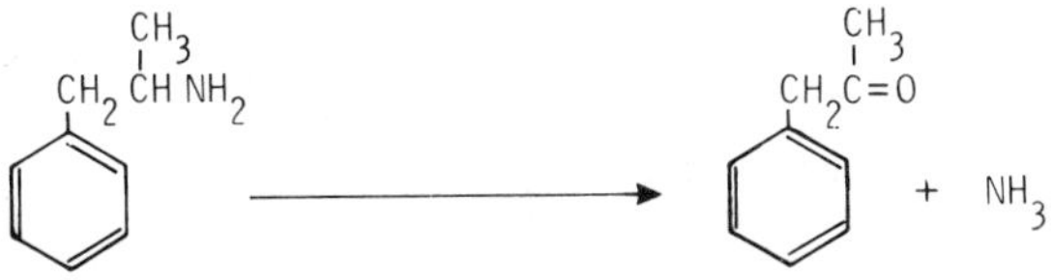

Amphetamine   Phenylacetone

FIGURE 4.27. Oxidative deamination of amphetamine.

formation. The phenylacetoneoxime is hydrolysed to phenylacetone. Also *N*-hydroxylation of amphetamine may take place and give rise to phenylacetone as a metabolite. This illustrates that there may be several routes to a particular metabolite.

OXIDATIVE DEHALOGENATION

Halogen atoms may be removed from xenobiotics in an oxidative reaction catalysed by cytochromes P-450. For example, the anaesthetic **halothane** is metabolized to trifluoroacetic acid via several steps, which involves the insertion of an oxygen atom and the loss of chlorine and bromine (figure 4.28). This is the major metabolic pathway in man and is believed to be involved in the hepatotoxicity of the drug. Trifluoroacetyl chloride is thought to be the reactive intermediate (see Chapter 7).

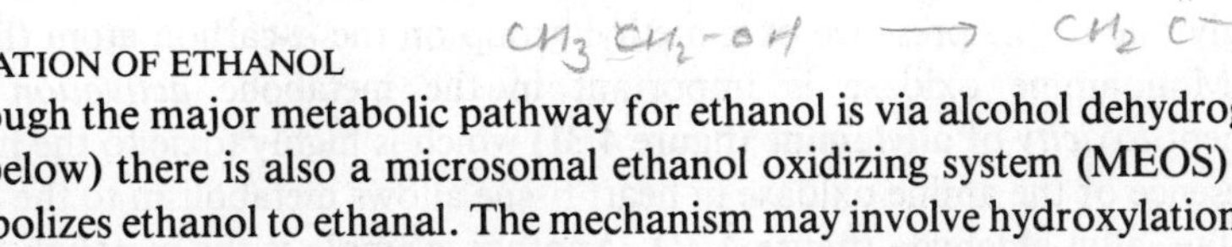

FIGURE 4.28. Oxidative metabolism of halothane.

OXIDATION OF ETHANOL

Although the major metabolic pathway for ethanol is via alcohol dehydrogenase (see below) there is also a microsomal ethanol oxidizing system (MEOS) which metabolizes ethanol to ethanal. The mechanism may involve hydroxylation at the carbon atom, although this is uncertain. Although this enzyme system is of minor importance in naive subjects, exposure to ethanol can induce the enzyme system such that it becomes the major enzyme system metabolizing ethanol.

DESATURATION OF ALKYL GROUPS

This novel reaction which converts a saturated alkyl compound into a substituted alkene and is catalysed by cytochromes P-450, has been described for the anti-epileptic drug, **valproic acid**. The mechanism proposed involves formation of a carbon-centred free radical which may form either a hydroxylated product (alcohol) or dehydrogenate to the unsaturated compound. It is believed to have a role in the hepatotoxicity of the drug (figure 4.29).

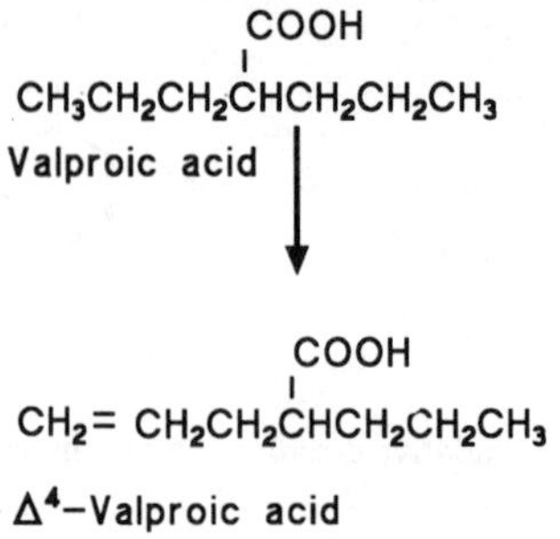

COOH
|
$CH_3CH_2CH_2CHCH_2CH_2CH_3$

Valproic acid

COOH
|
$CH_2= CH_2CH_2CHCH_2CH_2CH_3$

Δ^4-Valproic acid

FIGURE 4.29. Aliphatic desaturation of valproic acid.

Non-microsomal oxidation

AMINE OXIDATION

As well as the microsomal enzymes involved in the oxidation of amines, there are a number of other amine oxidase enzymes which have a different subcellular distribution. The most important are the monoamine oxidases and the diamine oxidases. The monoamine oxidases are located in the mitochondria within the cell and are found in the liver and also other organs such as the heart and central nervous system and in vascular tissue. They are a group of flavoprotein enzymes with overlapping substrate specificities. Although primarily of importance in the metabolism of endogenous compounds such as 5-hydroxytryptamine they may be involved in the metabolism of foreign compounds. The enzyme found in the liver will deaminate secondary and tertiary aliphatic amines as well as primary amines, although the latter are the preferred substrates and are deaminated faster. Secondary and tertiary amines are preferentially dealkylated to primary amines. For aromatic amines, such as **benzylamine**, electron withdrawing substituents on the ring will *increase* the reaction rate. The product of the reaction is an aldehyde (figure 4.30). Amines such as amphetamine are not substrates, seemingly due to the presence of a methyl group on the α-carbon atom (figure 4.27). Monoamine oxidase is important in the metabolic *activation* and subsequent *toxicity* of **allylamine** (figure 4.31) which is highly toxic to the heart. The presence of the amine oxidase in heart tissue allows metabolism to the toxic metabolite, allyl aldehyde (figure 4.31). Another example is the metabolism of MPTP to a toxic metabolite by monoamine oxidase in the central nervous system, which is discussed in more detail in Chapter 7.

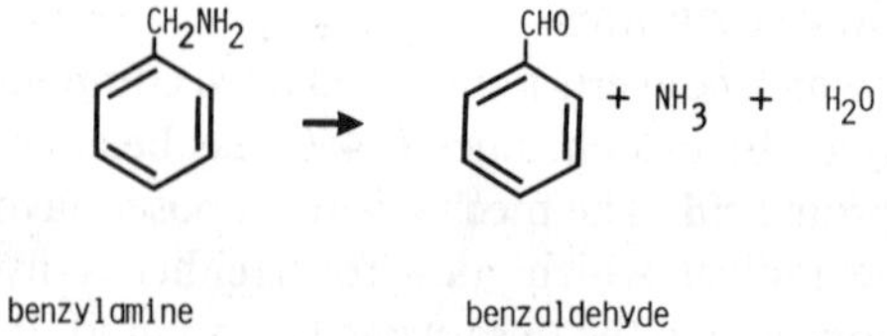

benzylamine   benzaldehyde

FIGURE 4.30. Oxidative deamination of benzylamine.

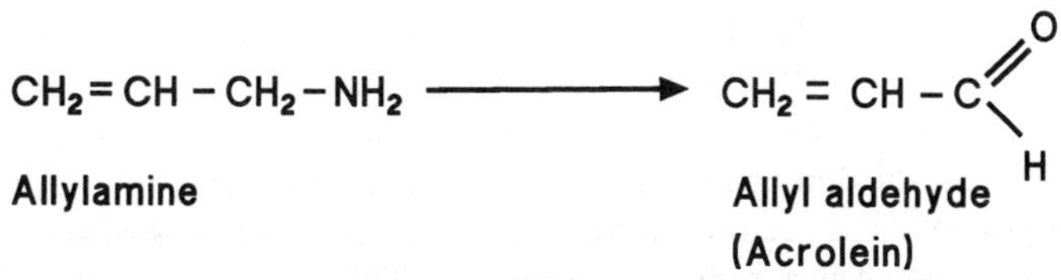

$$CH_2 = CH - CH_2 - NH_2 \longrightarrow CH_2 = CH - C{\overset{\displaystyle O}{\underset{\displaystyle H}{\diagdown}}}$$

Allylamine   **Allyl aldehyde**
                                                (Acrolein)

FIGURE 4.31. Oxidative deamination of allylamine.

Diamine oxidase, a soluble enzyme found in liver and other tissues, is mainly involved in the metabolism of endogenous compounds such as the aliphatic diamine putrescine (figure 7.26). This enzyme, which requires pyridoxal phosphate, does not metabolize secondary or tertiary amines or those with more than nine carbon atoms. The products of the reaction are aldehydes.

ALCOHOL AND ALDEHYDE OXIDATION

Although *in vitro* a microsomal enzyme system has been demonstrated which oxidizes ethanol (see above), probably the more important enzyme *in vivo* is alcohol dehydrogenase, which is found in the soluble fraction in various tissues. The coenzyme is normally NAD, and although NADP may be utilized, the rate of the reaction is slower. The enzyme is relatively non-specific and so accepts a wide variety of substrates including exogenous primary and secondary alcohols. Secondary alcohols are metabolized at a slower rate than primary alcohols and tertiary alcohols are not readily oxidized at all.

The product of the oxidation is the corresponding aldehyde if the substrate is a primary alcohol (figure 4.32) or a ketone if a secondary alcohol is oxidized. Secondary alcohols are oxidized much more slowly than primary alcohols, however. The aldehyde produced by this oxidation may be further oxidized by aldehyde dehydrogenase to the corresponding acid. This enzyme also requires NAD and is found in the soluble fraction. Alcohol dehydrogenase may have a

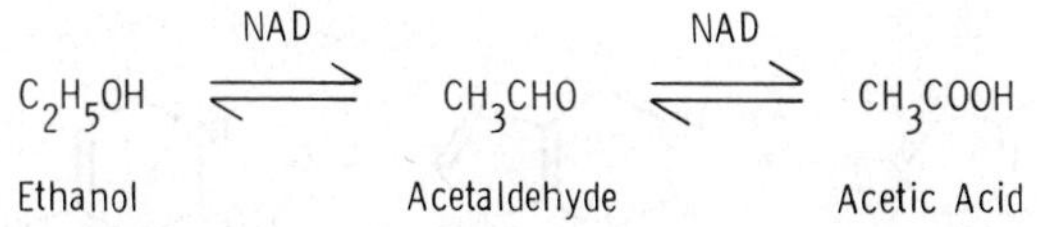

$$C_2H_5OH \overset{NAD}{\rightleftharpoons} CH_3CHO \overset{NAD}{\rightleftharpoons} CH_3COOH$$

Ethanol   Acetaldehyde   Acetic Acid

FIGURE 4.32. Oxidation of ethanol by alcohol dehydrogenase.

$$CH_2 = CH - CH_2OH \longrightarrow CH_2 = CH - C{\overset{\displaystyle O}{\underset{\displaystyle H\ (-^2H)}{\diagdown}}}$$
$$(-C^2H_2)$$

Allyl alcohol   **Allyl aldehyde**
(1,1-dideutero allyl alcohol)   **(1-deutero allyl aldehyde)**

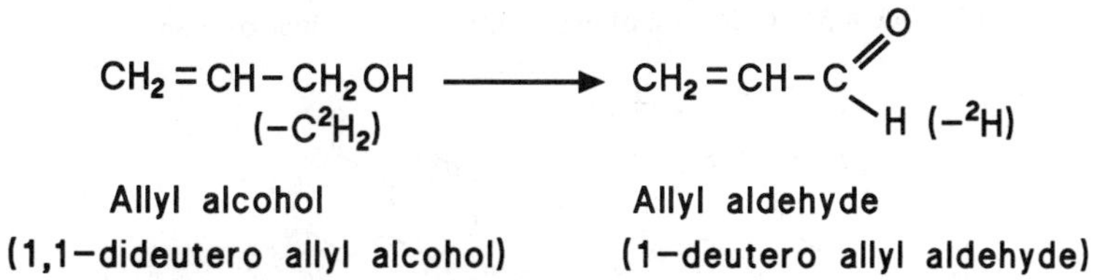

FIGURE 4.33. Oxidation of allyl alcohol by alcohol dehydrogenase. The figure also shows in brackets the position of deuterium labelling as discussed in the text.

role in the hepatotoxicity of allyl alcohol. This alcohol causes periportal necrosis (see page 200) in experimental animals; this is thought to be due to metabolism to allyl aldehyde (acrolein) (figure 4.33) in an analogous manner to allylamine (see above). The use of deuterium-labelled allyl alcohol showed that oxidation was necessary for toxicity: the replacement of the hydrogen atoms of the CH_2 group on the C^1 with deuterium reduced the toxicity. This was proposed to be due to an isotope effect as breakage of a carbon–deuterium bond is *more difficult* than breakage of a carbon–hydrogen bond. If this bond breakage is a *rate-limiting step* in the oxidation to allyl aldehyde, metabolism will be reduced.

Other enzymes may also be involved in the oxidation of aldehydes, particularly aldehyde oxidase and **xanthine oxidase**, which belong to the **molybdenum hydroxylases**. These enzymes are primarily cytosolic, although microsomal aldehyde oxidase activity has been detected. They are flavoproteins, containing FAD and also molybdenum, and the oxygen incorporated is derived from water *rather than* molecular oxygen. Aldehyde oxidase and xanthine oxidase in fact oxidize a wide variety of substrates, both aldehydes and nitrogen containing heterocycles such as caffeine and purines (see below). Aldehyde oxidase is found in highest concentrations in the liver, but xanthine oxidase is found in the small intestine, milk and the mammary gland.

PURINE OXIDATION

The oxidation of purines and purine derivatives is catalysed by xanthine oxidase. For example, the enzyme oxidizes hypoxanthine to xanthine and thence uric acid (figure 4.34). Xanthine oxidase also catalyses the oxidation of foreign compounds, such as the nitrogen heterocycle phthalazine (figure 4.35). This compound is also a substrate for aldehyde oxidase, giving the same product.

AROMATIZATION OF ALICYCLIC COMPOUNDS

Cyclohexane carboxylic acids may be metabolized by a mitochondrial enzyme system to an aromatic acid such as benzoic acid. This enzyme system requires

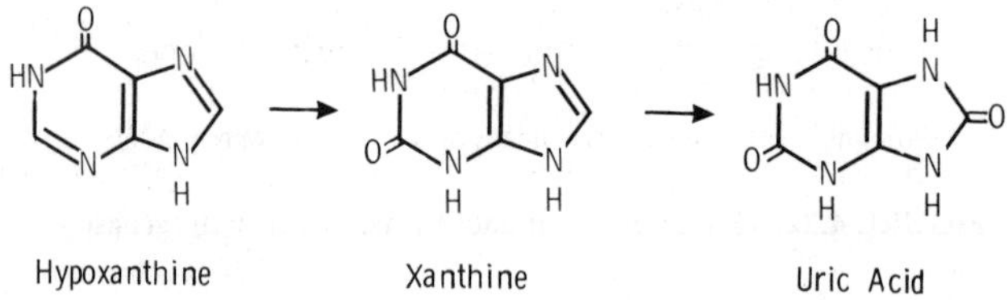

Hypoxanthine   Xanthine   Uric Acid

FIGURE 4.34. Oxidation of hypoxanthine by xanthine oxidase.

Phthalazine   Phthalazinone

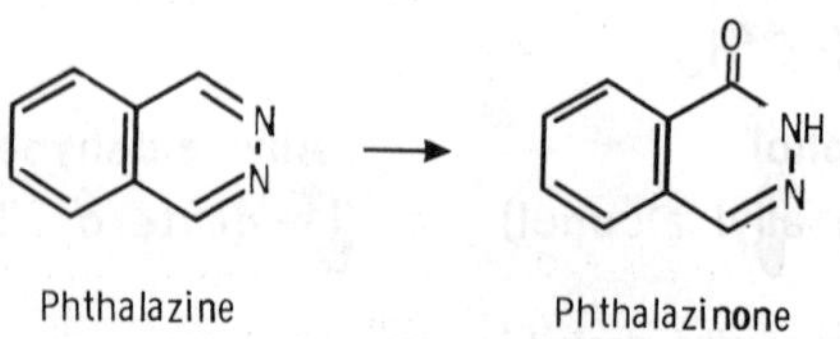

FIGURE 4.35. Oxidation of phthalazine by xanthine oxidase.

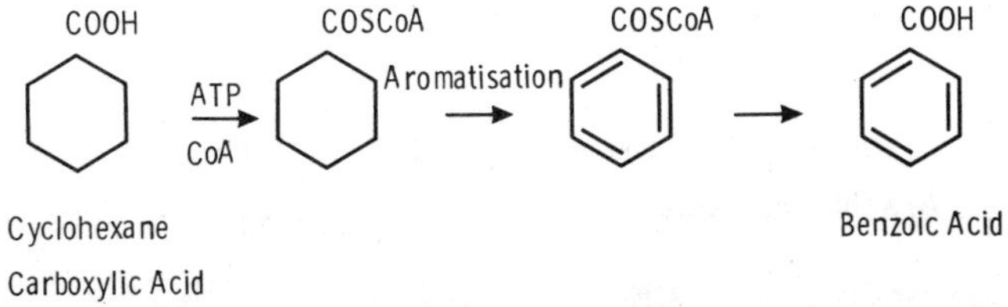

FIGURE 4.36. Aromatization of cyclohexane carboxylic acid.

CoA, ATP and oxygen and is thought to involve three sequential dehydrogenation steps after the initial formation of the cyclohexanoyl CoA (figure 4.36). The **aromatase** enzyme also requires the cofactor FAD.

PEROXIDASES

Another group of enzymes which are involved in the oxidation of xenobiotics are the peroxidases. There are a number of these enzymes in mammalian tissues: **prostaglandin synthase** found in many tissues, but especially seminal vesicles and also kidney, lung, intestine spleen and blood vessels; **lactoperoxidase** found in mammary glands; **myeloperoxidase** found in neutrophils, macrophages, liver Kupffer cells and bone marrow cells.

The overall peroxidase catalysed reaction may be summarized as follows:

$$\text{Peroxidase} + H_2O_2 \;\rightarrow\; \text{Compound I}$$
$$\text{Compound I} + RH_2 \;\rightarrow\; \text{Compound II} + \cdot RH_2{}^+$$
$$\text{Compound II} + RH_2 \rightarrow \text{Peroxidase} + \cdot RH_2{}^+$$

The haem iron in the peroxidase is oxidized by the peroxide from III^+ to V^+ in compound I. The compound I is reduced by two sequential one-electron transfer processes giving rise to the original enzyme. A substrate free radical is in turn generated. This may have toxicological implications. Thus the myeloperoxidase in the bone marrow may catalyse the metabolic activation of phenol or other metabolites of benzene. This may underlie the toxicity of **benzene** to the bone marrow which causes aplastic anaemia.

Similarly, uterine peroxidase has been suggested as being involved in the metabolic activation and toxicity of **diethylstilboestrol** (see Chapter 6 and figure 6.19). Probably the most important peroxidase enzyme system is prostaglandin synthase. This enzyme is found in many tissues, including kidney and seminal vesicles. It is a glycoprotein which is located in the endoplasmic reticulum. This enzyme system is involved in the oxygenation of polyunsaturated fatty acids and the biosynthesis of prostaglandins. The oxidation of arachidonic acid to prostaglandin H_2 is an important step in the latter (figure 4.37). The enzyme catalyses two steps, first the formation of a hydroperoxy endoperoxide, prostaglandin G_2, then metabolism to a hydroxy metabolite, prostaglandin H_2. In the second step, xenobiotics may be co-oxidized. There are a number of examples of xenobiotic metabolism catalysed by this system, such as the oxidation of *p*-**phenetidine**, a metabolite of the drug phenacetin (see Chapter 5

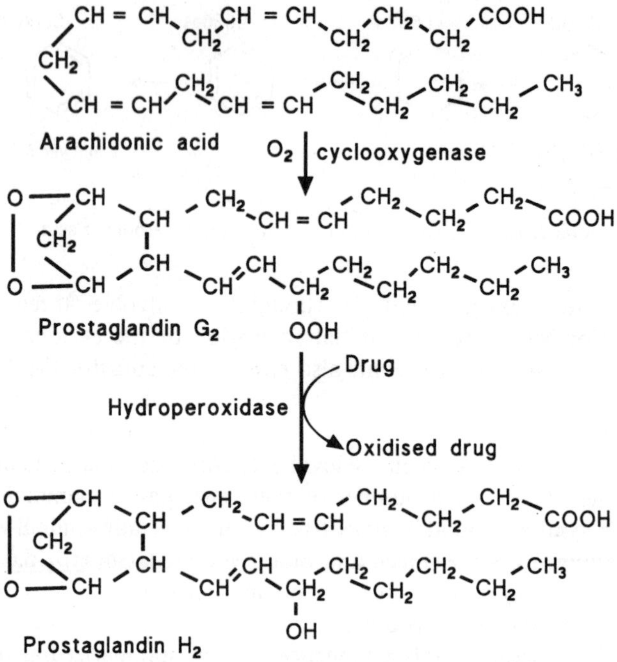

FIGURE 4.37. Co-oxidation of a drug by the prostaglandin synthetase enzyme system.

and figure 5.20), a process which may be involved in the *nephrotoxicity* of the drug. The prostaglandin synthase catalysed oxidation of this compound gives rise to free radicals which may be responsible for binding to DNA. Horseradish peroxidase will also catalyse the oxidation of *p*-phenetidine. **Paracetamol** can also be oxidized by prostaglandin synthase to a free radical intermediate, a semiquinone, which may be involved in the toxicity and yields a glutathione conjugate which is the same as that produced via the cytochromes P-450-mediated pathway (for more details see Chapter 7). Other examples of metabolic pathways catalysed by this pathway are *N*-demethylation of aminopyrine, formation of an aromatic epoxide of 7,8-dihydroxy, 7,8-dihydro,-benzo[*a*]pyrene and sulphoxidation of methyl phenyl sulphide. The exact mechanism of the co-oxidation is unclear at present, but may involve formation of free radicals via one-electron oxidation/hydrogen abstraction as with paracetamol (figure 7.11) or the direct insertion of oxygen as with benzo[*a*]pyrene. It is not yet clear how significant prostaglandin synthase catalysed routes of metabolism are in comparison with microsomal monooxygenase-mediated pathways, but they may be very important in tissues where the latter enzymes are not abundant.

The possible role of peroxidases in metabolic activation and cytotoxicity has only relatively recently attracted attention, but may prove to be of particular importance in underlying mechanisms of toxicity. It has been suggested, for example, that a number of adverse drug reactions are due to metabolism via peroxidases in leucocytes and other similar cells. Thus, leucocytes will metabolize the drug **procainamide** (figure 4.42) to a hydroxylamine metabolite. This is thought to be catalysed by a myeloperoxidase. In the presence of chloride ion, peroxidases such as myeloperoxidase also produce an oxidant similar to hypochlorite which may be involved in cytotoxicity and metabolism. Myeloperoxidase may thus give rise to chlorinated metabolites such as *N*-chloroprocainamide. For more details the reader is directed to the review by Uetrecht in the Bibliography.

Reduction

The enzymes responsible for reduction may be located in both the microsomal fraction and the soluble cell fraction. **Reductases** in the microflora present in the gastrointestinal tract may also have an important role in the reduction of xenobiotics. There are a number of different reductases which can catalyse the reduction of azo and nitro compounds. Thus, in the microsomal fraction, cytochromes P-450 and possibly a flavoprotein are capable of reductase activity. NADPH is required, but the reaction is inhibited by oxygen. FAD alone may also catalyse reduction by acting as an electron donor.

Reduction of the azo dye **prontosil** to produce the antibacterial drug sulphanilamide (figure 4.38) is a well-known example of azo reduction. This reaction is catalysed by cytochromes P-450 and is also carried out by the reductases in the gut bacteria. The reduction of azo groups in food colouring dyes such as **amaranth** is catalysed by several enzymes, including cytochromes P-450, NADPH cytochrome P-450 reductase and **DT-diaphorase**, a cytosolic enzyme.

The reduction of nitro groups may also be catalysed by microsomal reductases and gut bacterial enzymes. The reduction passes through several stages to yield the fully reduced primary amine, as illustrated for **nitrobenzene** (figure 4.38). The intermediates are nitrosobenzene and phenylhydroxylamine which are also

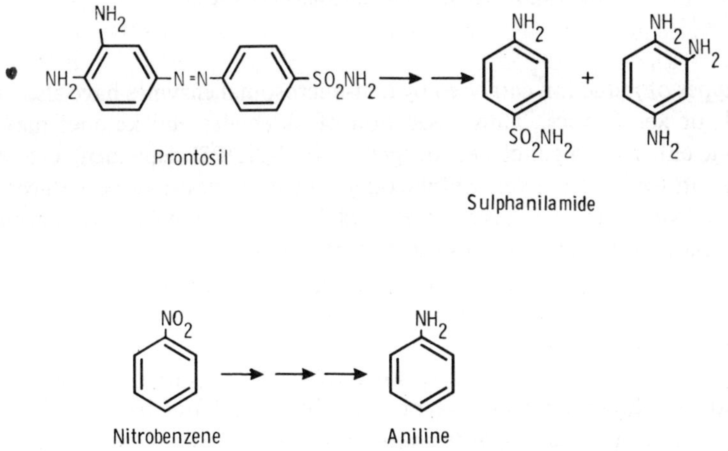

FIGURE 4.38. Reduction of the azo group in prontosil and the nitro group in nitrobenzene.

reduced in the microsomal system. These intermediates, which may also be produced by the oxidation of aromatic amines (figure 4.21), are involved in the toxicity of nitrobenzene to red blood cells after oral administration to rats. The importance of the gut bacterial reductases in this process is illustrated by the drastic reduction in nitrobenzene toxicity in animals devoid of gut bacteria, or when nitrobenzene is given by the intraperitoneal route.

Arylhydroxylamines, whether derived from nitro compounds by reduction or amines by *N*-hydroxylation, have been shown to be involved in the toxicity of a number of compounds. The possibility of reduction followed by oxidation in a cyclical fashion with the continual production of toxic metabolites has important toxicological implications.

Tertiary amine oxides and hydroxylamines are also reduced by cytochromes P-450. Hydroxylamines as well as being reduced by cytochromes P-450 are also reduced by a flavoprotein which is part of a system which requires NADH and includes NADH cytochrome b_5 reductase and cytochrome b_5.

Quinones such as the anti-cancer drug **adriamycin** can undergo one-electron reduction catalysed by NADPH cytochrome P-450 reductase. The semiquinone product may be oxidized back to the quinone with the concomitant production of superoxide anion radical, giving rise to **redox cycling** and potential cytotoxicity (see Chapter 6). An important example is the reduction of nitroquinoline *N*-oxide. This proceeds via the hydroxylamine, which is an extremely carcinogenic metabolite, probably the ultimate carcinogen (figure 4.39).

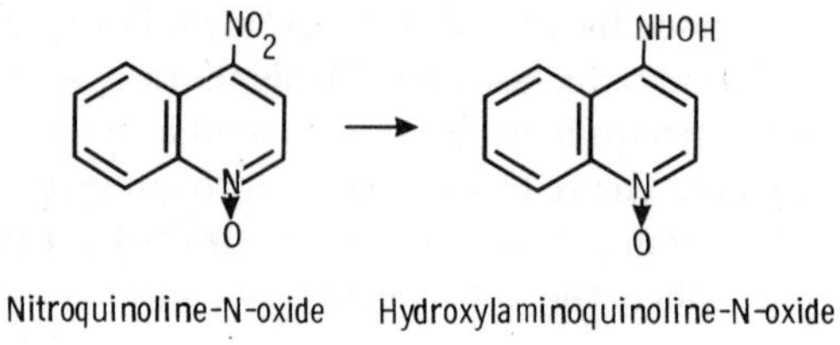

Nitroquinoline-N-oxide   Hydroxylaminoquinoline-N-oxide

FIGURE 4.39. Reduction of nitroquinoline *N*-oxide.

Other types of reduction catalysed by non-microsomal enzymes have also been described for xenobiotics. Thus, reduction of aldehydes and ketones may be carried out either by alcohol dehydrogenase or NADPH-dependent cytosolic reductases present in the liver. Sulphoxides and sulphides may be reduced by cytosolic enzymes, in the latter case involving glutathione and glutathione reductase. Double bonds in unsaturated compounds and epoxides may also be reduced.

REDUCTIVE DEHALOGENATION
The microsomal enzyme-mediated removal of a halogen atom from a foreign compound may be *either* reductive or oxidative. The latter has already been discussed with regard to the volatile anaesthetic **halothane** which undergoes both oxidative and reductive dehalogenation and which may be involved in the

hepatotoxicity (see figures 4.28 and 7.51). Reductive dehalogenation of halothane is catalysed by cytochromes P-450 under anaerobic conditions and may lead to reactive radical metabolites (figures 4.40 and 7.51). Another example of reductive dehalogenation of toxicological importance, also catalysed by cytochromes P-450, is the metabolic activation of **carbon tetrachloride** by dechlorination to yield a free radical (see figure 7.51). In both these examples the substrate binds to the cytochromes P-450 and then receives an electron from NADPH cytochrome P-450 reductase. The enzyme–substrate complex then loses a halogen ion and a free radical intermediate is generated. Alternatively, the enzyme–substrate complex may be further reduced by another electron from NADH via NADH cytochrome b_5 reductase and cytochrome b_5. The carbanion or carbene intermediates may rearrange with loss of a halogen ion (figure 7.51).

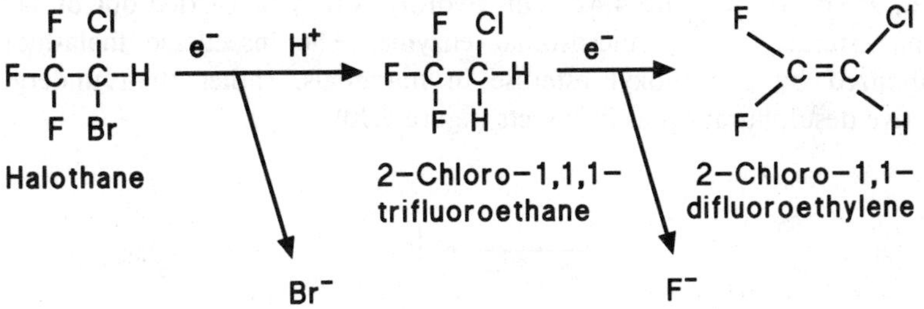

FIGURE 4.40. Reductive dehalogenation of halothane.

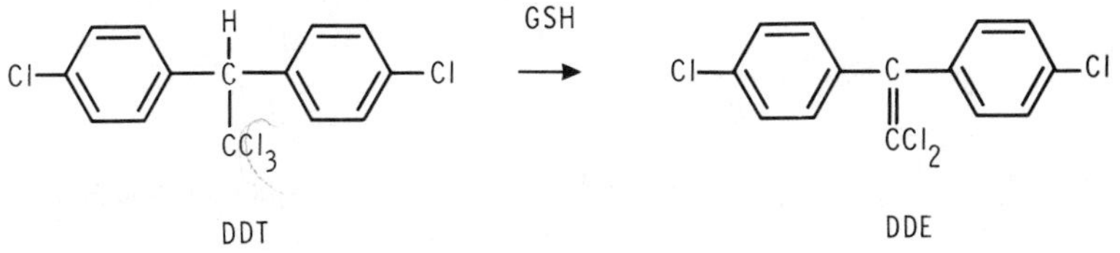

FIGURE 4.41. Dehydrohalogenation of DDT.

Dehalogenation of the insecticide **DDT** is catalysed by a soluble enzyme and requires glutathione. The overall reaction is a dehydrohalogenation and yields DDE (figure 4.41).

Hydrolysis

Esters, amides, hydrazides and carbamates can all be metabolized by hydrolysis. The enzymes which catalyse these hydrolytic reactions, carboxylesterases and amidases, are usually found in the cytosol, but microsomal esterases and amidases have been described and some are also found in the plasma. The various enzymes have different substrate specificities, but carboxylesterases have amidase activity and amidases have esterase activity. The two apparently different activities may therefore be part of the same overall activity.

Hydrolysis of esters

Various **esterases** exist in mammalian tissues, hydrolysing different types of esters. They have been classified as type A, B or C on the basis of activity towards phosphate triesters. A-Esterases, which include arylesterases, are not inhibited by phosphotriesters and will metabolize them by hydrolysis. B-Esterases are inhibited by paraoxon and have a serine group in the active site (see Chapter 7). Within this group are carboxylesterases, cholinesterases and arylamidases. C-Esterases are also not inhibited by paraoxon and the preferred substrates are acetyl esters, hence these are acetylesterases. Carboxythioesters are also hydrolysed by esterases. Other enzymes such as trypsin and chymotrypsin may also hydrolyse certain carboxyl esters.

Metabolism of the local anaesthetic **procaine** provides an example of esterase action, as shown in figure 4.42. This hydrolysis may be carried out by both a plasma esterase and a microsomal enzyme. The insecticide malathion is metabolized by a carboxyl esterase in mammals, rather than undergoing oxidative desulphuration as in insects (figure 5.10).

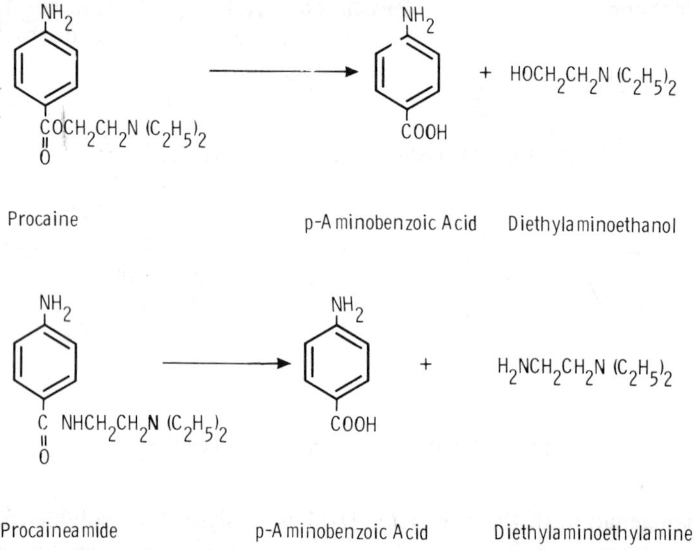

Procaine   p-Aminobenzoic Acid   Diethylaminoethanol

Procaineamide   p-Aminobenzoic Acid   Diethylaminoethylamine

FIGURE 4.42. Hydrolysis of the ester procaine and the amide procainamide.

Hydrolysis of amides

The amidase-catalysed hydrolysis of amides is rather slower than that of esters. Thus, unlike procaine, the analogue **procainamide** is not hydrolysed in the plasma at all, the hydrolysis *in vivo* being carried out by enzymes in other tissues (figure 4.42).

The hydrolysis of some amides may be catalysed by a liver microsomal carboxyl esterase, as is the case with **phenacetin** (figure 4.43). Hydrolysis of the acetylamino group resulting in deacetylation is known to be important in the

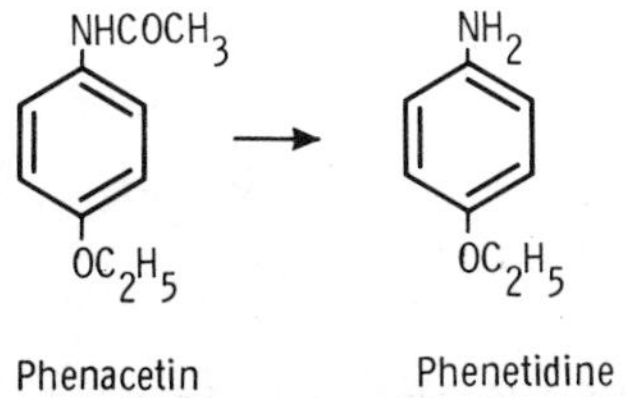

FIGURE 4.43. Deacetylation of phenacetin.

toxicity of a number of compounds. For example, the deacetylated metabolites of phenacetin are thought to be responsible for its toxicity, the oxidation of haemoglobin to methaemoglobin. This toxic effect occasionally occurs in subjects taking therapeutic doses of the drug who have a deficiency in the normal pathway of metabolism of phenacetin to paracetamol. Consequently, more phenacetin is metabolized by deacetylation and subsequent oxidation to toxic metabolites (figure 5.20).

Hydrolysis of hydrazides

The drug **isoniazid** (isonicotinic acid hydrazide) is hydrolysed *in vivo* to the corresponding acid and hydrazine, as shown in figure 4.44. However, in man, *in vivo*, hydrolysis of the acetylated metabolite acetylisoniazid is quantitatively more important and toxicologically more significant (figure 4.45 and see Chapter 7). This hydrolysis reaction accounts for about 45% of the acetylisoniazid produced. These hydrolysis reactions are probably catalysed by amidases and are inhibited by organophosphorus compounds such as **bis-*p*-nitrophenyl phosphate** (figure 5.32).

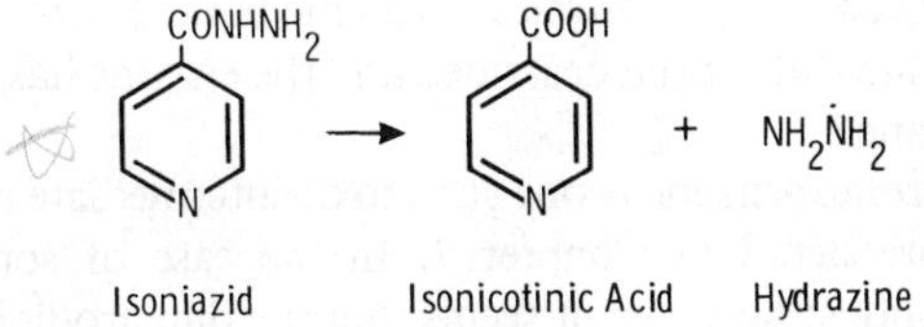

Isoniazid   Isonicotinic Acid   Hydrazine

FIGURE 4.44. Hydrolysis of isoniazid.

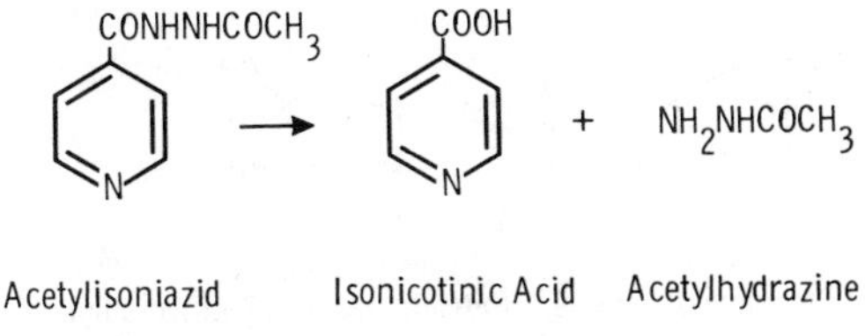

Acetylisoniazid   Isonicotinic Acid   Acetylhydrazine

FIGURE 4.45. Hydrolysis of acetylisoniazid.

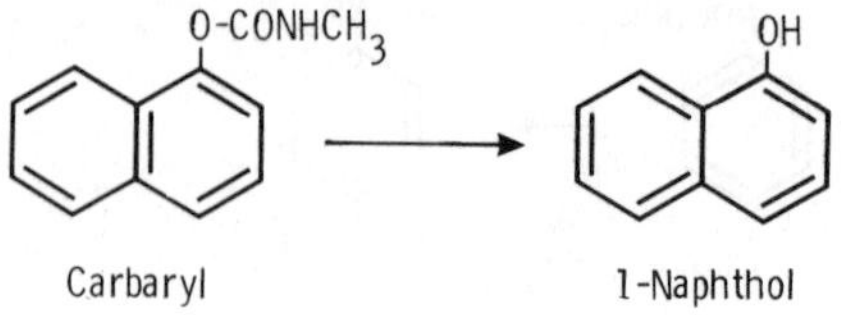

Carbaryl   1-Naphthol

FIGURE 4.46. Hydrolysis of carbaryl.

Hydrolysis of carbamates

The insecticide **carbaryl** is hydrolysed by liver enzymes to 1-naphthol (figure 4.46). This compound also undergoes extensive metabolism by other routes.

Hydration of epoxides

Epoxides, three-membered oxirane rings containing an oxygen atom, may be metabolized by the enzyme **epoxide hydrolase**. This enzyme adds water to the epoxide, probably by *nucleophilic attack* by OH^- on one of the carbon atoms of the oxirane ring, which may be regarded as electron deficient, to yield a dihydrodiol which is predominantly *trans* (figure 4.47) although the degree of stereospecificity is variable. Epoxides are often intermediates produced by the oxidation of unsaturated double bonds, aromatic, aliphatic or heterocyclic, as for example takes place during the hydroxylation of bromobenzene, the hepatotoxic solvent (figure 7.12).

The enzyme exists in multiple forms with different substrate specificities. It is found mainly in the endoplasmic reticulum in close proximity to cytochromes P-450, and like the latter is also present in greater amounts in the centrilobular areas of the liver. Epoxide hydrolase is therefore well placed to carry out its important role in detoxifying the chemically unstable and often toxic epoxide intermediates produced by cytochromes P-450 mediated hydroxylation. Soluble epoxide hydrolases have also been described and the enzyme has been detected in the nuclear membrane.

The epoxide of bromobenzene is one such toxic intermediate and this example is discussed in more detail in Chapter 7. In the case of some carcinogenic polycyclic hydrocarbons, however, it seems that the dihydrodiol products are in turn further metabolized to epoxide-diols, the ultimate carcinogens (see pages 288–90).

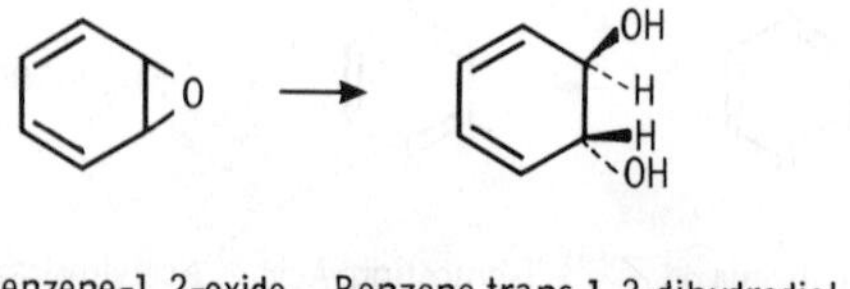

Benzene-1,2-oxide   Benzene trans 1,2-dihydrodiol

FIGURE 4.47. Hydration of benzene-1,2-oxide by epoxide hydrolase.

Phase 2 reactions

Conjugation

Conjugation reactions involve the addition to foreign compounds of endogenous groups which are generally polar and readily available *in vivo*. These groups are added to a suitable functional group present on the foreign molecule or introduced by phase 1 metabolism. This renders the whole molecule more polar and less lipid soluble, thus facilitating excretion and reducing the likelihood of toxicity. The endogenous groups donated in conjugation reactions include carbohydrate derivatives, amino acids, glutathione and sulphate. The mechanism commonly involves formation of a high energy intermediate, where either the endogenous metabolite or the foreign compound is activated (type 1 and type 2, respectively). The groups donated in conjugation reactions are often involved in intermediary metabolism.

Glucuronide formation

This is a major, type 1, conjugation reaction occurring in most species with a wide variety of substrates, including endogenous substances. It involves the transfer of glucuronic acid in an activated form as **uridine diphosphate glucuronic acid (UDPGA)** to hydroxyl, carboxyl, nitrogen sulphur and occasionally carbon atoms. The UDPGA is formed in the cytosol from glucose-1-phosphate in a two-step reaction (figure 4.48). The first step, addition of the UDP, is catalysed by UDP glucose pyrophosphorylase, the second step by UDP glucose dehydrogenase.

The enzyme catalysing the conjugation reaction, **UDP-glucuronosyl transferase** (glucuronyl transferase), exists in possibly four or more forms, each

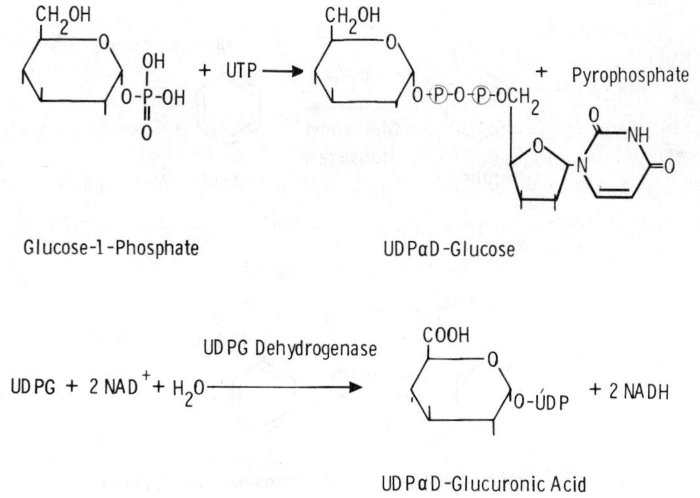

FIGURE 4.48. Formation of uridine diphosphate glucuronic acid (UDPGA).

with different substrate specificities. The following are examples of substrates for the different forms although some substrates may be glucuronidated by more than one form of the enzyme: form A 1-napthol; form B bilirubin; form C oestrone, form D morphine. The enzymes are located in the endoplasmic reticulum and are found in many tissues including the liver.

Conjugation with glucuronic acid involves nucleophilic attack by the oxygen, sulphur or nitrogen atom at the C-1 carbon atom of the glucuronic acid moiety. Glucuronides are therefore generally β in configuration. Conjugation with hydroxyl groups gives ether glucuronides and with carboxylic acids, ester glucuronides (figure 4.49). Amino groups may be conjugated directly, as in the case of aniline (figure 4.50) or through an oxygen atom as in the case of *N*-

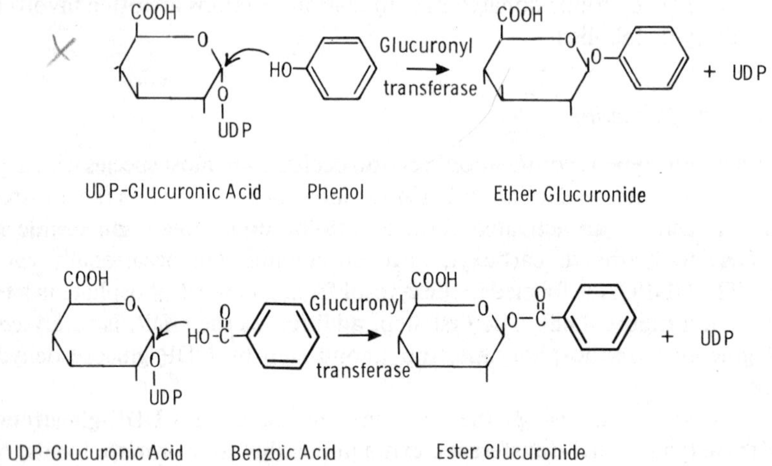

UDP-Glucuronic Acid   Phenol   Ether Glucuronide

UDP-Glucuronic Acid   Benzoic Acid   Ester Glucuronide

FIGURE 4.49. Formation of ether and ester glucuronides of phenol and benzoic acid respectively.

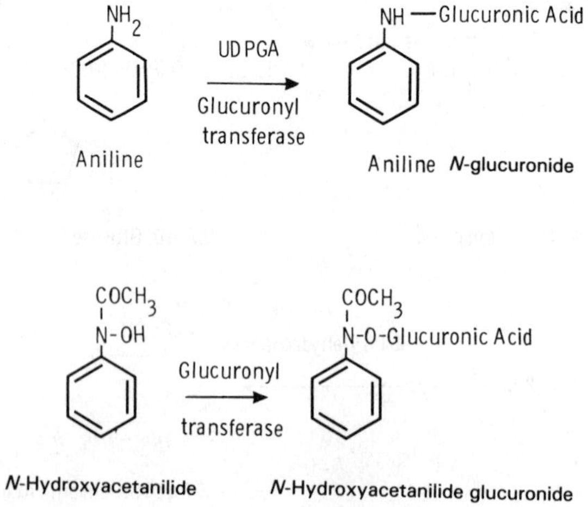

Aniline   Aniline *N*-glucuronide

N-Hydroxyacetanilide   *N*-Hydroxyacetanilide glucuronide

FIGURE 4.50. Glucuronidation of aniline and *N*-hydroxyacetanilide.

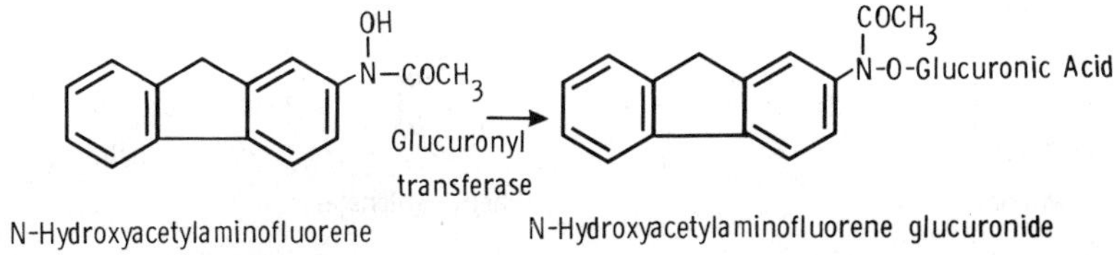

FIGURE 4.51. Formation of *N*-hydroxyacetylaminofluorene glucuronide.

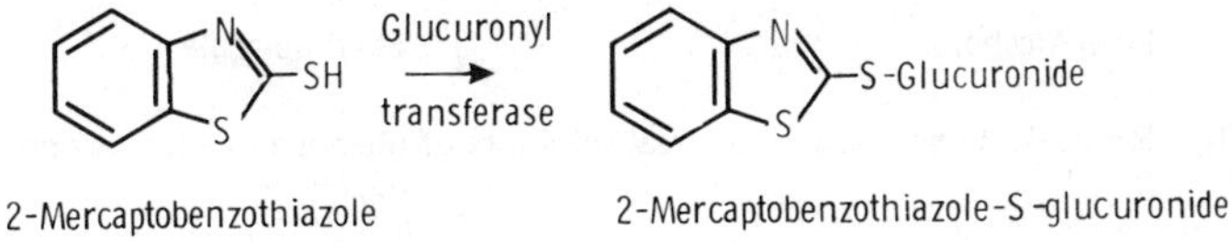

FIGURE 4.52. Formation of 2-mercaptobenzothiazole-*S*-glucuronide.

hydroxy compounds, such as the carcinogen *N*-hydroxyacetylaminofluorene (figure 4.51).

Certain thiols may be conjugated directly through the sulphur atom (figure 4.52). Glucuronic acid conjugated directly to carbon atoms has been reported such as with the drug phenylbutazone.

Although conjugation generally decreases biological activity, including toxicity, occasionally the latter is increased, as in the case of acetylamino-fluorene. The *N*-hydroxyglucuronide is a more potent carcinogen (figure 4.51).

Glucose conjugates (glucosides) may sometimes be formed, especially in insects, and the mechanism is analogous to that involved in the formation of glucuronides. Xylose and ribose are also sometimes utilized, for example, 2-hydroxynicotinic acid has been shown to form an *N*-ribose conjugate. Analogues of purines and pyrimidines may be conjugated with ribose or ribose phosphates to give ribonucleotides and ribonucleosides.

Sulphate conjugation

The formation of sulphate esters is a major route of conjugation for various types of hydroxyl group, and may also occur with amino groups. Thus, substrates include aliphatic alcohols, phenols, aromatic amines and also endogenous compounds such as steroids and carbohydrates (figure 4.53 and 4.54).

The sulphate donor for this type 1 reaction is in an activated form, as **3′-phosphoadenosine-5′-phosphosulphate (PAPS)** (figure 4.55), formed from inorganic sulphate and ATP:

$$SO_4^{2-} + ATP \xrightarrow{\text{ATP-sulphurylase}} \text{Adenosine-5′-phosphosulphate (APS)} + PPi$$

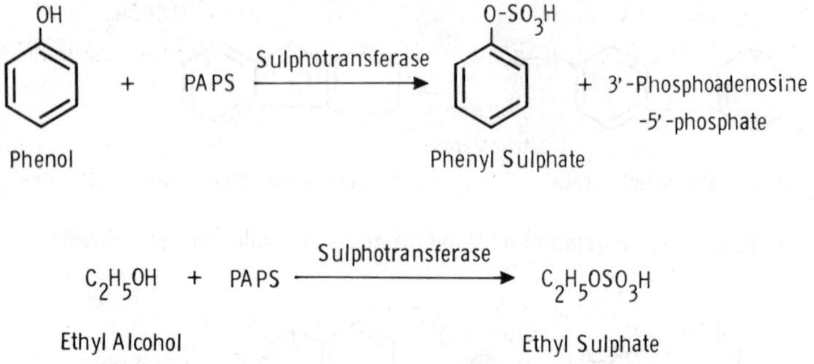

FIGURE 4.53. Formation of ethereal sulphates of phenol and ethyl alcohol.

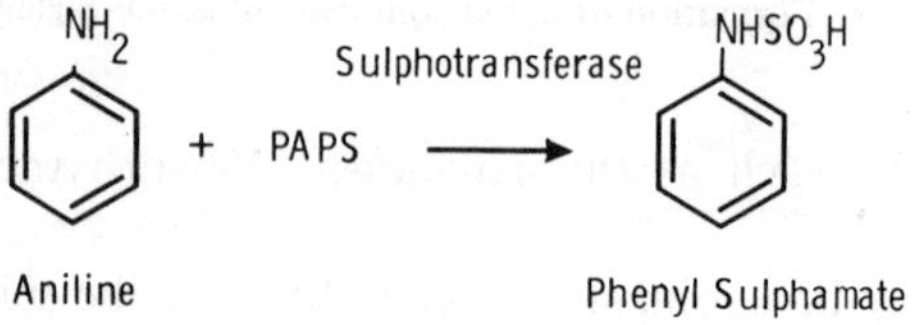

FIGURE 4.54. Sulphate conjugation of aniline.

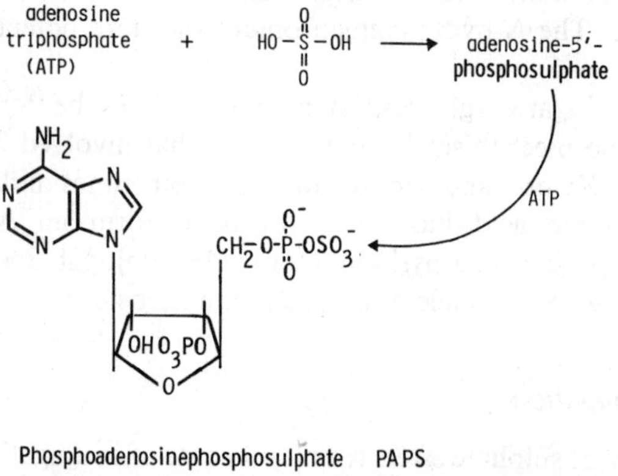

Phosphoadenosinephosphosulphate   PAPS

FIGURE 4.55. Formation of 3-phosphoadenosine-5′-phosphosulphate (PAPS).

$$\text{APS} + \text{ATP} \xrightarrow{\text{APS-kinase}} 3'\text{-Phosphoadenosine-5}'\text{-phosphosulphate (PAPS)}$$
$$+ \text{ADP}$$

The conjugation is catalysed by a **sulphotransferase** enzyme which is located in the cytosol and is found particularly in the liver, gastrointestinal mucosa and kidney. There are several different sulphotransferases, classified by the particular substrate type: arylsulphotransferase, hydroxysteroid sulphotransferase, estrone sulphotransferase and bile salt sulphotransferase. Some of these have been separated into different forms which catalyse the sulphate conjugation of various different substrates such as phenols, arylamines and alcohols.

The inorganic sulphate precursor of PAPS may become depleted when large amounts of a foreign compound conjugated with sulphate, such as paracetamol, are administered. Sulphate conjugation may increase toxicity in certain rare cases as with the conjugation of *N*-hydroxyacetylaminofluorene, the carcinogen (see Chapter 7, figure 7.1).

Glutathione conjugation

Glutathione is one of the most important molecules in the *cellular defence* against toxic compounds. This *protective function* is due in part to its involvement in conjugation reactions, and a number of toxicological examples of this such as bromobenzene and paracetamol hepatotoxicity are discussed later (see Chapter 7). The other protective functions of glutathione are discussed in Chapter 6. Glutathione is a tripeptide (figure 4.56), composed of glutamic acid, cysteine and glycine (glu–cys–gly). It is found in most cells, but is especially abundant in the liver where it reaches a concentration of 5 mM or more in mammals. The presence of cysteine provides a sulphydryl group, which is nucleophilic and so glutathione will react, probably as the thiolate ion, GS^-, with electrophiles. These electrophiles may be chemically reactive, metabolic products of a phase 1 reaction, or they may be more stable foreign compounds which have been ingested. Thus, glutathione protects cells by removing reactive metabolites. Unlike glucuronic acid or sulphate conjugation, however (type 1 conjugation reactions), the conjugating moiety (glutathione) is not activated in some high energy form. Rather the substrate is often in an activated form. Glutathione conjugation *may* be an enzyme-catalysed reaction or *simply* a chemical reaction. The glutathione conjugate produced by the reaction may then either be excreted, usually into the bile rather than the urine, or the conjugate may be further metabolized. This involves several steps: removal of the glutamyl and glycinyl groups and acetylation of the cysteine amino group to yield a mercapturic acid or *N*-acetylcysteine conjugate. This is illustrated for the compound **naphthalene** which is metabolized by cytochromes P-450 to a reactive epoxide intermediate, then conjugated with glutathione and eventually excreted as an *N*-acetylcysteine

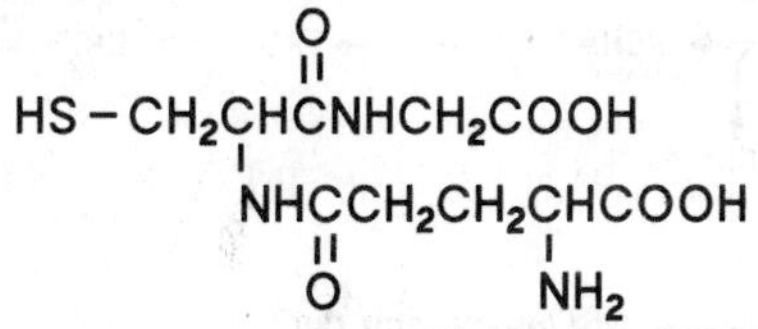

$$
\begin{array}{c}
\quad\quad\quad\quad O \\
\quad\quad\quad\quad \| \\
HS-CH_2CHCNHCH_2COOH \\
\quad\quad\quad | \\
\quad\quad\quad NHCCH_2CH_2CHCOOH \\
\quad\quad\quad \|\quad\quad\quad\quad | \\
\quad\quad\quad O\quad\quad\quad\quad NH_2
\end{array}
$$

Glutathione (γ–glutamylcysteinylglycine)

FIGURE 4.56. Structure of glutathione.

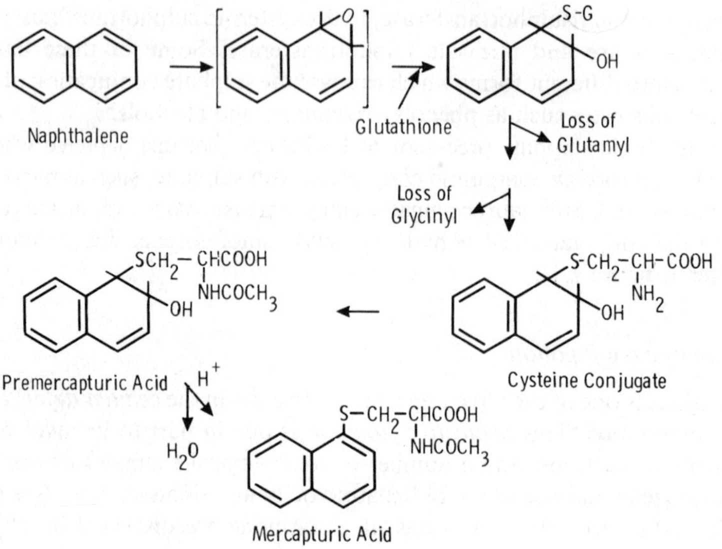

FIGURE 4.57. Conjugation of naphthalene-1,2-oxide with glutathione and formation of naphthalene mercapturic acid.

conjugate (figure 4.57). This sequence of further catabolic steps has been termed **phase** 3 metabolism. Naphthalene is toxic to the lung and these metabolic pathways are important in this toxicity (see below).

There are many types of substrates for glutathione conjugation including aromatic, aliphatic, heterocyclic and alicyclic epoxides, halogenated aliphatic and aromatic compounds, aromatic nitro compounds, unsaturated aliphatic

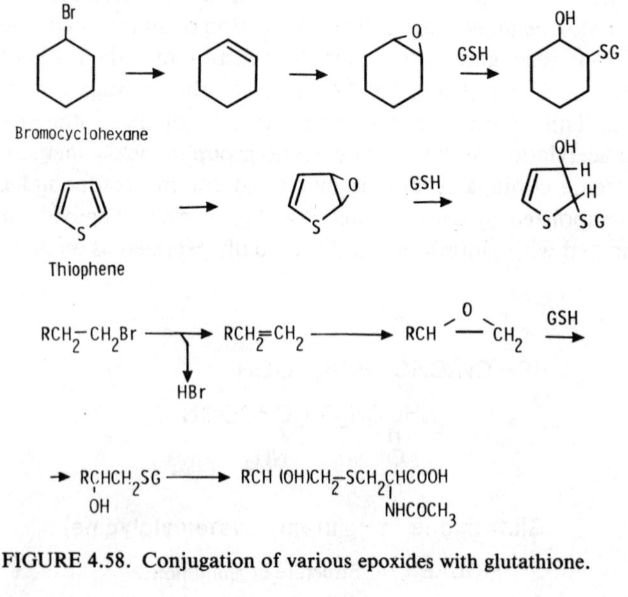

FIGURE 4.58. Conjugation of various epoxides with glutathione.

compounds and alkyl halides (figures 4.58–4.60). In each case the glutathione is reacting with an electrophilic carbon atom in an addition or substitution reaction. With the reactive epoxides the two carbon atoms of the oxirane ring will be electrophilic and suitable for reaction with glutathione. Such a reaction occurs with bromobenzene and the polycyclic hydrocarbon benzo[*a*]pyrene (see Chapter 7). With aromatic and aliphatic halogen compounds and aromatic nitro compounds, the nucleophilic sulphydryl group of the glutathione attacks the electrophilic carbon atom to which the electron withdrawing halogen or group is attached, and the latter is replaced by glutathione (figure 4.59). With unsaturated compounds such as **diethylmaleate**, the electron withdrawing substituents allow nucleophilic attack on one of the unsaturated carbon atoms and addition of a proton to the other, leading to an addition reaction (figure 4.60). Diethylmaleate reacts readily with glutathione *in vivo* and has been used to deplete tissue levels of the tripeptide. The reaction may be catalysed by one of the **glutathione-*S*-transferases**. These are cytosolic enzymes, although also detectable in the microsomal fraction, and are found in many tissues but particularly liver, kidney, gut, testis and adrenal gland. There are at least six isoenzymes of glutathione transferase, each having specificity for a particular substrate type and there are

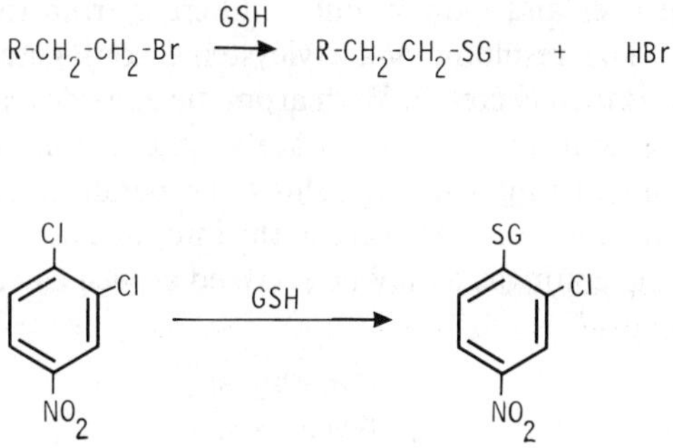

$$R\text{-}CH_2\text{-}CH_2\text{-}Br \xrightarrow{\text{GSH}} R\text{-}CH_2\text{-}CH_2\text{-}SG \quad + \quad HBr$$

3,4-Dichloronitrobenzene

FIGURE 4.59. Displacement of aliphatic and aromatic halogens by glutathione.

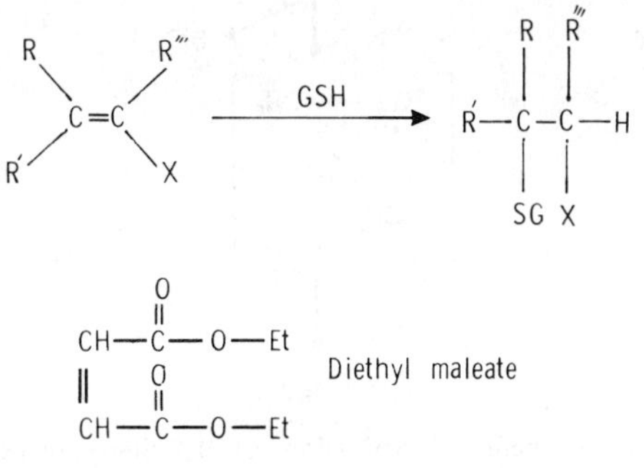

Diethyl maleate

FIGURE 4.60. Conjugation of an unsaturated aliphatic compound with glutathione and structure of diethylmaleate, a typical example.

three or more non-identical sub-units arranged as dimers, various combinations of which make up the functional isoenzymes. Although a wide variety of substrates may be accepted, there is absolute specificity for glutathione. However, the substrates have certain characteristics; namely they are hydrophobic, contain an electrophilic carbon atom and react non-enzymatically with glutathione to some extent. It appears that as well as catalysing conjugation reactions some of the transferases have a binding function. Thus, transferase B is also known as **ligandin** and will bind both endogenous substances such as bilirubin and such exogenous compounds as the drugs penicillin and tetracycline. This binding facility is associated with one particular sub-unit but is not directly related to catalytic activity. Thus some of the compounds bound to glutathione transferase B (ligandin) are not substrates for the transferase activity. Thus, the enzyme also has a transport or storage function.

After the conjugation reaction, the first catabolic step, removal of the glutamyl residue, is catalysed by the enzyme γ-**glutamyltranspeptidase (glutamyl-transferase)**. This is a membrane-bound enzyme found in high concentrations in the kidney. In the second step the glycine moiety is removed by the action of a peptidase, **cysteinyl glycinase**. The final step is acetylation of the amino group of cysteine by an *N*-acetyl transferase which utilizes acetyl CoA and is a microsomal enzyme found in liver and kidney, but is different from the cytosolic enzyme described below. The resultant *N*-acetylcysteine conjugate, also known as a mercapturic acid, is then excreted. With aromatic epoxides, as in the example of naphthalene shown in figure 4.57, the *N*-acetylcysteine conjugate may lose water and regain the aromatic ring structure. This will generally not occur in other types of glutathione conjugation reaction. Also the intermediate such as the cysteinyl-glycine and cysteine conjugates may be excreted as well as being metabolized to

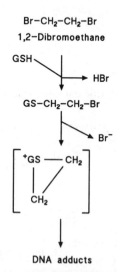

FIGURE 4.61. Glutathione mediated activation of 1,2-dibromoethane. The addition of glutathione is catalysed by glutathione transferase. Loss of bromide from the glutathione conjugate gives rise to an episulphonium ion. This can react with bases such as guanine in DNA.

the *N*-acetylcysteine derivative. It should be noted, however, that there are now examples of glutathione conjugates being involved in toxicity. For example **1,2-dibromoethane** forms a glutathione derivative catalysed by glutathione transferase which loses a halogen atom and becomes a charged episulphonium ion (figure 4.61). This reacts with DNA to give a guanine adduct which is believed to be responsible for the mutagenicity of the 1,2-dibromoethane. The diglutathione conjugate of **bromobenzene** is believed to be involved in the nephrotoxicity of bromobenzene after further metabolic activation (figure 7.14)

Cysteine conjugate β-lyase

Cysteine conjugates resulting from initial glutathione conjugation as described above may undergo further catabolism to give the thiol compound, pyruvate and ammonia (figure 4.62). The enzyme which catalyses this reaction, cysteine conjugate β-lyase (or CS lyase), requires pyridoxal phosphate, is cytosolic and will not accept the acetylated cysteine derivative. The thiol conjugate produced as a result of the action of β-lyase may be further metabolized by methylation and then *S*-oxidation (see below). The cleavage of cysteine conjugates can occur in the gut or kidney and is known to be involved in the nephrotoxicity of a number of compounds such as *S*-**(1,2-dichlorovinyl)**-L-**cysteine** and **hexachlorobutadiene** (see Chapter 7). Thus, although the initial glutathione conjugation may be a detoxication step, the final product of this phase 3 reaction may prove to be toxic.

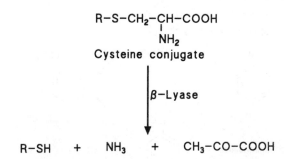

FIGURE 4.62. Metabolism of a cysteine conjugate by CS lyase (β-lyase).

Acetylation

Acetylation is an important route of metabolism for aromatic amines, sulphonamides, hydrazines and hydrazides and there is a wide variety of substrates. This metabolic reaction is one of two types of acylation reaction and involves an activated conjugating agent, acetyl CoA. It is hence a type 1 reaction. Acetylation is notable in that the product may be *less water soluble* than the parent compound. This fact gave rise to problems with **sulphonamides** when these were administered in high doses. The acetylated metabolites, being less soluble in urine, crystallized out in the kidney tubules, causing tubular necrosis (table 4.1). The enzymes which catalyse the acetylation reaction, acetyl-transferases, are cytosolic and are found in the liver, in both hepatocytes and Kupffer cells, in the gastrointestinal mucosa and in white blood cells. The enzyme has been purified and its mechanism of action extensively studied and is now well understood. This involves first acetylation of the enzyme by **acetyl CoA** (figure 4.63), followed by addition of the substrate and then transfer of the acetyl group to the substrate. With loss of the acetylated product the enzyme is regenerated.

A typical substrate is sulphanilamide, which may be acetylated or either the N^4, amino nitrogen or the N^1, sulphonamido nitrogen (figure 4.64).

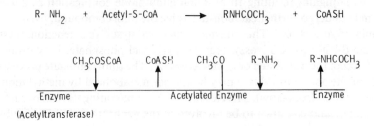

FIGURE 4.63. Reaction sequence for *N*-acetyltransferase.

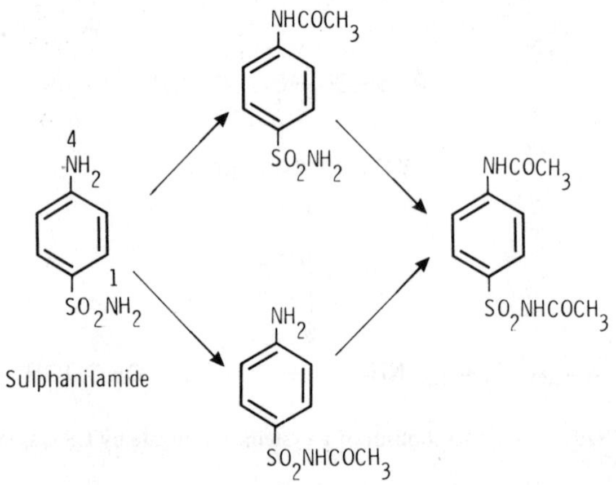

NHCOCH$_3$

^{4}-NH$_2$

SO$_2$NH$_2$

NHCOCH$_3$

^{1} SO$_2$NH$_2$

NH$_2$

SO$_2$NHCOCH$_3$

Sulphanilamide

SO$_2$NHCOCH$_3$

FIGURE 4.64. Acetylation of sulphanilamide on the N^1-sulphonamido or N^4-amino nitrogen to give the N^1 and N^4-acetyl and N^1,N^4-diacetyl derivatives.

It has been found, however, that the acetylation of certain compounds in man and in the rabbit shows wide interindividual variation. This variation in acetylation has a genetic basis and shows a bimodal distribution, the two phenotypes being termed rapid and slow acetylators. It is probable that in man this polymorphism reflects different forms of the acetyltransferase, as has been shown to be the case in the rabbit.

The acetylation polymorphism has a number of toxicological consequences which will be discussed more fully in Chapters 5 and 7. Only certain substrates are polymorphically acetylated. Some compounds, notably **sulphanilamide, *p*-aminosalicylic acid** and ***p*-aminobenzoic acid**, are monomorphically acetylated (figures 4.64 and 5.18).

As well as *N*-acetylation a related reaction, *N,O*-transacetylation, may also occur. This reaction applies to arylamines which first undergo *N*-hydroxylation (see figure 4.21) and then the hydroxylamine group is acetylated to yield an arylhydroxamic acid (see figure 7.1). This may then transfer the acetyl group to another amine or to the hydroxy group to yield a highly reactive acyloxy arylamine which is capable of reacting with proteins and nucleic acids. The enzyme which catalyses this, ***N,O*-acyltransferase** is a cytosolic enzyme.

Conjugation with amino acids

This is the second type of acylation reaction. However, in this type the xenobiotic *itself* is activated, and it is therefore a type 2 reaction. Organic acids, either aromatic such as salicylic acid (see Chapter 7), or aliphatic such as **2-methoxyacetic acid**, are the usual substrates for this reaction which involves conjugation with an endogenous amino acid. The particular amino acid depends on the species of animal exposed although species *within* a similar evolutionary group tend to utilize the same amino acid. Glycine is the most commonly used amino acid, but taurine, glutamine, arginine and ornithine can also be utilized. The mechanism involves first activation of the xenobiotic carboxylic acid group by reaction with acetyl CoA (figure 4.65). This reaction requires ATP and is catalysed by a **ligase** or **acyl Co A synthetase** which is a mitochondrial enzyme.

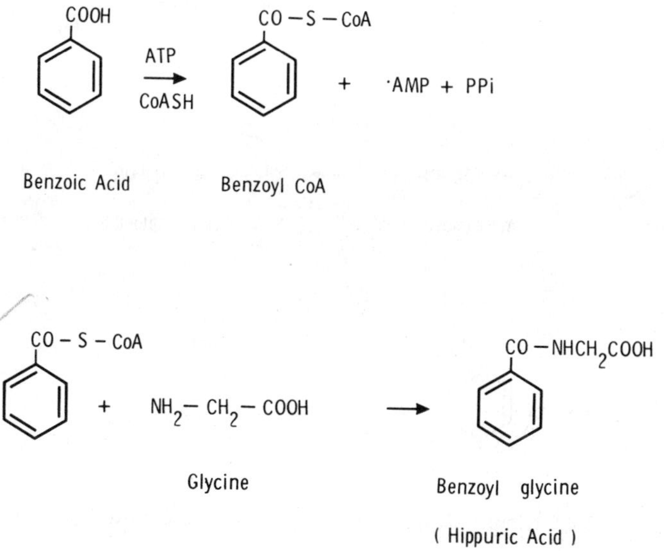

FIGURE 4.65. Conjugation of benzoic acid with glycine.

The S-CoA derivative then acylates the amino group of the particular amino acid in an analogous way to the acetylation of amine groups described above, yielding a peptide conjugate. This is catalysed by an amino acid *N*-acyltransferase which is located in the mitochondria. Two such enzymes have been purified, each utilizing a different group of CoA derivatives.

Bile acids are also conjugated with amino acids in a similar manner but different enzymes are involved.

Methylation

Amino, hydroxyl and thiol groups in foreign compounds may undergo methylation *in vivo* (figure 4.66) in the same manner as a number of endogenous compounds. The methyl donor is *S*-adenosyl methionine formed from methionine and ATP (see Chapter 7, figure 7.41) and so again it is a type 1 reaction with an activated donor. The reactions are catalysed by various **methyltransferase** enzymes which are mainly cytosolic although some are located in the endoplasmic reticulum. Some of these enzymes are highly specific for endogenous compounds such as the *N*-methyltransferase which methylates histamine. However, a non-specific *N*-methyltransferase exists in lung and other tissues which methylates both endogenous compounds and foreign compounds such as the ring nitrogen in nornicotine and the secondary amine group in

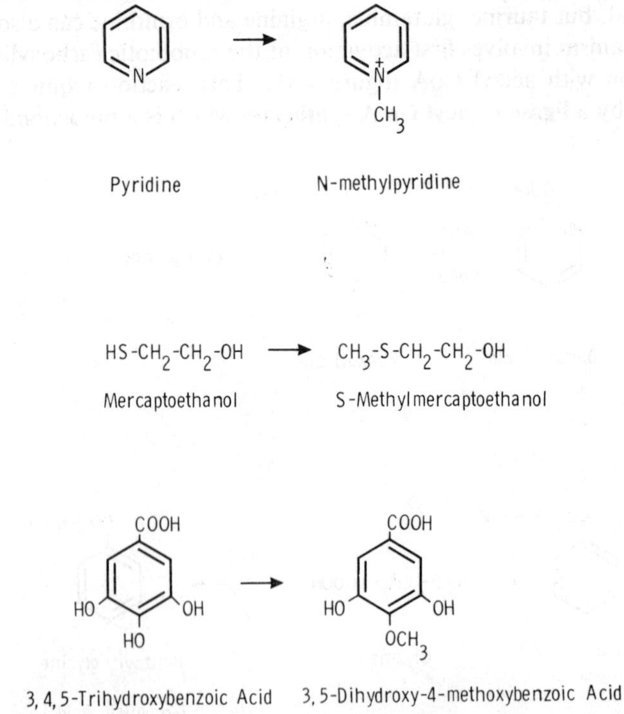

Pyridine　　　　　　　　N-methylpyridine

$$HS-CH_2-CH_2-OH \longrightarrow CH_3-S-CH_2-CH_2-OH$$

Mercaptoethanol　　　　　S-Methylmercaptoethanol

3, 4, 5-Trihydroxybenzoic Acid　　3, 5-Dihydroxy-4-methoxybenzoic Acid

FIGURE 4.66. *N*-, *O*- and *S*-methylation.

desmethylimipramine. Catechol-*O*-methyltransferase is a cytosolic enzyme which exists in multiple forms and catalyses the methylation of both *endogenous* and *exogenous* catechols and trihydric phenols. A microsomal *O*-methyltransferase is also found in mammalian liver which methylates phenols such as paracetamol. Thiol *S*-methyltransferase is a microsomal enzyme found in liver, lung and kidney, which will catalyse the methylation of a wide variety of foreign compounds. Thus **hydrogen sulphide**, H_2S, is detoxified by methylation first to methanthiol (CH_3SH) which is highly toxic, but is then further methylated to dimethylsulphide (CH_3-S-CH_3). The thiol products of β-lyase may also be methylated by this enzyme.

Metals may also undergo methylation reactions. Thus **mercury** is methylated by micro-organisms:

$$Hg^{2+} \rightarrow CH_3 Hg^+ \rightarrow (CH_3)_2 Hg$$

In this reaction the physicochemical properties, toxicity and environmental behaviour are altered. Thus the products methylmercury and dimethylmercury are more lipophilic, neurotoxic and persistent in the environment than elemental or inorganic mercury (see Chapter 7). Other metals such as **tin**, **lead** and **thallium** may also be methylated as can the elements **arsenic** and **selenium**. Thus, the methylation reaction, like acetylation, tends to reduce the water solubility of a foreign compound rather than increase it.

The foregoing discussion has by no means covered *all* of the possible metabolic transformations that a foreign compound may undergo and for more information the reader should consult the Bibliography. However, some general points should be made at this stage.

1. Generally the enzymes involved in xenobiotic metabolism are less specific than the enzymes involved in intermediary, endogenous substrate metabolism.

2. However, apart from absolute specificity, foreign compounds may also be substrates for enzymes involved in endogenous pathways, often with toxicological consequences. Thus, for example, with valproic acid (see above), fluoroacetate and galactosamine (see Chapter 7) involvement in endogenous metabolic pathways is a crucial aspect of the toxicity. The chemical structure and physicochemical characteristics will determine whether this occurs or not.

3. Foreign compounds do not necessarily only undergo one metabolic transformation. It is obvious from the preceding discussion that there are many possible routes of metabolism for many foreign compounds. What determines which will prevail is not always clear. However, in many cases several routes will operate at once giving rise to a variety of metabolites each with different biological activity. The balance and competition between these various routes will therefore be important in determining toxicity. This is well illustrated by bromobenzene and isoniazid (see below and Chapter 7).

4. Metabolism may involve many sequential steps, not just one phase 1 followed by one phase 2 reaction. Phase 1 reactions can sometimes follow phase 2 reactions, one molecule can undergo several phase 1 reactions, and cyclical or reversible metabolic schemes may operate. Thus, further metabolic transformations, sometimes termed phase 3 reactions, can convert a detoxified metabolite into a toxic product. An example is the metabolism of glutathione conjugates to toxic thiols previously mentioned. Further metabolism may occur as a result of biliary excretion for example. Thus glucuronic acid conjugates excreted via the bile into the intestine can be catabolized by bacterial β-**glucuronidase** to the aglycone, which may be either reabsorbed or further metabolized by the gut flora and then reabsorbed. Similarly, glutathione conjugates can follow the same type of pathway by the action of intestinal and bacterial enzymes to yield thiol conjugates as already described.

5. The rates of the various reactions will vary. This may be due to the availability of cofactors, concentration of enzyme in a particular tissue, competition with other, possibly endogenous, substrates or to intrinsic factors within the enzymes involved. This variation in rates will clearly affect the concentrations of metabolites in tissues, and the half-life of parent compound and metabolites. It may lead to accumulation of intermediate metabolites.

6. Metabolism of foreign compounds is not necessarily detoxication. This has already been indicated in examples and will become more apparent later in this book. This may involve activation by a phase 1 or phase 2 pathway or transport to a particular site followed by metabolism. Thus, sulphate conjugation and acetylation may be involved in the metabolic activation of N-hydroxy aromatic amines, glutathione conjugation may be important in the nephrotoxicity of compounds, methylation in metal toxicity, glucuronidation in the carcinogenicity of β-naphthylamine and 3,2'-dimethyl-4-aminobiphenyl.

Control of metabolism

The metabolic pathways which have been discussed in this chapter are *influenced* by many factors, some of which are discussed in Chapter 5. Such factors may have an effect on the toxicity of a compound as indicated below, and these are discussed in more detail in Chapter 5. It is important to appreciate that the metabolism of foreign compounds is not completely separate from intermediary metabolism, but linked to it. Consequently this will exert a controlling influence on the metabolism of foreign compounds. Thus some of the important factors controlling xenobiotic metabolism are:

(a) the availability of cofactors such as NADPH
(b) the availability of co-substrates such as oxygen and glutathione
(c) the level of particular enzymes.

The NADPH level is clearly important for phase 1 reactions, yet many biochemical processes, such as fatty acid biosynthesis, utilize this coenzyme. It is derived from either the pentose phosphate shunt or isocitrate dehydrogenase. Consequently, the *overall metabolic condition* of the organism will have an influence on the NADPH supply as there will be competition for its use. This may be important in paraquat toxicity where NADPH may be depleted (see Chapter 7).

Oxygen is normally readily available to all reasonably well-perfused tissues, but deep inside organs such as the liver, especially the centrilobular area (see Chapter 6), there will be a reduction in the oxygen concentration. This is clearly important when both oxidative and reductive pathways are available for a particular substrate. Thus, as conditions in a particular tissue become more anaerobic, reductive pathways will become more important. This is well illustrated by the metabolism of halothane where, in the rat, hypoxia will increase reductive metabolism and hepatotoxicity (see Chapter 7). Glutathione is an extremely important cofactor, involved in both protection and conjugation. It may be depleted by both of these processes or under certain circumstances such as hereditary glucose-6-phosphate deficiency in man, supply may be reduced (see Chapter 5). This will clearly influence toxicity and there are a number of examples discussed in Chapter 7 in which it is important.

Other co-substrates possibly limited in supply are inorganic sulphate and glycine for conjugation; these may be important factors in paracetamol hepatotoxicity and salicylate poisoning, respectively (Chapter 7).

The level of a particular enzyme involved in xenobiotic metabolism can obviously affect the extent of metabolism by that enzyme. Again competition may play a part if endogenous and exogenous substrates are both metabolized by an enzyme, as is the case with some of the forms of cytochromes P-450 which metabolize steroids, or NADPH cytochrome P-450 reductase and cytochrome b_5 reductase which are also involved in haem catabolism and fatty acid metabolism, respectively. The synthesis and degradation of enzymes such as cytochromes P-450 are therefore important factors and as discussed in detail in the next chapter, can be modified by exogenous factors.

Toxication versus detoxication

Although the biotransformation of foreign compounds is often regarded as a *detoxication* process, this is not always the case. Metabolites or intermediates may sometimes be produced which are more toxic than the parent compound. These may be the result of phase 1, 2 or 3 reactions although phase 1 reactions are the most commonly involved. The intermediates or metabolites responsible for the toxicity may be chemically reactive or stable. When the metabolic process produces a metabolite which is chemically reactive, this process is known as **metabolic activation** or **bioactivation**. The exact chemical reactivity may indeed be crucial, and there may be an optimum level for this reactivity. Thus, metabolism may underlie the toxicity of a compound and there will be many examples given throughout this book. However, a discussion of the general principles is appropriate here.

It is often the case that there are several metabolic pathways available for a foreign compound. Some of these pathways may be detoxication pathways while others may lead to toxicity (figure 4.67). When this situation arises there is the possibility of competition between pathways and, as indicated above, various factors may influence the balance between them. Furthermore, biological systems often have protective mechanisms for the removal of such reactive intermediates. However, these systems may sometimes be overloaded, suffer failure or be absent in some tissues.

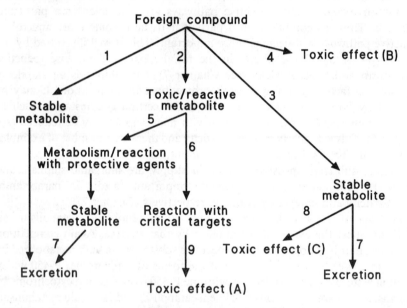

FIGURE 4.67. The various possible consequences of metabolism of a foreign compound. The compound may undergo detoxication (1); metabolic activation (2); formation of a stable metabolite (3) which may cause a toxic effect (C) or (4) cause a direct toxic effect (b). The rective metabolite may be detoxified (5/7) or (6/9) cause a toxic effect (A).

Thus the balance between detoxication and toxication will be affected by many factors such as:

(1) relative rates of the toxication and detoxication pathways: these will be influenced by the availability of enzymes and the kinetic parameters of the enzymes involved
(2) availability of cofactors
(3) availability of protective agents
(4) dose and saturability of metabolic pathways
(5) genetic variation in enzymes catalysing various pathways
(6) induction or inhibition of the enzymes involved
(7) species or strain of animal

(8) tissue differences in enzyme and isoenzyme patterns
(9) diet
(10) age
(11) disease
(12) sex.

Many of these factors will operate *independently* and toxicity will often only result when several conditions apply together. These various factors are discussed in more detail in Chapter 5 and are also apparent in some of the examples in Chapter 7.

Bibliography

ALVARES, A. P. and PRATT, W. B. (1990) Pathways of drug metabolism. In *Principles of Drug Action*, edited by W. B. Pratt and P. Taylor (New York: Churchill Livingstone).

BEEDHAM, C. (1988) Molybdenum hydroxylases. In *Metabolism of Xenobiotics*, edited by J. W. Gorrod, H. Oelschlager and J. Caldwell (London: Taylor & Francis).

ELING, T., BOYD, J., REED, G., MASON, R. and SIVARAJOH, K. (1983) Xenobiotic metabolism by prostaglandin endoperoxide synthetase. *Drug Metab. Rev.*, **14**, 1023.

GIBSON, G. G. and SKETT, P. (1986) *Introduction to Drug Metabolism* (London: Chapman & Hall).

GONZALEZ, F. J. (1988) The molecular biology of cytochrome P-450s. *Pharmacol. Rev.*, **40**, 243.

GONZALEZ, F. J., MATSUNAGA, T. and NAGATA, K. (1989) Structure and regulation of P-450s in the rat P450IIA gene subfamily. *Drug Metab. Rev.*, **20**, 827.

GORROD, J. W., OELSCHLAGER, H. and CALDWELL, J. (editors) (1988) *Metabolism of Xenobiotics* (London: Taylor & Francis).

GRAM, T. E. (editor) (1980) *Extrahepatic Metabolism of Drugs and Other Foreign Compounds* (Jamaica, New York: Spectrum Publications). (This text contains chapters on various aspects of drug metabolism, both phase 1 and phase 2.)

HATHWAY, D. E. BROWN, S. S., CHASSEAUD, L. F. and HUTSON, D. H. (reporters) (1970–1981) *Foreign Compound Metabolism in Mammals*, Vols. 1–6 (London: The Chemical Society).

HAWKINS, D. R. (editor) (1988–1990) *Biotransformations*, Vols 1 and 2 (London: Royal Society of Chemistry).

HODGSON, E. and GUTHRIE, F. E. (editors) (1980) *Introduction to Biochemical Toxicology*, Chapters 5 and 6 (New York: Elsevier-North Holland).

HUCKER, H. B. (1973) Intermediates in drug metabolism reactions. *Drug Metab. Rev.*, **2**, 33.

JAKOBY, W. R. (editor) (1980) *Enzymic Basis of Detoxication*, Vols 1 and 2 (New York: Academic Press).

JAKOBY, W. R., BEND, J. R. and CALDWELL, J. (editors) (1982) *Metabolic Basis of Detoxication* (New York: Academic Press).

JENNER, P. and TESTA, B. (1978) Novel pathways in drug metabolism. *Xenobiotica*, **8**, 1.

JERINA, D. M. and DALY, J. W. (1974) Arene oxides: A new aspect of drug metabolism. *Science*, **185**, 573.

JERINA, D. M., DALY, J. W. and WITKOP, B. (1971) The 'NIH Shift' and a mechanism of enzymatic oxygenation. In *Biogenic Amines and Physiological Membranes in Drug Therapy*, edited by I. H. Biel and L. G. Abood (New York: Marcel Dekker).

KAO, J. and CARVER, M. P. (1990) Cutaneous metabolism of xenobiotics. *Drug Metab. Rev.*, **22**, 363.

KETTERER, B. and TAYLOR, J. B. (1990) Glutathione transferases, In *Frontiers of Biotransformation*, Vol. II, *Principles, Mechanisms and Biological Consequences of Induction*, edited by K. Ruckpaul and H. Rein (London: Taylor & Francis).

MACKENZIE, P. I. (1990) Structure and regulation of UDP glucuronosyltransferase. In *Frontiers of Biotransformation*, Vol. II, *Principles, Mechanisms and Biological Consequences of Induction*, edited by K. Ruckpaul and H. Rein (London: Taylor & Francis).

MUKHTAR, H. and KHAN, W. A. (1989) Cutaneous cytochrome P-450. *Drug Metab. Rev.*, **20**, 657.

NEBERT, D. W. and GONZALEZ, F. J. (1987) P-450 genes: structure, evolution and regulation. *Annu. Rev. Biochem.*, **56**, 943.

NEBERT, D. W., NELSON, D. R., COON, M. J., ESTABROOK, R. W., FEYERSEISEN, R., FUJII-KURIYAMA, Y., GONZALEZ, F. J., GUENGERICH, F. P., GUNSALUS, I. C., JOHNSON, E. F., LOPER, J. C., SATO, R., WATERMAN, M. R. and WAXMAN, D. J. (1991) The P450 superfamily: update on new sequences, gene mapping and recommended nomenclature. *DNA Cell Biol.*, **10**, 1.

NEBERT, D. W. and GONZALES, F. (1990) The P450 gene superfamily. In *Frontiers of Biotransformation*, Vol. II, *Principles, Mechanisms and Biological Consequences of Induction*, edited by K. Ruckpaul and H. Rein (London: Taylor & Francis).

ORTIZ DE MONTELLANO, P. R. (editor) (1986) *Cytochrome P-450: Structure, Mechanism, and Biochemistry* (New York: Plenum Press).

ORTIZ DE MONTELLANO, P. R. and STEARNS, R. A. (1989) Radical intermediates in the cytochrome P-450 catalysed oxidation of aliphatic hydrocarbons. *Drug Metab. Rev.*, **20**, 183.

PARKE, D. V. (1968) *The Biochemistry of Foreign Compounds* (Oxford: Pergamon).

RUCKPAUL, K. and REIN, H. (editors) (1989) *Frontiers of Biotransformation*, Vol, I, *Basis and Mechanism of Regulation of Cytochrome P450* (London: Taylor & Francis).

SIES, H. and KETTERER, B. (editors) (1989) *Glutathione Conjugation. Mechanisms and Biological Significance.* (London: Academic Press).

TARLOFF, J. B., GOLDSTEIN, R. S. and HOOK, J. B. (1990) Xenobiotic biotransformation by the kidney: pharmacological and toxicological aspects. In *Progress in Drug Metabolism*, Vol. 12, edited by G. G. Gibson (London: Taylor & Francis).

THOMAS, H., TIMMS, C. W. and OESCH, F. (1990) Epoxide hydrolases: Molecular properties, induction, polymorphisms and function, In *Frontiers of Biotransformation*, Vol. II, *Principles, Mechanisms and Biological Consequences of Induction*, edited by K. Ruckpaul and H. Rein (London: Taylor & Francis).

TRAGER, W. F. (1989) Stereochemistry of cytochrome P-450 reactions. *Drug Metab. Rev.*, **20**, 489.

UETRECHT, J. (1990) Drug metabolism by leukocytes and its role in drug induced lupus and other idiosyncratic drug reactions. *Crit. Rev. Toxicol.*, **20**, 213.

WILLIAMS, R. T. (1959) *Detoxication Mechanisms* (London: Chapman & Hall).

ZEIGLER, D. M. (1988) Flavin-containing monooxygenases: catalytic mechanism and substrate specificities. *Drug Metab. Rev.*, **19**, 1.

Chapter 5

Factors affecting metabolism and disposition

Introduction

In the preceding two chapters, the disposition and metabolism of foreign compounds, as determinants of their toxic responses, were discussed. In this chapter, the influence of various chemical and biological factors on these determinants will be considered.

It is becoming increasingly apparent that the toxicity of a foreign compound and its mode of expression are dependent on many variables. Apart from large variations in susceptibility between species, within the same species many factors may be involved. The genetic constitution of a particular organism is known to be a major factor in conferring susceptibility to toxicity in some cases. The age of the animal and certain characteristics of its organ systems may also be important internal factors.

External factors such as the dose of the compound or the manner in which it is given, the diet of the animal and other foreign compounds to which it is exposed, are also important for the eventual toxic response. Although some of these factors may be controlled in experimental animals, in the human population they remain and may be extremely important.

For a logical use of experimental animals as models for man in toxicity testing, therefore, these factors must be appreciated and utilized for the fullest possible exploration of potential toxicity.

The factors affecting the disposition and toxicity of a foreign compound may be divided into chemical and biological factors:

Chemical factors: lipophilicity, size, structure, pK_a, ionization, chirality.
Biological factors: species, strain, sex, genetic factors, disease and pathological conditions, hormonal influences, age, stress, diet, tissue and organ specificity, dose, enzyme induction and inhibition.

Chemical factors

The importance of the physicochemical characteristics of compounds has already been alluded to in the previous two chapters. Thus *lipophilicity* is a factor of major importance for the absorption, distribution, metabolism and excretion of foreign compounds. Lipophilic compounds are more readily absorbed, metabolized and distributed, but more poorly excreted, than hydrophilic compounds.

The distribution of compounds is profoundly affected by lipophilicity and this may in turn influence the biological activity. Lipophilic compounds are more readily able to distribute into body tissues than hydrophilic compounds and there exert toxic effects or be sequestered and be redistributed later to other tissues. As already discussed (see above pages 33 and 55) comparison of the two drugs thiopental and pentobarbital illustrates the importance of lipophilicity. Similarly, the influence of size, structure and ionization have been mentioned. For example, there is a correlation between the nephrotoxicity of the aminoglycoside antibiotics, such as **gentamycin**, and structure, although other factors are also involved. The more ionizable amino groups on the aminoglycoside molecule, the greater the nephrotoxicity. The underlying reason for this is binding of the drug to anionic phospholipids on the brush border of the proximal tubular cells in the kidney and subsequent accumulation (see Chapters 6 and 7). The selective uptake of paraquat is another excellent example of the importance of size and shape in the disposition and toxicity of foreign compounds (see page 46 and Chapter 7). Similarly, the chemical similarities between carbon monoxide and oxygen are important in the toxic effects caused as are discussed in detail in Chapter 7.

Metabolism is also affected by the physico-chemical characteristics of a compound as discussed by Hansch (1972) for example (see Bibliography). Thus, with the mono-oxygenases there is a correlation between the lipophilicity of a compound, as measured by the partition coefficient, and metabolism by certain routes, such as N-demethylation for example. This correlation is not always clear-cut, however, as other factors may be involved. Ionization is another factor which may inhibit the ability of compounds to be metabolized. Size and molecular structure are clearly important in metabolism. Very large molecules may not be readily metabolized because of their inability to fit into the active site of an enzyme. As will already be apparent from Chapter 4, the molecular structure will determine what types of metabolic transformation will take place.

Chiral factors

The importance of chiral factors in disposition and toxicity has been fully recognized only relatively recently, although important examples have been known for some time. For instance the S (−) enantiomer of **thalidomide** is known to have greater embryotoxicity than the R (+) enantiomer (see Chapter 7). Ariens (see Bibliography) has been in the vanguard of those trying to highlight the importance of chirality, and particularly its implications for drugs.

The presence of a chiral centre in a molecule, giving rise to **isomers**, may influence the disposition of a compound and therefore its toxicity or other biological activity. It is clear from a consideration of the biochemistry of endogenous compounds where only one isomer may be metabolized or active, that biological systems are intrinsically chiral. Therefore it is hardly surprising that these considerations should apply also to foreign compounds.

All four phases of disposition may be influenced by chirality. Absorption is not often directly affected by the presence of a chiral centre except when an active transport process is involved. Thus **L-DOPA** is more readily absorbed from the gastrointestinal tract than the D-isomer. Various aspects of distribution may be affected by chirality such as protein and tissue binding. Thus for the drug **ibuprofen** the ratio of plasma protein binding for the (+) and (−) enantiomers is 1·5. It has been shown that the *S* enantiomer of **propranolol** undergoes selective storage in adrenergic nerve endings in certain tissues such as the heart.

The renal excretion of compounds can also be affected by the presence of a chiral centre, probably as a result of active secretion. For example, with the drug **terbutyline** the ratio for the excretion of the (+) to (−) enantiomers was found to be 1·8. Similarly biliary excretion of compounds may show stereoselectivity.

Probably of more significance are chiral effects in metabolism which can be divided into (i) **substrate stereoselectivity** (the effect of a pre-existing chiral centre in a molecule); (ii) **product stereoselectivity** (the production of metabolic products with chiral centres); (iii) **inversion of configuration**; and (iv) **loss of chirality** as a result of metabolism.

When **racemic mixtures** of drugs or foreign compounds are administered to animals, as is currently often the case, **stereoselective metabolism** means that either two or more different isomeric products are formed or only one isomer is metabolized. Alternatively there may be differences in rates of metabolism for the different isomers. All of these factors may have significant implications with regard to the biological activity of the molecule in question. Thus differences in rates or routes of metabolism for different isomers may underlie species, organ or genetic differences in metabolism and toxicity. Stereoselectivity in metabolism occurs with cytochromes P-450 and also with other enzymes such as epoxide hydrolase and glutathione transferases.

As an example of the first type of chiral effect, metabolism of the drug **bufuralol** may be considered. Hydroxylation in the 1′ position only occurs with the (+) isomer whereas for hydroxylation in the 4 and 6 position the (−) isomer is the preferred substrate (figure 5.1). Glucuronidation of the side chain hydroxyl group is specific for the (+) isomer. A further complication in human subjects is

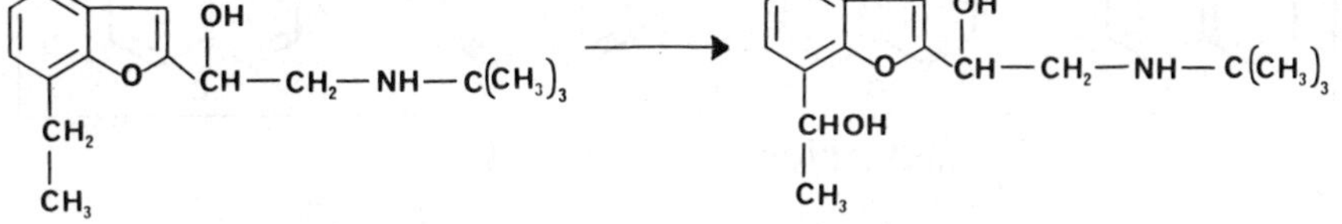

FIGURE 5.1. Aliphatic hydroxylation of bufuralol (1′ position).

that the 1-hydroxylation is under genetic control, being dependent on the **debrisoquine hydroxylator status** (see below). The selectivity for the isomers for the hydroxylations is virtually abolished in poor metabolizers.

There are other examples in which chirality is a factor in determining which particular metabolic route occurs, such as the metabolism of the drugs propranolol, metoprolol and warfarin.

An example of the second type of chiral effect in metabolism is afforded by **benzo[a]pyrene**, also discussed in more detail in Chapter 7. This carcinogenic polycyclic hydrocarbon is metabolized stereoselectively by a particular cytochrome P-450 isozyme, CYP 1A1, to the (+)-7R, 8S oxide (figure 7.2) which in turn is metabolized by epoxide hydrolase to the (−)-7R, 8R dihydrodiol. This metabolite is further metabolized to (+)-benzo[a]pyrene, 7R, 8S dihydrodiol, 9S, 10R epoxide in which the hydroxyl group and epoxide are *trans* and which is more mutagenic than other enantiomers. The (−)-7R, 8R dihydrodiol of benzo[a]pyrene is ten times more tumorigenic than the (+)-7S, 8S enantiomer. It was reported that in this case the configuration was more important for tumorigenicity than the chemical reactivity.

Another example is the metabolism of naphthalene which may cause lung damage in certain species. Thus **naphthalene** is metabolized first to an epoxide as previously discussed (see Chapter 4). However, this is *stereospecific* giving rise predominantly to the 1R, 2S-naphthalene oxide (figure 5.2) in the susceptible species (mouse) compared with ratios of 1R, 2S : 1S, 2R of one or less in non-susceptible species (rat and hamster). The 1R, 2S enantiomer was found to be a better substrate for epoxide hydrolase than the 1S, 2R enantiomer. The relationship of the stereospecificity to the lung toxicity is not yet clear, but differences in cytotoxicity between the two enantiomers were found in isolated hepatocytes. A further complication is the production of two chiral centres and hence **diastereoisomers** from metabolism of a chiral compound. For example,

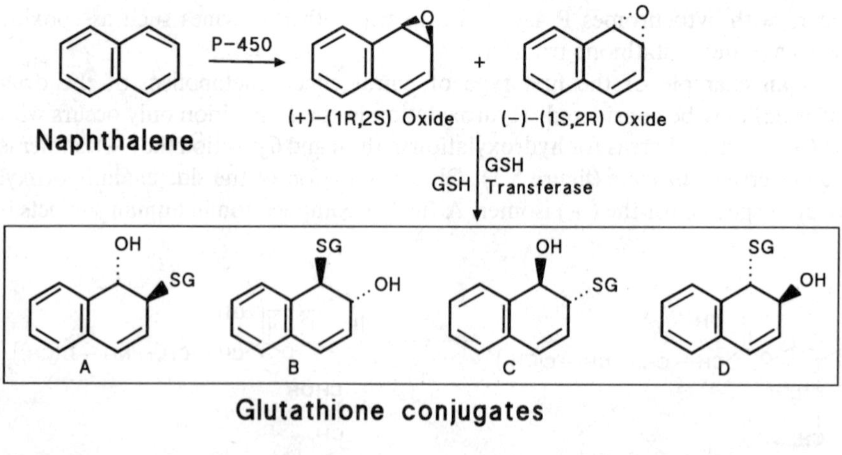

Glutathione conjugates

FIGURE 5.2. The metabolism of naphthalene showing the various possible isomeric glutathione conjugates.

this may occur when conjugation of isomers occurs with other chiral molecules such as β-D-glucuronic acid or L-glutathione. Thus, using the example of naphthalene, the oxirane ring of the two enantiomeric epoxides may be opened by attack by glutathione at either the 1- or 2-position to give four different diastereoisomeric glutathione conjugates (figure 5.2). The formation of the various diastereoisomers was found to show considerable species differences *in vitro*.

The third type of chiral effect, inversion of configuration, has been shown to occur with a number of compounds. For example, the anti-inflammatory drug **ibuprofen**, an arylproprionic acid, undergoes inversion from the R- to the pharmacologically active S-isomer. Furthermore, stereoselective uptake of the R-ibuprofen into fat tissue occurs as a result of selective formation of the coenzyme A thioester of the R-isomer. This thioester may then undergo inversion to the S-thioester. Both thioester isomers are incorporated into **triglycerides** forming hybrid products (figure 5.3). Thus, after S-ibuprofen is administered to rats, only a fraction is found in fat tissue in comparison with the incorporation after R-ibuprofen or the racemate is administered. Although the fate of these hybrid triglycerides is currently unknown, they might potentially interfere in lipid metabolism with possible toxicological consequences. This would be especially likely after chronic administration when accumulation could occur. There are also various factors which may affect the inversion *in vivo*, such as the species and reduction of renal excretion.

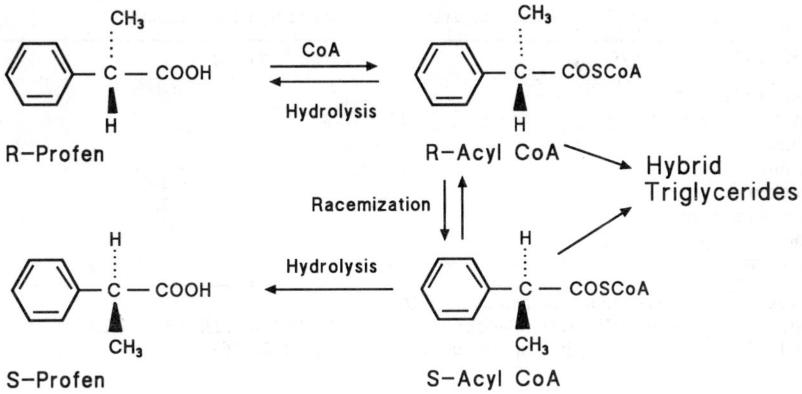

FIGURE 5.3. The mechanism of chiral inversion of ibuprofen and formation of hydrid triglycerides.

In conclusion, it cannot be stressed too strongly that the physicochemical characteristics of a foreign compound are factors of paramount importance in determining its toxicity.

Biological factors

Species

There are many different examples of species differences in the toxicity of foreign compounds, some of which are commercially useful to man, as in the case of pesticides and antibiotic drugs where there is exploitation of **selective toxicity**. Species differences in toxicity are often related to differences in the metabolism and disposition of a compound, and an understanding of such differences is extremely important in the **safety evaluation** of compounds in relation to the extrapolation of toxicity from animals to man and hence **risk assessment**.

Some species differences are due to differences in the response of the organism to insult or in the repair mechanisms available. There may be very simple differences; for example, rats are susceptible to certain rodenticides which they ingest by mouth, as unlike most other mammals, they are unable to vomit. There may be differences in receptor sensitivity such as for the organophosphorus compound **paraoxon** (see figure 4.25). Thus the cholinesterase enzyme in the mouse is more sensitive to inhibition than that in the frog ($79 \times$) and the frog is correspondingly less sensitive to the toxicity of the compound ($22 \times$). Another example is the very big species difference in the acute toxicity of 2, 3, 7, 8-tetrachlorodibenzdioxin (**TCDD** or **dioxin**) as can be seen from table 5.1. The difference seems to be due at least in part to the different sensitivities of the thymus to TCDD.

Table 5.1.   Species differences in the acute toxicity of dioxin*.

Species	LD_{50} (μg/kg body wt)
Guinea-pig	0·5–2
Rat	22–100
Mouse	114–284
Rabbit	10–115
Chicken	25–50
Rhesus monkey	<70
Dog	>30–300
Hamster	5051†

*Dioxin: 2,3,7,8-tetrachlorodibenzdioxin;TCDD.
Data from Reggiani (1978) *Arch. Toxicol.*, **40**, 161 and †EPA (1984) *Health Assessment Document for PCDDs*, Parts 1 and 2, *External Reviewers Draft*; PB 84–220268.

Species differences in disposition

ABSORPTION
Absorption of foreign compounds from various sites is dependent on the physiological and physical conditions at these sites. These, of course, may be subject to species variations. Absorption of compounds through the skin shows considerable species variation. Table 5.2 gives an example of this and shows the species differences in toxicity of an organophosphorus compound absorbed percutaneously. Human skin is generally less permeable to chemicals than that of

Table 5.2. Species differences in the relative percutaneous toxicity and skin penetration of organophosphorus compounds.

Species	Rate (μg/cm^2)/min	Compound 1[†]	Compound 2[†]	2/1
Pig	0·3	10·0	80·0	8·0
Dog	2·7	1·9	10·8	5·7
Monkey	4·2	4·4	13·0	3·0
Goat	4·4	3·0	4·0	1·3
Cat	4·4	0·9	2·4	2·7
Rabbit	9·3	1·0	5·0	5·0
Rat	9·3	17·0	20·0	1·2
Mouse	—	6·0	9·2	1·5
Guinea-pig	6·0	—	—	—

[†]Values are expressed as the ratio of the LD_{50} of that compound to the rabbit LD_{50} of compound 1. Data from McCreesh (1965) *Toxic. Appl. Pharmac.*, Suppl. 2,7,20, and Adamson and Davies (1973). In *International Encyclopaedia of Pharmacology and Therapeutics*, Section 85, Chapter 9 (Oxford: Pergamon).

rabbits, mice and rats, although there is variation. For some compounds rat skin has similar permeability to human skin and seems to be less permeable than that of the rabbit.

Oral absorption depends partially on the pH of the gastrointestinal tract which is known to vary between species, as shown in table 5.3. Clearly, therefore, considerable differences in the absorption of weak acids from the stomach may occur between species. Similarly, differences might be seen in compounds which are susceptible to the acidic conditions of the stomach, such that a foreign compound would be more stable in the gastric juice of a sheep than that of a rat. For example there is a difference in the acute toxicity in **pyrvinium chloride**, a rodenticide, between rats and mice due to differences in absorption from the gastrointestinal tract. Thus, after intraperitoneal dosing the toxicity is similar in the two species (LD_{50} 3–4 mg/kg) whereas after an oral dose it is very different, being more toxic in the mouse (LD_{50} 15 mg/kg) than in the rat (LD_{50}

Table 5.3. Species differences in pH of saliva and gastric juices.

Species	pH (Saliva)	pH (Gastric juice)
Man	6·75	1·5–2·5
Dog	7·5	1·5–2·0[1]
		4·5[2]
Cat	7·5	—
Rat	8·2–8·9	2·0–4·0
Horse	7·3–8·6	4·46
Cattle	8·1–8·8	5·5–6·5
Sheep	8·4–8·7	7·6–8·2
Chicken		4·2
Frog		2·2–3·7

[1]Fasting; [2]Fed.
Data from Altman and Dittmer (1961) *Blood and Other Body Fluids* (Washington D.C.: Federation of American Societies for Experimental Biology); Dobson (1967) *Fed. Proc.*, **26**, 994; Levine (1965) *Life Sci.*, **4**, 959; Prosser and Brown (1961) *Comparative Animal Physiology* (Philadelphia: W. B. Saunders); Bishop *et al.* (1950) *Comparative Animal Physiology* (Philadelphia: W. B. Saunders).

1550 mg/kg). Species differences in lung physiology may be important in considerations of absorption of compounds by inhalation. Small animals such as rats and mice and birds have a more rapid respiration rate than larger animals such as humans. Consequently, for compounds with high solubility in the blood where absorption is ventilation limited, exposure will be greater for these small animals when exposed to the same concentration of a compound. This is the basis of the use of canaries in mines to warn of the build up of dangerous gases.

DISTRIBUTION

The distribution of foreign compounds may vary between species because of differences in a number of factors such as proportion and distribution of body fat, rates of metabolism and excretion and hence elimination and the presence of specific uptake systems in organs. For instance, differences in localization of **methylglyoxal-bis-guanyl hydrazone** (figure 5.4) in the liver accounts for its greater hepatotoxicity in rats than in mice. The hepatic concentration in mice is only $0 \cdot 3 - 0 \cdot 5\%$ of the dose after 48 h, compared with 2–8% in the rat.

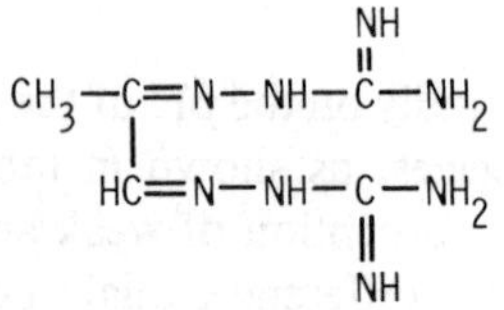

FIGURE 5.4. Structure of methylglyoxal-bis-guanyl hydrazone.

The plasma protein concentration is a species-dependent variable, and the proportions and types of proteins may also vary. The concentration may vary from about 20 g/l in certain fish to 83 g/l in cattle. Thus, foreign compounds may bind to plasma proteins to very different extents in different species (table 5.4). Because the extent of binding may be a very important determinant of the free concentration of a compound in the plasma and the tissues, this species difference may be an important determinant of toxicity. The free form of the compound is the important moiety as far as toxicity is concerned.

Table 5.4. Binding of various sulphonamides to plasma of various species.

| Sulphonamide | Percent bound at concentration of 100 μg/ml | | | | | | | |
	Human	Monkey	Dog	Cat	Mouse	Chicken	Bovine plasma	Bovine albumin
Sulphadiazine	33	35	17	13	7	16	24	24
Sulphamethoxypyridazine	83	81	60	49	28	14	66	60
Sulphisoxazole	84	86	68	43	31	5	76	76
Sulphaethylthiadiazole	95	90	86	76	38	48	87	87

Data from Adamson and Davies (1973). In *International Encyclopaedia of Pharmacology and Therapeutics*, Section 85, Chapter 9 (Oxford: Pergamon).

EXCRETION

Renal excretion. Although most mammals have similar kidneys, there are functional differences between species and urine pH, and volume and rate of production may vary considerably (table 5.5). Thus, the rate of urine production in the rat is an order of magnitude greater than the rate in man. Although the pH ranges for the urine of a number of mammals may overlap (table 5.5), a small change in pH may markedly change the solubility of a foreign compound and therefore its excretion. For instance, some of the sulphonamides and their acetylated metabolites show marked changes in solubility for a pH change of one unit (table 4.1), and renal toxicity due to crystallization of the drug or its metabolites in the renal tubules has been known to occur when high doses are used. The species differences in renal excretion for an unmetabolized compound, **methylglyoxal-bis-guanylhydrazone** (figure 5.4), are shown in table 5.6. It can be seen that rats and mice excrete *twice* the amount excreted by man in 24 h. This may be at least partially due to the greater rate of urine flow in the rodent. The

Table 5.5. Variation in urinary volume and pH with species.

Species	Volume (ml/kg)/day)	pH
Man	9–29	6·3(4·8–7·8)
Monkey	70–80	—
Dog	20–100	5·0–7·0
Cat	10–20	5·0–7·0
Rabbit	50–75	—
Rat	150–300	—
Horse	3–18	7–8
Cattle	17–45	7–8
Sheep	10–40	7–8
Swine	5–30	Acid or Alkaline

Data from Altman and Dittmer (1961) *Blood and Other Body Fluids* (Washington D.C.: Federation of American Societies for Experimental Biology); Bloom (1960) *The Urine of the Dog and Cat* (New York: Gamma Publications); Cornelius and Kaneko (1963) *Clinical Biochemistry of Domestic Animals* (New York: Academic Press).

Table 5.6. Urinary excretion of methylglyoxal-bis-guanylhydrazone in mammalian species.

Species	Dose (mg/kg)	Percent excreted	Time period for excretion (h)
Mouse	20 (i.v.)	51	24
Rat	20 (i.p.)	65	24
Dog	20 (i.v.)	26	24
		52	48
		66	96
Monkey	25 (i.v.)	47	24
Man	4 (i.v.)	25	24
		42	118
		49	166

Data from Oliverio *et al.* (1963) *J. Pharmac. Exp. Ther.*, **141**, 149; Adamson and Davies (1973). In *International Encyclopaedia of Pharmacology and Therapeutics*, Section 85, Chapter 9 (Oxford: Pergamon).

Table 5.7.　Biliary excretion of compounds of molecular size 300–500 in various species.

	Percent dose excreted in bile				
	Methylenedisalicylic acid (mol. wt. 288, 10 mg/kg, i.v., 6h)	Succinylsulphathiazole (mol. wt. 355, 20 mg/kg, i.v., 6h)	Stilboestrol glucuronide (mol. wt. 445, 10 mg/kg, i.v., 3h)	Sulphadimethoxine-N^1-glucuronide (mol. wt. 487, 15 mg/kg, i.v., 3h)	Phenolphthalein glucuronide (mol. wt. 495, 10 mg/kg, i.v., 3h)
Rhesus monkey	—	0·2	—	—	9
Rat	54	29	95	43	54
Hen	—	25	—	—	—
Dog	65	20	65	43	81
Cat	—	7	77	—	34
Sheep	—	7	—	—	38
Rabbit	5	1	32	10	13
Guinea-pig	4	—	20	12	6
Pig	—	0·2	—	—	—

Data from Abou-El-Makarem *et al.* (1967) *Biochem. J.*, **105**, 1289, and Davison and Williams (1968) *J. Pharm. Pharmac.*, **20**, 12.

observation that renal tubular atrophy is caused by certain diuretics such as furosemide (figure 3.33) in the dog, but not in the rat or monkey, has been explained in terms of differences in the vascular system of the dog kidney from that in rats, monkeys and humans. For compounds which are not actively secreted in the kidney species differences in plasma-protein binding may indirectly lead to differences in urinary excretion.

Biliary excretion. The extent of excretion of foreign compounds via the bile is influenced by a number of factors, the molecular weight of the compound being the major one. However, the molecular weight threshold for biliary excretion may show considerable species differences. Little biliary excretion (5–10% of the dose) occurs for compounds of molecular weight of less than 300. Above this value, however, the bile may become a major route of elimination, and it is probably around this value that species variations are most noticeable. Thus, for methylene di-salicylic acid (mol. wt. 288), the dog excretes 65% in the bile, whereas the guinea-pig excretes only 4% (table 5.7). Similarly, the biliary excretion of succinyl sulphathiazole (mol. wt. 355) shows more than a ten-fold variation between the rhesus monkey and the rat (table 5.7). Thus the approximate molecular weight thresholds are 500–700 in humans, 475 in rabbits, 400 in guinea-pigs and 325 in rats.

The species pattern of the rabbit and the guinea-pig being poor biliary excretors and the rat being an extensive biliary excretor is maintained with many other compounds. With compounds of higher molecular weight, however, species differences are less, as illustrated by the compound **indocyanine green** (table 5.8). The metabolism of a compound obviously influences the extent of biliary excretion, and therefore species differences in metabolism may also be a factor.

The rate of bile secretion and the pH of the bile may also be determinants of the extent of biliary excretion of a foreign compound, and these also show species variations. The fate of compounds excreted in the bile may also depend on the species, as differences in intestinal pH and flora occur. A particularly important consequence of biliary excretion is metabolism by the gut flora and reabsorption. This enterohepatic circulation prolongs the length of time the animal is exposed to the foreign compound, and may introduce novel toxic metabolites. This could therefore result in marked species differences in toxicity.

Table 5.8. Biliary excretion of indocyanine green in various species.

Species	Dose (mg/kg, i.v.)	% Dose in bile†
Rat	0·5	60
Rat	2·5	82
Dog	1–7	97
Rabbit	2·5	94
Man	0·5	High
Man	2·0	High

†Excreted unchanged.
Data from Caesar *et al.* (1961) *Clin. Sci.*, **21**, 43; Cherrick *et al.* (1960) *J. Clin. Invest.*, **39**, 592; Delaney *et al.* (1969) unpublished data; Levine *et al.* (1970) *Biochem. Pharmac.*, **19**, 235; Wheeler *et al.* (1958) *Proc. Soc. Exp. Bio. Med. N.Y.*, **99**, 11.

Species differences in metabolism

Differences in metabolism between species may be either *quantitative* or *qualitative*, but quantitative differences are more common. In general small animals such as mice metabolize foreign compounds at a faster rate than larger animals such as humans, consistent with differences in overall metabolic rate. An extreme example of a difference in rates of metabolism is afforded by the drug **oxyphenbutazone**. In the dog it is rapidly metabolized and has a half-life of around 30 min, in several other species such as the rat, rabbit and monkey the half-life is between 3 and 6 h, whereas in humans metabolism is very slow and therefore the drug has a half-life of about 3 days. Quantitative differences also exist although with a few exceptions, it is generally difficult to discern useful patterns. Even the simplest organisms such as bacteria seem to be able to carry out many different types of reaction. The differences which are clear and fall within taxonomic groups are mainly found with phase 2 reactions. Differences in some cases are related to diet, and so herbivores and carnivores may show differences.

Examples of toxicologically important species differences in metabolism will therefore be dealt with by considering the different types of metabolic reactions.

PHASE 1 REACTIONS

Oxidation. Although most of the common mammals used as experimental animals carry out oxidation reactions, there may be large variations in the extent to which some of these are carried out. The most common species differences are in the rate at which a particular compound is oxidized rather than the particular pathway through which it is metabolized. Most species are able to hydroxylate aromatic compounds, but there is no apparent species pattern in the ability to carry out this metabolic transformation.

Fish have a relatively poor ability for oxidative metabolism compared with the commonly used laboratory animals such as rats and mice. Insects such as flies have microsomal enzymes, and these are involved in the metabolism of the insecticide parathion to the more toxic paraoxon as discussed in the previous chapter (figure 4.25).

However, there are known instances of differences in the preferred route of metabolism which are important in toxicity, as well as simple differences in the route of a particular oxidation. For example, the oxidative metabolism of ethylene glycol gives rise to either carbon dioxide or oxalic acid (figure 5.5). The relative importance of these two pathways is reflected in the toxicity. Thus, the production of oxalic acid is in the order: cat > rat > rabbit, and this is also the order of increasing toxicity (figure 5.6). The aromatic hydroxylation of aniline (figure 5.7) shows marked species differences in the position of substitution, as shown in table 5.9. Thus carnivores such as the ferret, cat and dog excrete mainly *o*-aminophenol whereas herbivores such as the rabbit and guinea-pig excrete mainly *p*-aminophenol. The rat, an omnivore, is intermediate.

The preferred route of hydroxylation also correlates with the toxicity, such that those species to which aniline is particularly toxic, such as the cat and dog,

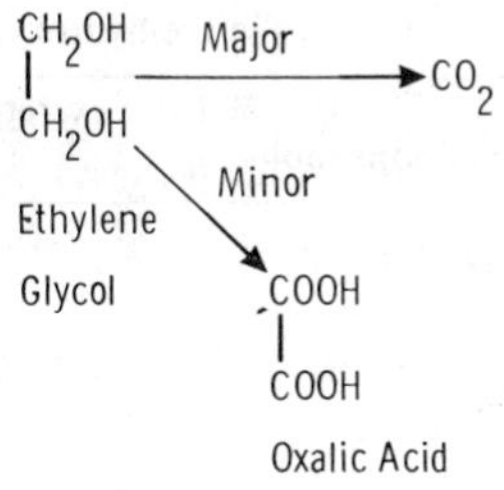

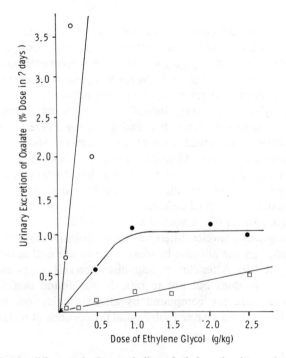

FIGURE 5.5. Metabolism of ethylene glycol.

FIGURE 5.6. Species differences in the metabolism of ethylene glycol to oxalate after increasing doses. Species used were cats (○), rats (●) and rabbits (□).
Data from Gessner *et al.* (1961) *Biochem. J.*, **79**, 482.
Adapted from Parke, D.V. (1968) *The Biochemistry of Foreign Compounds* (Oxford: Pergamon).

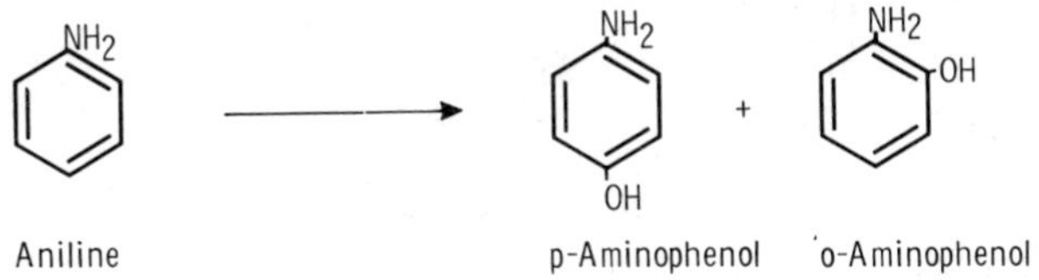

Aniline   p-Aminophenol   o-Aminophenol

FIGURE 5.7. Aromatic hydroxylation of aniline.

Table 5.9.   Species differences in the hydroxylation of aniline.

Species	% Dose excreted	
	o-Aminophenol	*p*-Aminophenol
Gerbil	3	48
Guinea-pig	4	46
Golden hamster	6	53
Chicken	11	44
Rat	19	48
Ferret	26	28
Dog	18	9
Cat	32	14

Data from Parke, D. V. (1968) *The Biochemistry of Foreign Compounds* (Oxford: Pergamon).

produce mainly *o*-aminophenol, whereas those producing *p*-aminophenol, such as the rat and hamster, seem less susceptible. Conversely, the hydroxylation of coumarin at the seven position (figure 4.15) is an important pathway in the rabbit and also the hamster and cat, but not in the rat or mouse. It is clear that, even with aromatic hydroxylation, species cannot be readily grouped.

The *N*-hydroxylation of acetylaminofluorene and paracetamol are two toxicologically important examples illustrating species differences (see Chapter 7). Another example is the metabolism of **amphetamine**, which reveals marked species differences in the preferred route, as shown in figure 5.8.

Species differences in the rate of metabolism of **hexobarbital** *in vitro* correlate with the plasma half-life and duration of action *in vivo* as shown in table 5.10. This data shows that the marked differences in enzyme activity between species is the major determinant of the biological activity in this case.

A recent example of a species difference in metabolism causing a difference in toxicity is afforded by the alicyclic hydroxylation of the oral anti-allergy drug, **proxicromil** (figure 5.9). After chronic administration this compound was found to be hepatotoxic in dogs but not in rats. It was found that dogs did not significantly metabolize the compound by alicyclic oxidation, whereas rats, hamsters, rabbits and man excreted substantial proportions of metabolites in the

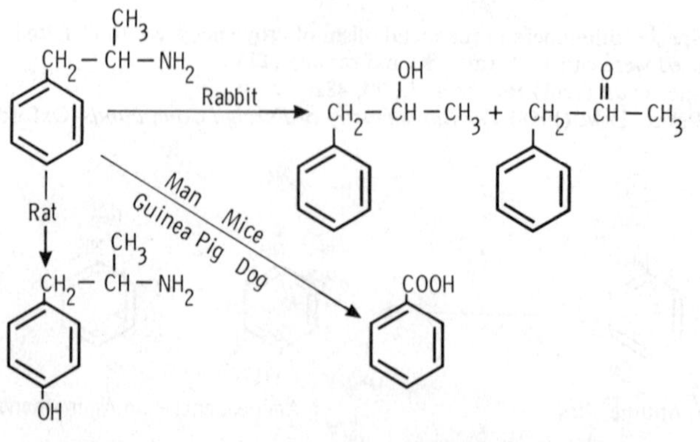

FIGURE 5.8. Species differences in the metabolism of amphetamine.

Table 5.10. Species differences in the duration of action and metabolism of hexobarbital (dose of barbiturate 100 mg/kg (50 mg/kg in dogs)).

Species	Duration of action (min)	Plasma half-life (min)	Relative enzyme activity ((μg/g)/h)	Plasma level on awakening (μg/ml)
Mouse	12	19	598	89
Rabbit	49	60	196	57
Rat	90	140	135	64
Dog	315	260	36	19

Data from Quinn *et al.* (1958) *Biochem. Pharmac.*, **1**, 152.

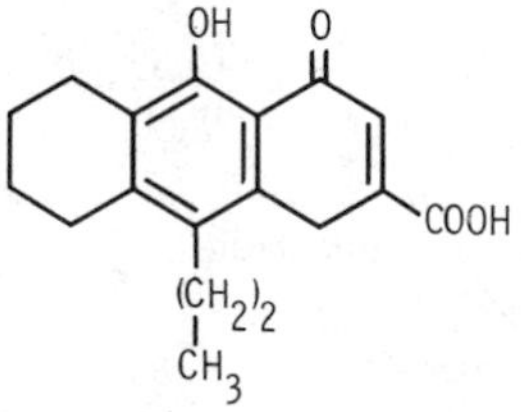

FIGURE 5.9. The structure of proxicromil.

urine. In the dog, biliary excretion was the route of elimination of the unchanged compound, and after toxic doses were administered, this route was saturated. Hence, the toxicity was probably due to the accumulation of high levels of the unchanged compound.

Hydrolytic reactions. There are numerous different esterases responsible for the hydrolysis of esters and amides, and they occur in most species. However, the activity may vary considerably between species. For example, the insecticide **malathion** owes its selective toxicity to this difference. In mammals, the major route of metabolism is hydrolysis to the dicarboxylic acid, whereas in insects it is oxidation to malaoxon (figure 5.10). Malaoxon is a very potent cholinesterase

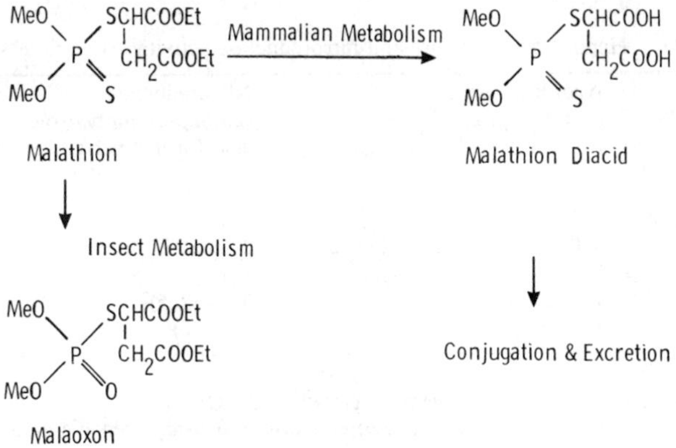

FIGURE 5.10. Metabolism of malathion.

inhibitor, and its insecticidal action is probably due to this property. The hydrolysis product has a low mammalian toxicity (see Chapter 7).

Another example is **dimethoate**, the toxicity of which is related to its rate of hydrolysis. Those species which are capable of metabolizing the insecticide are less susceptible than those species which are poor metabolizers. The metabolism of dimethoate is shown in figure 5.11. Studies on the metabolism *in vitro* of dimethoate have shown that sheep liver produces only the first metabolite, whereas guinea-pigs produce only the final product (figure 5.11). Rats and mice metabolize dimethoate to both products. The toxicity is in the descending order—sheep > dog > rat > cattle > guinea-pig > mouse.

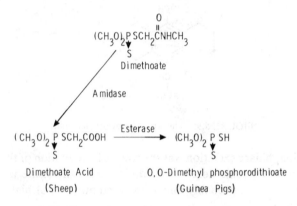

FIGURE 5.11. Metabolism of dimethoate.

Reduction. The activity of azo and nitroreductase varies between different species, as shown by the *in vitro* data in table 5.11. Thus, azoreductase activity is particularly high in the guinea-pig, relative to the other species studied, whereas nitroreductase activity is greatest in the mouse liver.

Table 5.11. Hepatic azoreductase and nitroreductase activities of various species.

Species (male)	Azoreductase (μmol sulphanilamide formed per g liver per h)	Nitroreductase (μmol p-aminobenzoic acid formed per g liver per h)
Mouse	6·7–9·6[1]	2·1–3·2[1]
Rat	5·9	2·1
Guinea-pig	9·0	2·0
Pigeon	7·1	1·1
Turtle	1·4 (0·5)[2]	0·15 (2·5)[2]
Frog	1·2 (0·6)[2]	0 (0)[2]

[1]According to strain.
[2]Temperature of incubation 21°C (temperature elsewhere 37°C).
Substrates used were neoprontosil for the azoreductase and *p*-nitrobenzoic acid for the nitro-reductase.
Data from Adamson *et al.* (1965) *Proc. Natn. Acad. Sci.*, **54**, 1386.

PHASE 2 REACTIONS

Species vary considerably in the extent to which they conjugate foreign compounds, but this is generally a quantitative rather than a qualitative difference. Thus, most species have a preferred route of conjugation but other routes are still available and utilized.

Glucuronide conjugation. Conjugation of foreign compounds with glucuronic acid is an important route of metabolism in most animals, namely mammals (table 5.12) birds, amphibians and reptiles, but not fish. In insects, glucoside conjugates utilizing glucose rather than glucuronic acid are formed. The major exception with regard to glycoside conjugation is the **cat** which is virtually unable to form glucuronic acid conjugates with certain foreign compounds, in particular phenols. However, bilirubin, thyroxine and certain steroids are conjugated with glucuronic acid in the cat. This may be explained by the presence of multiple forms of the enzyme UDP-glucuronosyl transferase shown in the rat for example, which catalyse conjugation with different types of substrate (see Chapter 4). Presumably the cat lacks the isoenzyme which catalyses the glucuronidation of phenols. The cat is therefore more susceptible to the toxic effects of phenols than species able to detoxify them by glucuronide conjugation.

Table 5.12.   Conjugation of phenol with glucuronic acid and sulphate.

Species	Phenol conjugation (% total)	
	Glucuronide	Sulphate
Cat	0	87
Gerbil	15	69
Man	23	71
Rat	25	68
Rhesus monkey	35	65
Ferret	41	32
Rabbit	46	45
Hamster	50	25
Squirrel monkey	70	10
Guinea-pig	78	17
Indian fruit bat	90	10
Pig	100	0

Data from Hodgson, E. and Levi, P.E. (1987) *A Textbook of Modern Toxicology* (New York: Elsevier).

Sulphate conjugation. Conjugation of foreign compounds with sulphate occurs in most mammals (table 5.12), amphibians, birds, reptiles and insects, but, as with glucuronidation, not in fish. The **pig**, however, has a reduced ability to form certain ethereal sulphate conjugates, such as with phenol, whereas l-naphthol is excreted as a sulphate conjugate. As there are several forms of sulphotransferase, specific for different substrates, the inability of the pig to form a sulphate conjugate with phenol may be due to the lack of one particular form of the enzyme. The inability of the **guinea-pig** to form a sulphate conjugate of *N*-hydroxyacetylaminofluorene helps confer resistance to the tumorigenicity of **acetylaminofluorene** on this species (see Chapter 7).

It can be seen from table 5.12 that the relative proportions of glucuronide and sulphate conjugates vary between the species with the cat and pig being at the opposite extremes. Interestingly humans and rats excrete similar proportions of conjugates, but the squirrel monkey, also a primate, is quite different.

Conjugation with amino acids. Considerable species differences exist in the conjugation of aromatic carboxylic acids with amino acids. A number of amino acids may be utilized, although conjugation with glycine is the most common route (figure 4.65) and occurs in most species except some birds, where ornithine is the preferred amino acid. Humans and Old World monkeys utilize glutamine for conjugation of **arylacetic acids** and in the pigeon and ferret taurine is used. Reptiles may excrete ornithine conjugates as well as glycine conjugates, and some insects utilize mainly arginine.

Aromatic acids may also be excreted as glucuronic acid conjugates, and the relative importance of glucuronic acid conjugation versus amino acid conjugation depends on the particular species and the structure of the compound. Herbivores generally favour amino acid conjugation, carnivores favour glucuronide formation, and omnivores, such as man, utilize both routes of metabolism.

There are also species differences in the site of conjugation; this usually occurs in both liver and kidney, but dogs and chickens carry out this conjugation only in the kidney.

Glutathione conjugation. Conjugation with glutathione, which results in the urinary excretion of *N*-acetylcysteine or cysteine derivatives (figure 4.57) occurs in man, rats, hamsters, mice, dogs, cats, rabbits and guinea-pigs. Guinea-pigs are unusual, however, in generally not excreting *N*-acetylcysteine conjugates, as the enzyme responsible for the acetylation of cysteine is lacking.

Insects are also capable of forming glutathione conjugates, this being probably involved in the dehydrochlorination of the insecticide DDT, a reaction at least some insects, such as flies, are able to carry out (figure 4.41).

Methylation. Methylation of oxygen, sulphur and nitrogen atoms seems to occur in most species of mammal and in those birds, amphibia and insects which have been studied.

Acetylation. Most mammalian species are able to acetylate aromatic amino compounds, the major exception being the **dog**. Thus, for a number of amino compounds such as procainamide (figure 4.42), sulphadimethoxine, sulpha-methomidine, sulphasomizole and the N^4 amino group of sulphanilamide (figure 4.64), the dog does not excrete the acetylated product. However, the dog does have a high level of deacetylase in the liver and also seems to have an acetyltransferase inhibitor in the liver and kidney. Consequently, acetylation may not be absent in the dog, but rather the products may be hydrolysed or the reaction effectively inhibited.

The dog does, however, acetylate the N^1, sulphonamido group of sulphanilamide (figure 4.64), and also acetylates aliphatic amino groups. The guinea-pig is unable to acetylate aliphatic amino groups such as that in cysteine. Consequently, it excretes cysteine rather than *N*-acetylcysteine conjugates or mercapturic acids. Birds, some amphibia and insects are also able to acetylate aromatic amines, but reptiles do not utilize this reaction.

CONCLUDING REMARKS

Thus most species differences in metabolism are quantitative rather than qualitative; only occasionally does a particular single species show an inability to carry out a particular reaction, or to be its sole exponent. The more common quantitative differences depend on species differences in the enzyme concentration or its kinetic parameters, the availability of cofactors, the presence of reversing enzymes or inhibitors and the concentration of substrate in the tissue.

These quantitative differences may often mean, however, that different metabolic routes are favoured in different species, with a consequent difference in pharmacological or toxicological activity.

In general, man is able to carry out all the metabolic transformations found in other mammals and does not show any particular differences in the presence or absence of an enzymatic pathway.

However, it would be difficult at this time to pick a single species as the best model for man on *metabolic grounds alone*. It may be possible to do this after a consideration of the structure of the foreign compound in question however.

Strain

Differences in the disposition of foreign compounds between different strains of the same species have been well documented. For example, mice and rats of various strains show marked differences in the duration of action of **hexobarbital**, whereas within any one strain the response is uniform (table 5.13). It is noteworthy that the variation as indicated by the standard deviation is greatest for the outbred group.

The metabolism of antipyrine in rats varies widely between different strains. A well known example of a strain difference is that of the Gunn rat, which is unable to form *o*-glucuronides of bilirubin and most foreign compounds. This defect is due to a deficiency in glucuronyl transferase. *N*-acetyltransferase activity has been found to vary with the strain of rats and mice. The hydrolysis of

Table 5.13. Strain differences in the duration of action of hexobarbital in mice (dose of barbiturate 125 mg/kg body weight).

Strain	Numbers of animals	Mean sleeping time (min) ± S.D.
A/NL	25	48 ± 4
BALB/cAnN	63	41 ± 2
C57L/HeN	29	33 ± 3
C3HfB/HeN	30	22 ± 3
SWR/HeN	38	18 ± 4
Swiss (non-inbred)	47	43 ± 15

Data from Jay (1955) *Proc. Soc. Exp. Biol. Med.*, **90**, 378.

acetylisoniazid to acetylhydrazine (figure 4.45) was found to be significantly different between two strains of rats, as was the hepatotoxicity of the acetylisoniazid, being greater in the strain with the greater acylamidase activity (see Chapter 7). Humans also show such genetically based variations and this is discussed later in this chapter. Strain or genetic differences in *responsiveness* may also be important, such as in the response to enzyme inducers discussed below.

Sex

There are a number of documented differences in the disposition and toxicity of foreign compounds which are related to the sex of the animal. For instance the organophosphorus compound, **parathion,** is twice as toxic to female as compared with male rats. One of the first sex differences to be noted was the fact that **hexobarbital**-induced sleeping time is longer in female than in male rats. This is in accord with the view that, in general, male rats metabolize foreign compounds more rapidly than females. Thus, the biological half-life of hexobarbital is considerably longer in female than in male rats, and *in vitro*, the liver microsomal fraction metabolizes both hexobarbital and aminopyrine more rapidly when derived from male rather than female rats.

As well as phase 1 reactions, phase 2 reactions also show sex differences. Thus glucuronic acid conjugation of **1-naphthol** (figure 4.4) is greater in male than female rats and this difference is also found in microsomes *in vitro*. The acetylation of **sulphanilamide** (figure 4.64) is also greater in male than female rats. Sex differences in metabolism depend on the substrate, however. For example, the hydroxylation of aniline or zoxazolamine shows little difference

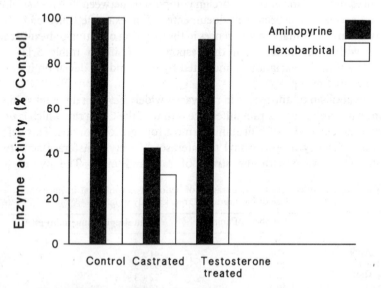

FIGURE 5.12. The effect of castration and androgen replacement on the metabolism of aminopyrine and hexobarbital in the rat. Data from Kato and Onoda (1970) *Biochem. Pharmacol.,* **19**, 1649.

between the sexes, in contrast to the three-fold greater metabolism of hexobarbital or aminopyrine in male compared with female rats.

Sex differences in metabolism are less pronounced in species other than the rat. In humans, differences, when noted are similar to those in the rat, whereas in the mouse they are often the reverse of those in the rat. The differences in metabolism between males and female animals are due to the influence of hormones and genetic factors. Thus, the differences appear at puberty and the administration of androgens to female animals abolishes the differences from the males. The sex hormones appear to have a powerful influence on the metabolism of foreign compounds. For example, study of the metabolism of **aminopyrine** and **hexobarbital** in male rats showed that castration markedly reduced metabolism of both drugs, but this was restored by the administration of androgens (figure 5.12).

However, the **control of metabolism** appears to be more complicated than this as it also involves the **hypothalamus** and **pituitary** gland. It seems that the male hypothalamus produces a factor which inhibits the release of a hormone and which therefore leaves the liver in a particular, *male state*. In the female the hypothalamus is inactive, and therefore produces no factor and hence the pituitary releases a feminizing factor (possibly growth hormone) which changes the liver to the *female state*.

This activity of the hypothalamus does not depend on the genotypic sex of the animal, but on events during the perinatal period. For example, male rats have very low activity for the hydroxylation of steroid sulphate conjugates (15β-hydroxylase). This applies to both normal and castrated male animals. However, if male animals are castrated immediately after birth, they show a greater, female level of activity. This effect can be reversed by treatment of castrated males with testosterone during the period immediately after birth, but not by treatment of the adult animal. This has been interpreted as indicating that it is the absence or presence of androgen in the period *immediately after birth* which determines the maleness or femaleness of the animal as regards the enzymes involved in the metabolism of foreign compounds. The phenomenon is known as imprinting.

As already mentioned, the pharmacological activity of certain drugs is lengthened in the female rat, and also the toxicity of certain compounds may be increased. **Procaine** is hydrolysed more in male rats (figure 4.42), with consequently a lower toxicity in the sex, whereas the insecticides **aldrin** and **heptachlor** are metabolized more rapidly to the more toxic epoxides in males and are therefore less toxic to females (figure 5.13).

Probably the best example of a sex difference in toxicity is that of the renal toxicity of **chloroform** in mice. The males are markedly more sensitive than the females, and this difference can be removed by castration of the male animals, and subsequently restored by administration of androgens.

In vitro, chloroform is metabolized ten times faster by kidney microsomes from male compared with female mice. In contrast to rats, male mice given hexobarbital metabolize it more slowly, and the pharmacological effect is more prolonged than in females. The excretion of the food additive butylated hydroxytoluene (figure 5.14) shows an interesting six difference in rats, being

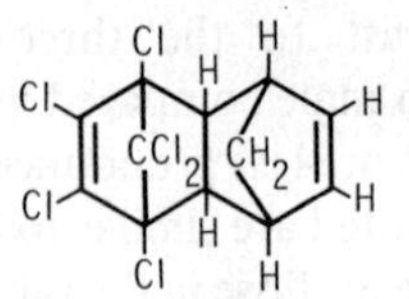

Aldrin

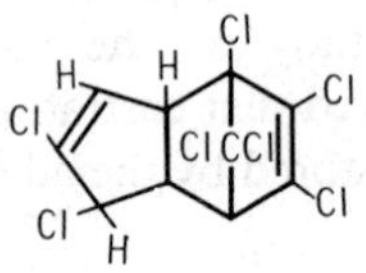

Heptachlor

FIGURE 5.13. Structures of the insecticides aldrin and heptachlor.

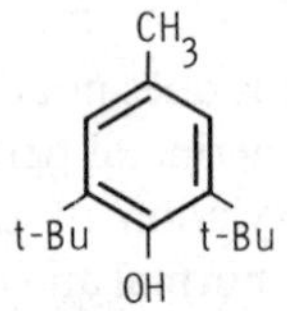

FIGURE 5.14. Structure of the food additive butylated hydroxytoluene.

mainly via the urine for males but predominantly via faecal excretion in females, which probably reflects differences in the rates of production of glucuronide and mercapturate conjugates between the sexes.

A similar but particularly important example which results in a difference in toxicity is afforded by **2,4-dinitrotoluene** (figure 5.15). This compound, used in industry, has been found to be *hepatocarcinogenic* in rats, but more so in males than females. The compound is first metabolized by hydroxylation of the methyl group followed by conjugation with glucuronic acid. The *greater susceptibility* to hepatocarcinogenicity of the male animals has been shown to be due to the greater biliary excretion of the glucuronide conjugate in the males. This is followed by breakdown in the gut to release the aglycone (1-hydroxymethyl,2,4-dinitrotoluene), reduction of the aglycone and then reabsorption of a reduced metabolite. It is this reduced metabolite which is believed to be responsible for the liver tumours.

Kidney tumours caused by several different compounds including 1,4-dichlorobenzene, isophorone and unleaded petrol have been found to be both sex and species dependent. Thus, only male rats suffer from $\alpha 2\mu$-globulin nephropathy and renal tubular adenocarcinoma as a result of the accumulation of a compound–protein complex in the epithelial cells of renal proximal tubules

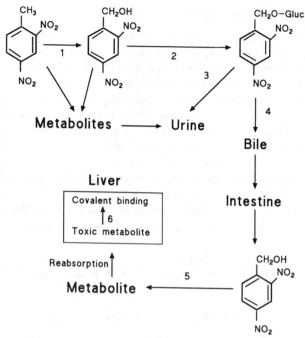

FIGURE 5.15. The role of metabolism and enterohepatic recirculation in the sex difference in toxicity of 2,4-dinitrotoluene in rats. In the female animal the glucuronide conjugate is excreted mainly in the urine (3) whereas in the male it undergoes enterohepatic recirculation following biliary excretion (4). Metabolism by the gut flora releases the aglycone which is metabolized (5) and reabsorbed. In the liver it is metabolically activated to a reactive metabolite which interacts with liver macromolecules (6) and causes liver tumours.

(see Chapter 6). The synthesis of the protein involved, $\alpha 2\mu$-globulin, is under *androgenic control* in the male rat.

Genetic factors

Inherited differences in the metabolism and disposition of foreign compounds which may be seen as strain differences in animals, and as racial and interindividual variability in man, are of great importance in toxicology.

There are many examples of human subjects showing idiosyncratic reactions to the pharmacological or toxicological actions of drugs. In some cases, the genetic basis of such reactions has been established and these cases underline the need for an understanding and appreciation of genetic factors and their role in the causation of toxicity.

Genetic factors may affect the toxicity of a foreign compound in one of two ways:

(a) by influencing the *response* to the compound;
(b) by affecting the *disposition* of the compound.

Glucose-6-phosphate dehydrogenase deficiency

A well known example of the first type in humans is deficiency of the enzyme glucose-6-phosphate dehydrogenase which is associated with susceptibility to drug-induced **haemolytic anaemia**. This is a *genetically determined trait*, carried on the X chromosome and so it is sex-linked, but the inheritance is not simple. Overall 5–10% of Negro males suffer the deficiency, but it is particularly common in the Mediterranean area and in some ethnic groups, such as male Sephardic Jews from Kurdistan; its incidence may be as high as 53%. The biochemical basis for this increased sensitivity or response, is the result of variants in the glucose-6-phosphate dehydrogenase enzyme rather than a complete absence. This in turn gives rise to a deficiency in the concentration of reduced glutathione in the red blood cell, as shown in figure 5.16. The reaction, catalysed by glucose-6-phosphate dehydrogenase and carried out in the red blood cell, is the first step in the **pentose-phosphate shunt**, or hexose monophosphate pathway.

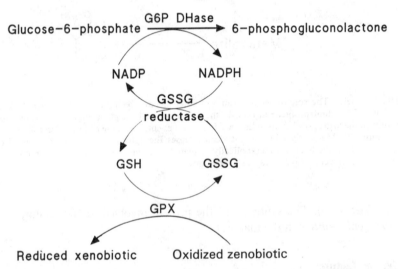

FIGURE 5.16. The interrelationship between glucose-6-phosphate dehydrogenase (G6PDHase), NADPH, GSH and the reduction of oxidized metabolites. GSSG reductase: glutathione reductase; GPX: glutathione peroxidase.

The enzyme maintains the level of NADPH in the red cell which in turn maintains the level of reduced glutathione. This functions to protect the haemoglobin from damage caused by oxidizing agents which might cause haemolysis. Thus, if the enzyme glucose-6-phosphate dehydrogenase has reduced activity, the levels of NADPH, and therefore of reduced glutathione (GSH), are low. This situation leaves the haemoglobin unprotected, and consequently when the afflicted patient takes a drug such as **primaquine** or is exposed to a chemical such as **phenylhydrazine**, haemolytic anaemia ensues. Another cause of this type of haemolytic anaemia is eating Fava beans which contain a substance able to damage haemoglobin in a similar way, hence the syndrome is also called **Favism**.

The enzyme variants are intrinsic to the red cell, and so exposure of red cells to a suitable foreign compound will lead to lysis *in vitro*.

Other similar genetic factors which lead to a heightened response are those which affect haemoglobin more directly. Thus, there are genetic traits in which the haemoglobin itself is sensitive to drugs such as **sulphonamides**. Exposure to certain drugs such as primaquine and **dapsone** can oxidize haemoglobin to methaemoglobin which is unable to carry oxygen. Individuals who have a genetically determined deficiency in **methaemoglobin reductase** are unable to remove this methaemoglobin and hence suffer cyanosis.

Heightened sensitivity to **alcohol** in individuals of Oriental origin is another example of a genetically determined increased response.

The second type of genetic factor, where the disposition is affected, is probably the more important in terms of the toxicity of foreign compounds. Variability in human populations makes drug therapy unpredictable and risk assessment for chemicals difficult. An indication of the scale of variability is afforded by the drug **paracetamol** which is discussed in more detail in Chapter 7. Thus, rates of metabolic oxidation for this drug were found to vary over a ten-fold range in human volunteers. The highest rate of oxidation, occurring in about 5% of the population studied, was comparable to that in the hamster, the species most susceptible to the hepatotoxicity of paracetamol. However, those human individuals with the lowest rates would probably be relatively resistant to the hepatotoxicity, similar to the rat. This example highlights the problems of using a single, inbred animal model to try to predict toxicity and risk to heterogeneous human populations. The variation seen may be only partly due to genetic influences, there being many factors in the human environment which may also affect drug disposition, as discussed later in this chapter. Genetically determined enzyme deficiencies which affect the disposition of foreign compounds occur in both man and animals, and the example of the Gunn rat, which is deficient in glucuronosyl transferase, has already been mentioned.

Similar defects have been described in man; for example **Gilbert's syndrome** and the **Crigler–Najjar syndrome** are both associated with reduced glucuronyl transferase and the consequent reduced ability to conjugate bilirubin. The administration of drugs also conjugated as glucuronides and therefore in competition with the enzyme may lead to blood levels of unconjugated bilirubin sufficient to result in brain damage.

ACETYLATOR PHENOTYPE

Perhaps one of the best known and fully described genetic factors in drug disposition and metabolism is the acetylator phenotype. It has been known for more than 30 years that for certain drugs which are acetylated, and **isoniazid** (figure 7.15) was the prototype, in human populations there is variation in the amount of this acetylation. This variation was found to have a genetic basis and did not show a normal Gaussian distribution, but one interpreted as **bimodal**, suggesting a **genetic polymorphism** (figure 5.17). The two groups of individuals, were termed **rapid** and **slow acetylators** because of the difference in both the amount and rate of acetylation. There are now a number of drugs which show

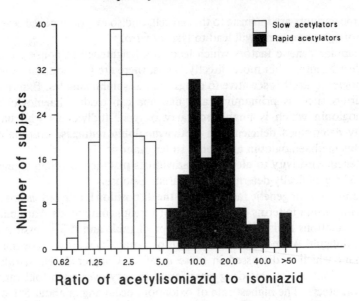

FIGURE 5.17. Frequency distribution for isoniazid acetylation. The acetylated metabolite (acetylisoniazid) and total isoniazid (acid-labile isoniazid) were measured in the urine. Data from Ellard *et al.*, (1975) *Tubercle*, **56**, 203.

this polymorphism in humans including isoniazid, **hydralazine, procainamide, dapsone, phenelzine, aminoglutethimide, sulphamethazine,** products of **caffeine** metabolism and probably aromatic amines such as the carcinogen **benzidine**. The acetylation polymorphism is an important genetic factor in a number of toxic reactions (table 5.14), and both isoniazid and hydralazine are discussed in more detail in Chapter 7. The trait has now been observed in other species, including the rabbit, mouse, hamster and rat.

Studies in human populations suggest that the acetylator phenotype is a *single gene trait*, with two **alleles** at a single autosomal genetic locus with slow acetylation a simple Mendelian *recessive* trait. The dominant–recessive relationship is unclear however. Genetic studies in the rabbit, mouse and hamster indicate that there is *codominant* expression of these alleles. Thus there are three possible genotypes with the following arrangement of the slow (r) and rapid (R) alleles: rr, rR and RR.

These would be manifested as slow, intermediate and rapid acetylators. However the ability to distinguish the three phenotypes in human populations is dependent on the method used and the particular drug administered. The *heterozygous* group, rR, may therefore be indistinguishable from the *homozygous* rapid group. The genetic trait governs the forms of the *N*-acetyltransferase enzyme, which catalyses the acetylation of amine, hydrazine and sulphonamide groups (see Chapter 4). In rabbits the activity of liver *N*-acetyltransferase was studied *in vitro* and found to show trimodal or bimodal distributions.

However, some aromatic amino compounds are not polymorphically

Table 5.14. Some toxicities and diseases proposed to be associated with acetylator phenotype.

Foreign compound	Adverse effect	Higher incidence in
Isoniazid	Peripheral neuropathy	Slow acetylators
Isoniazid	Hepatic damage	Slow acetylators
Hydralazine	Lupus erythematosus	Slow acetylators
Procainamide	Lupus erythematosus	Slow acetylators
Isoniazid in combination with phenytoin	Central nervous system toxicity	Slow acetylator
Sulphasalazine	Haemolytic anaemia	Slow acetylators
Aromatic amines	Bladder cancer	Slow acetylators
Aromatic amines	Mutagenesis/carcinogenesis	Rapid acetylators
Food pyrolysis products*	Colo-rectal cancer	Rapid acetylators

*Has yet to be fully substantiated.

acetylated, such as **p-aminobenzoic acid** and **p-aminosalicylic acid**. This is clearly illustrated by procainamide (figure 5.18) which is itself *polymorphically* acetylated (figure 5.19a) whereas the hydrolysis product p-aminobenzoic acid (figure 5.18) is *monomorphically* acetylated in the same individuals (figure 5.19b). Studies in rabbits showed that there were only small differences in acetylation *in vitro* between liver from rapid and slow acetylator phenotypes for monomorphic

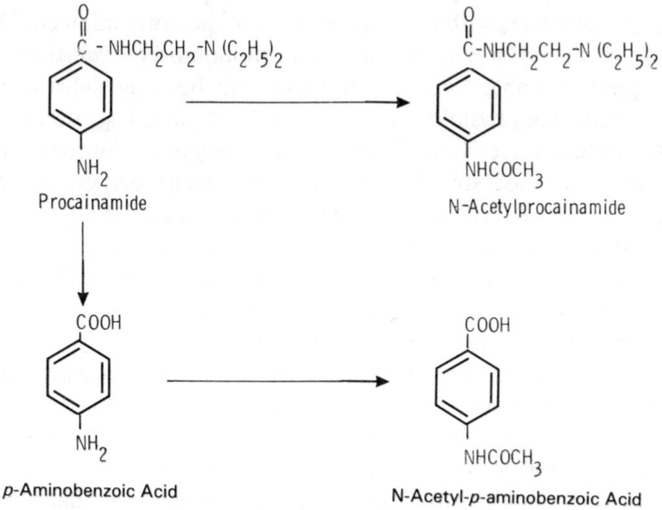

FIGURE 5.18. Metabolism of procainamide. Procainamide and the hydrolysis product p-aminobenzoic acid are both acetylated.

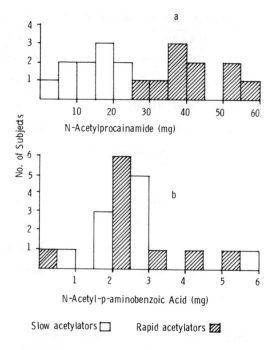

FIGURE 5.19. Frequency distribution for the acetylation of procainamide (a) and *p*-aminobenzoic acid (b) in human subjects. Data represents excretion of acetylated product in the urine 6 h after dosing.
Data from De Souich and Erill (1976) *Eur. J. Clin. Pharm.*, **10**, 283.

substrates such as *p*-aminobenzoic acid (2–2.5-fold) compared with large differences for polymorphic substrates (10–100-fold). However, using polymorphic substrates with the isolated enzyme, no correlation could be found between K_m and acetylator status; only when monomorphic substrates were used was this apparent (table 5.15). It is notable that both the substrates used are negatively charged at physiological pH. Thus the K_m and V_{max} data obtained for the isolated rabbit *N*-acetyltransferase enzyme using *p*-aminobenzoic acid and *p*-aminosalicylic acid show striking differences between genotypes, and support the hypothesis that the enzymes from slow and rapid acetylators are structurally different (table 5.15). The enzyme from heterozygous animals was intermediate between the homozygous values. The slow acetylator enzyme thus has a much greater affinity for these particular substrates, but a lower capacity than the rapid acetylator enzyme. It seems that in the rabbit the *N*-acetyltransferase from the slow phenotype is probably saturated *in vivo* at normal concentrations of drug and is therefore operating at maximum efficiency, whereas the rapid enzyme has such a high K_m that it would remain unsaturated and would be below maximum capacity. For monomorphic substrates therefore the overall rates of metabolism *in vivo* would be similar. For polymorphic substrates (sulphamethazine and procainamide, for example), the K_m values are much greater and more similar in the two phenotypes but the V_{max} values are markedly different and therefore the

Table 5.15. Characteristics of rabbit N-acetyltransferase with various substrates.

Substrate*	Apparent K_m (μM)		Apparent V_{max} (nmol/min/mg)	
	rr	RR	rr	RR
PABA	<5	105	0·24	9·3
PAS	<5	74	0·31	5·0
PA	200	67	0·35	4·4
SMZ	160	90	0·38	4·8

*PABA, p-aminobenzoic acid; PAS, p-aminosalicylic acid; PA, procainamide; SMZ, sulphameth-azine.
Data from Weber (1987) *The Acetylator Genes and Drug Response* (New York: Oxford University Press).

metabolism of such compounds *in vivo* is different in the two phenotypes (table 5.15).

It should be noted that in the hamster the situation is the reverse, with *p*-aminobenzoic acid and *p*-aminosalicylic acid being polymorphically acetylated, and sulphamethazine and procainamide monomorphically acetylated.

Thus, in the rabbit, hamster and mice the basis for the acetylator polymorphism is a qualitative difference in the *N*-acetyltransferase enzyme. In mice recent studies have indicated that other factors, such as **modifier genes**, may also affect the expression of acetyltransferase activity. In the human there are two N-acetyltransferases, NAT1 and NAT2. NAT1 is an enzyme with a wide tissue distribution which has a high affinity for monomorphic substrates such as *p*-aminobenzoic acid. NAT2 is mainly found in liver and has a high affinity for polymorphic substrates. However, there is post translational modification of the primary gene product. This has been related to several mutant alleles which have been identified in human slow acetylator populations. Thus, the human slow acetylators mainly have two mutations which results in less functional NAT2.

As can be seen from table 5.14, the acetylator phenotype is a factor in a number of toxic effects due to foreign compounds including the carcinogenicity of aromatic amines. Generally the slow acetylators are more at risk, probably because acetylation protects the amino or hydrazine group from metabolic activation. Thus, in the case of carcinogenic aromatic amines such as benzidine, slow acetylators are more susceptible to bladder cancer (relative risk, slow : rapid (S:R) 1.36), especially those individuals exposed to aromatic amines in industry (relative risk S:R, 1·7). Conversely evidence is accumulating suggesting rapid acetylators may be more at risk from colo-rectal cancer, possibly as a result of exposure to aromatic amines. Pyrolysis of food during cooking produces various mutagenic and carcinogenic amines which are known to be acetylated. Studies have shown that rapid acetylator mice have greater DNA adduct formation with **2-aminofluorene** than slow acetylator mice. The role of acetylation in carcinogenicity is discussed in more detail in Chapter 7 with regard to aminofluorene derivatives.

There are examples where several factors, including the acetylator phenotype, operate together. Hydralazine toxicity is one such example which is discussed in

Table 5.16.   Acetylator phenotype distribution in various ethnic groups
(INH, Isoniazid; SMZ, Sulphamethazine).

Ethnic Group	Rapid acetylators (%)	Drug
Eskimos	95–100	INH
Japanese	88	INH
Latin Americans	70	INH
Black Americans	52	INH
White Americans	48	INH
Africans	43	SMZ
South Indians	39	INH
Britons	38	SMZ
Egyptians	18	INH

Data from Lunde *et al.* (1977) *Clin. Pharmacokin.*, **2**, 182.

detail in Chapter 7. Another is the haemolytic anaemia caused by the drug **thiozalsulphone (Promizole)** which occurs particularly in those individuals who are both glucose-6-phosphate dehydrogenase deficient *and* slow acetylators. Promizole is acetylated, and studies in rapid and slow acetylator mice confirmed that acetylation was a factor as well as extent of hydroxylation. The latter may also be another factor in humans as is discussed below.

The acetylator polymorphism also exhibits an interesting **ethnic distribution** in humans, as shown in table 5.16, which may have important implications for the use of drugs in different parts of the world. As well as being a factor in the toxicity of several drugs, the acetylation polymorphism may influence the *efficacy* of treatment. For example, the plasma half-life of isoniazid is two to three times longer and the concentration higher in slow acetylators compared with rapid acetylators. Therefore the therapeutic effect will tend to be greater in the slow acetylators as the target microorganism is exposed to higher concentrations of drug.

HYDROXYLATOR STATUS

More recently it has become apparent that genetic factors also affect phase 1, oxidation pathways. Early reports of the defective metabolism of diphenylhydantoin in three families and of the defective de-ethylation of phenacetin in certain members of one family indicated a possible genetic component in microsomal enzyme-mediated reactions. Both these cases resulted in enhanced toxicity. Thus, diphenylhydantoin, a commonly used anticonvulsant, normally undergoes aromatic hydroxylation and the corresponding phenolic metabolite is excreted as a glucuronide (figure 7.35). Deficient hydroxylation results in prolonged high blood levels of diphenylhydantoin and the development of toxic effects, such as nystagmus, ataxia and dysarthria. The deficiency in the ability to hydroxylate diphenylhydantoin is inherited with dominant transmission.

The defective de-ethylation of phenacetin was discovered in a patient suffering methaemoglobinaemia after a reasonably small dose of the drug. This toxic effect was observed in a sister of the patient but not in other members of the family. The metabolism of phenacetin in the patient and in the sister was found to involve the

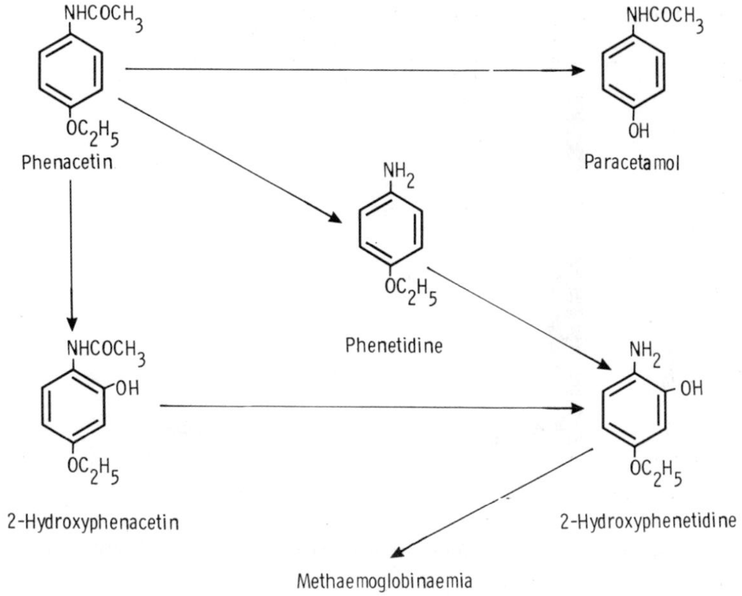

FIGURE 5.20. Metabolism of phenacetin.

production of large amounts of the normally minor metabolites 2-hydroxyphen-acetin and 2-hydroxyphenetidine, with a concomitant reduction in the excretion of paracetamol, the major metabolic product of de-ethylation in normal individuals (figure 5.20). It was suggested that autosomal recessive inheritance was involved, with the 2-hydroxylated metabolites probably responsible for the methaemoglobinaemia.

These early observations suggesting that genetic factors affect the oxidation of foreign compounds were confirmed by studies on the metabolism of the anti-hy-pertensive drug **debrisoquine**. The benzylic oxidation in the 4-position of the alicyclic ring (figure 5.21) has been found to be defective in 5–10% of the white population of Europe and North America. This is detected as a bimodal distribution when the metabolic ratio, urinary 4-hydroxydebroso-

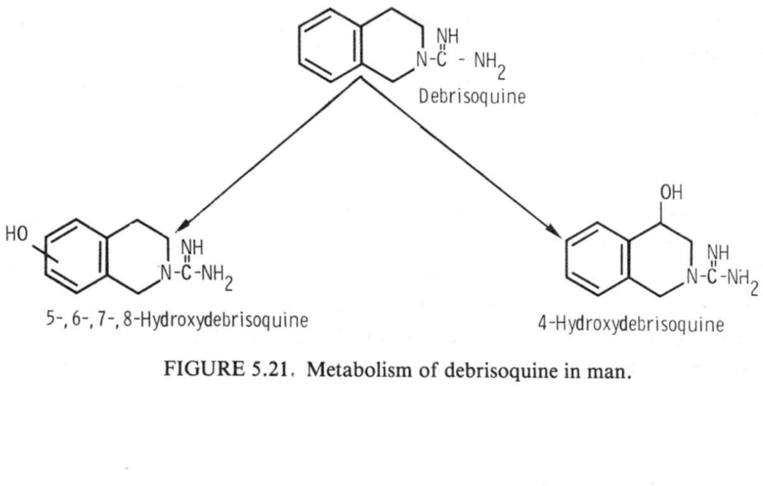

FIGURE 5.21. Metabolism of debrisoquine in man.

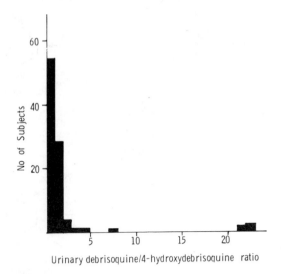

FIGURE 5.22. Frequency distribution for the ratio urinary debrisoquine: 4-hydroxydebrisoquine in human subjects.
Data from Magoub *et al.* (1977) *Lancet*, **ii**, 584.

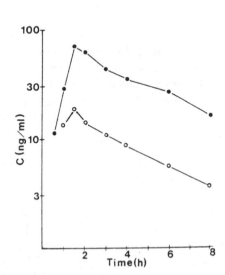

FIGURE 5.23. The plasma concentration of debrisoquine after a single oral dose (10 mg) in human subjects of the extensive (○) and poor (●) metabolizer phenotypes. Data from Sloan *et al.*, (1983) *Br. J. Clin. Pharmacol.*, **15**, 443.

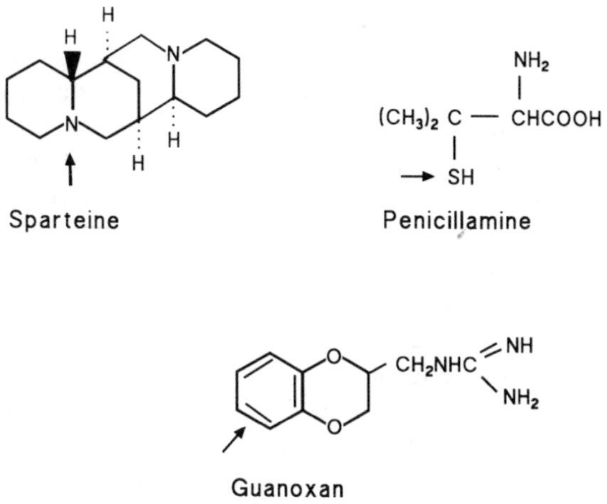

Sparteine

Penicillamine

Guanoxan

FIGURE 5.24. Compounds known to show the same oxidation polymorphism in humans as debrisoquine. The arrows indicate the site of oxidation.

quine : debrisoquine, is plotted against frequency of occurrence in the population (figure 5.22). The two phenotypes detectable are known as *poor (PM)* and *extensive (EM) metabolizers* and the poor metabolizer phenotype behaves as an autosomal recessive trait. Thus the extensive metabolizers are either homozygous (DD) or heterozygous (Dd) and the poor metabolizers are homozygous (dd). Poor metabolizers suffer an *exaggerated pharmacological effect* after a therapeutic dose of the drug as a result of a higher plasma level of the unchanged drug (figure 5.23). Extensive metabolizers excrete 10–200 times more 4-hydroxydebrisoquine than poor metabolizers. The deficiency extends to more than 20 different drugs, and various types of metabolic oxidation reaction. For example, the aromatic hydroxylation of **guanoxon**, *S*-oxidation of **penicillamine**, oxidation of **sparteine** (figure 5.24) and the hydroxylation of **bufuralol** (figure 5.1) have all been shown to be polymorphic. There are now a number of **adverse drug reactions** which are associated with the poor metabolizer status. For example perhexiline may cause hepatic damage and peripheral neuropathy in poor metabolizers in whom the half-life is significantly extended; penicillamine causes skin rashes, haematuria and thrombocytopenia in poor metabolizers; lactic acidosis may be associated with the use of phenformin in poor metabolizers.

The biochemical basis for the trait is an almost complete absence of one form of cytochrome P-450, CYP 2D1. It seems that there are several mutations which give rise to the poor metabolizer phenotype. These mutations produce incorrectly spliced, variant mRNAs in the liver from poor metabolizers, and two mutant alleles have been identified which are linked to the poor metabolizer phenotype. However, for some of the substrates of the polymorphism, such as bufuralol, **biphasic kinetics** has been demonstrated *in vitro*. This data and data from studies using inhibitors such as **quinidine**, which is specific for the debrisoquine hydroxylase form of cytochromes P-450, has indicated that two functionally different forms of cytochrome P-450 isoenzymes, are involved. These two forms differ in that one has low affinity whereas the other has high affinity. Only one of these may be under polymorphic control. The metabolism of **bufuralol** is further complicated by the fact that it is stereospecific. The aliphatic hydroxylation (1-hydroxylation) is selective for the $(+)$ isomer, whereas the aromatic 4- and 6-hydroxylation are selective for the $(-)$ isomer. Both aliphatic and aromatic hydroxylation of bufuralol are under the same genetic control, and the selectivity is virtually abolished in the poor metabolizer. Only the high affinity, polymorphic enzyme is stereospecific; the low affinity enzyme is not.

Studies with **sparteine** have also indicated that there are two isoenzymes involved in metabolism to 2- and 5-dehydrosparteine. Thus, there is a high affinity, quinidine sensitive form and a low affinity, quinidine insensitive form. The formation of the metabolites by the two isoenzymes is, however, quantitatively and qualitatively different. Both isoenzymes exhibit large interindividual differences, but it is believed that the low affinity enzyme is not controlled by the debrisoquine polymorphism. The variation in the high affinity isoenzyme is suggested as being due to **allozymes** of cytochrome P-450; that is, different enzymes from the alleles comprising one gene.

The importance of such polymorphisms in human susceptibility to diseases, such as cancer, is now increasingly being recognized. For example, studies in humans with **aminobiphenyl**, a liver and bladder carcinogen, have shown that both the acetylator phenotype and hydroxylator status are important in the formation of adducts (see Chapter 6).

Another similar, but *distinct*, genetic polymorphism concerns the aromatic 4-hydroxylation of the drug **mephenytoin** (figure 5.25). This hydroxylation deficiency occurs in 2–5% of Caucasians and in 20% of Japanese. Again it is an autosomal recessive trait. The enzyme, cytochrome P-450, form CYP 2C, from poor metabolizers has a high K_m and low V_{max}. As with the hydroxylation of bufuralol, the hydroxylation is stereoselective. Thus, only S-mephenytoin undergoes aromatic 4-hydroxylation and only this route is affected by the polymorphism.

ALCOHOL METABOLISM

Another oxidation reaction which shows variation in human populations is the oxidation of ethanol. This has been shown to be significantly lower in Canadian Indians compared with Caucasians, and thus the Indians are more susceptible to the effects of alcoholic drinks. The rate of metabolism *in vivo* in Indians is $0 \cdot 101$ g/kg/h compared with $0 \cdot 145$ g/kg/h in Caucasians.

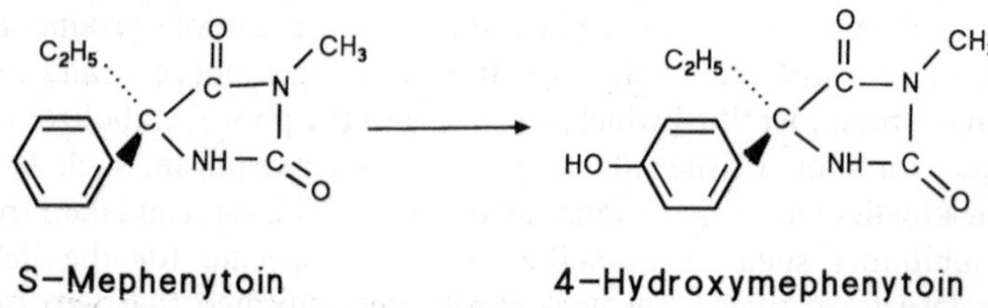

FIGURE 5.25. Aromatic hydroxylation of mephenytoin.

This seems to be due to variants in alcohol dehydrogenase, although differences in aldehyde dehydrogenase may also be involved. Variants of alcohol dehydrogenase resulting in increased metabolism have also been described within Caucasian and Japanese populations.

ESTERASE ACTIVITY

Hydrolysis reactions can also exhibit genetic influences such as the plasma enzyme, a **pseudocholinesterase**, which is responsible for the metabolism and inactivation of the drug **succinylcholine** (figure 7.36). Certain individuals may be defective in the ability to hydrolyse, and therefore inactivate, this drug. Such individuals have a form of the enzyme with a decreased hydrolytic activity, and the half-life of the drug is dramatically increased. Consequently, such individuals may be affected by the drug for 2–3 *hours* instead of the more normal 2–3 *minutes*. In extreme cases apnoea may result from the prolonged neuromuscular blockade and muscular relaxation caused by the drug. Family studies are consistent with this deficiency being a recessive trait governed by an autosomal autonomous gene with two alleles. The three genotypes are manifested as two phenotypes: rapid and slow hydrolysis of succinyl choline, although careful analysis may reveal three. Thus, the atypical pseudocholinesterase occurs in the individuals homozygous for the abnormal gene, the individuals homozygous for the normal gene have the normal enzyme and the heterozygotes produce a mixture of enzymes. The trait has a frequency of about 2% in some populations studied (British, Greek, Portuguese, North African), but it is absent from other populations such as Japanese and Eskimo, for example. There are in fact a number of variants of the atypical enzyme which can be distinguished by using specific inhibitors.

Environmental factors

There are many factors in the environment which may influence drug disposition, metabolism and toxicity to a greater or lesser extent. However, as the influence of certain foreign compounds, both drugs and those in the environment, on microsomal enzymes has been well studied, this will constitute a separate section (page 172).

Stress

Adverse environmental conditions or stimuli which create stress in an animal may influence drug metabolism and disposition. Cold stress, for instance, increases aromatic hydroxylation, as does stress due to excessive noise. It should be noted that the microsomal mono-oxygenases show a diurnal rhythm in both rats and mice with the greatest activity at the beginning of the dark phase.

Diet

The influence of diet on drug metabolism, disposition and toxicity consists of many constituent factors. Food additives and naturally occurring contaminants in food may influence the activities of various enzymes by induction or inhibition. However, these factors are discussed in a later section (page 172). The factors with which this section will be concerned are the nutritional aspects of diet.

The multitude of factors contained within the environment which may influence drug disposition and metabolism are difficult to separate. The finding that race and diet may affect the clearance of drugs such as antipyrine is such an example. Thus, meat-eating Caucasians have a significantly greater clearance and a shorter plasma half-life for antipyrine than Asian vegetarians. However, the relative importance of race versus diet as contributions to those differences is not clear.

The nutritional status of an animal is well recognized as having an important influence on drug metabolism, disposition and toxicity. The lack of various nutrients may affect drug metabolism, though not always causing a depression of metabolic activity. Lack of protein has been particularly well studied in this respect, and shows a marked influence on drug metabolism. Thus, rats fed on low-protein diets (5%) show a marked *loss* of microsomal enzyme activity when compared with those animals fed a 20% protein diet (table 5.17). The decline in activity is accompanied by a decline in the level of liver microsomal protein. Both cytochrome P-450 content and cytochrome P-450 reductase activity are reduced by a 5% protein diet.

Table 5.17. Effect of reduced protein diet on hepatic microsomal enzyme activity in rats.

	Ethylmorphine *N*-demethylation (nmol HCHO/100 g body wt/10min)
20% protein diet: control	10·5
5% protein diet: 4 days	6
5% protein diet: 8 days	<1

Data from Gibson and Skett (1986) *Introduction to Drug Metabolism* (London: Chapman & Hall).

Measurements *in vivo*, such as barbiturate sleeping times, are in agreement with these findings, sleeping times being longer in protein-deficient animals. Toxicity may also be influenced by such factors as a low-protein diet. For example, the hepatotoxicity of carbon tetrachloride is markedly less in protein-deficient rats than in normal animals, and this correlates with the reduced ability to metabolize the hepatotoxin in the protein-deficient animals. However, the reverse is the case with the hepatotoxicity of paracetamol, which is increased after a low-protein diet. This may be due to the reduced levels of glutathione in rats fed low-protein diets, which offsets the reduced amount of cytochromes P-450 caused by protein deficiency.

The carcinogenicity of **aflatoxin** is reduced by protein deficiency, presumably because of reduced metabolic activation to the epoxide intermediate which may be the ultimate carcinogen which binds to DNA (figure 5.26). A deficiency in dietary fatty acids also decreases the activity of the microsomal enzymes. Thus, ethylmorphine, hexobarbital and aniline metabolism are decreased, possibly because lipid is required for cytochromes P-450. Thus, a deficiency of essential

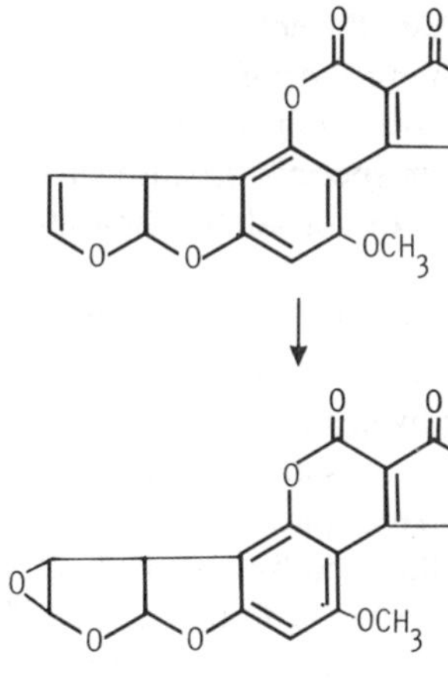

FIGURE 5.26. Epoxidation of aflatoxin B_1.

fatty acids leads to a decline in both cytochromes P-450 levels and activity *in vivo*.

Mineral and **vitamin deficiencies** also tend to reduce the metabolism of foreign compounds. Carbohydrates, however, do not seem to have major direct effects on the metabolism of foreign compounds. **Starvation** or changes in diet may also reduce supplies of essential cofactors such as sulphate which is required for phase 2 conjugation reactions and may be readily depleted (see Chapter 7). Overnight fasting of animals may also have significant effects on metabolism and toxicity. Thus the dealkylation of **dimethylnitrosamine** (figure 7.3) is increased by this short-term food deprivation and the hepatotoxicity is consequently increased. The hepatotoxicity of **paracetamol** and **bromobenzene** are also both increased by overnight fasting of animals, probably because this results in the depletion of glutathione to 50% of normal levels. There is thus less glutathione available for the detoxication of these compounds (see Chapter 7).

The nutritional status of an animal may affect the disposition of a foreign compound *in vivo* as well as the metabolism. Many drugs are protein-bound in the plasma, and alteration of the extent of binding for compounds extensively bound may have important toxicological implications. Thus, the decreased plasma levels of **albumin** after low-protein diets, such as occur in the human deficiency disease Kwashiorkor, might lead to significantly increased plasma levels of the free drug and therefore the possibility of increased toxicity.

Age effects

It is well known that the drug-metabolizing capacity and various other metabolic and physiological functions in man and other animals are influenced by age. Furthermore, sensitivity to the toxic and pharmacological effects of drugs and other foreign compounds is often different in young and geriatric animals. As might be expected, at the extremes of age, drug-metabolizing activity is often impaired, plasma-protein binding capacity is altered and the clearance of foreign compounds from the body may be less efficient. These differences generally lead to exposure to a higher level of unchanged drug for longer periods in the young and geriatric animals than in adults, with all the implications for toxicity that this has. The case of the foetus is rather special, because of the influence of the maternal organism, and will therefore be considered separately at the end of this section.

Although the toxicological significance has yet to be studied, age-related differences in the absorption of foreign compounds are demonstrable. Neonatal and geriatric human subjects have low gastric acid secretion, and consequently the absorption of some foreign compounds may be altered. Thus, in the neonate **penicillin** absorption is enhanced whereas paracetamol absorption is decreased.

Intestinal motility may also be influenced by age, with various effects on the absorption of foreign compounds dependent upon the site of such absorption. The absence of gut flora in the neonate may have as yet unknown influences on the disposition of foreign compounds in the gastro-intestinal tract.

Once absorbed, foreign compounds may react with plasma proteins and distribute into various body compartments. In both neonates and elderly human subjects both total plasma-protein and plasma-albumin levels are decreased. In the neonate the plasma proteins may also show certain differences which decrease the binding of foreign compounds, as will the reduced level of protein. For example, the drug lidocaine is only 20% bound to plasma proteins in the newborn compared with 70% in adult humans. The reduced plasma pH seen in neonates will also affect protein binding of some compounds as well as the distribution and excretion. Distribution of compounds into particular compartments may vary with age resulting in differences in toxicity. For example, **morphine** is between three and ten times more toxic to newborn rats than adults because of increased permeability of the brain in the newborn. Similarly this difference in the **blood–brain barrier** underlies the increased neurotoxicity of lead in newborn rats.

Total body water, particularly extracellular water, has been found to be greater in neonates than in adults and to decrease with age. The distribution of water-soluble drugs could clearly be influenced by this; lower plasma levels being one possible result. Both glomerular filtration and renal tubular secretion are lower in neonates and geriatrics, and human infants achieve adult levels of glomerular filtration by one year of age at the earliest. The consequences of this are reduced excretion, and hence reduced body clearance of foreign compounds. Consequently, toxicity may be increased by prolongation of exposure or accumulation if chronic dosing is involved.

This was an important factor in the development of adverse reactions to the anti-arthritis drug **benoxaprofen (Opren)**, which caused serious toxicity in some elderly patients, necessitating withdrawal of the drug from the market.

Many drug-metabolizing enzyme systems show marked changes around the time of birth, being generally reduced in the foetus and neonate. In some cases, adult activity may be achieved within a few days, whereas for some enzymes, such as cytochromes P-450, several weeks may be necessary to obtain optimum levels. The development of these enzyme systems does however depend on the species. For example, in the rat, microsomal mono-oxygenase activity is *negligible* at birth whereas in humans at 6 months of gestation, enzyme activity is not only detectable, but may be between 20 and 50% of adult levels. In the rat development to adult levels of activity takes about 30 days after which levels decline towards old age. In humans, however, hydroxylase activity increases up to the age of 6 years, reaching levels greater than those in the adult, which only decrease after sexual maturation. Thus the elimination of antipyrine and theophylline was found to be greater in children than in adults. It should be noted, however, that proportions of isoenzymes may be very different in neonates from the adult animal, and the development of the isoenzymes may be different. Thus, in the rat there seem to be four types of development for phase 1 metabolizing enzymes: linear increase from birth to adulthood, type A (aniline 4-hydroxylation); low levels until weaning, then an increase to adult levels, type B (N-demethylation); rapid development after birth followed by rapid decline to low levels in adulthood, type C (hydroxylation of 4-methylcoumarin); rapid increase after birth to a maximum and then decline to adult levels, type D. Patterns of development may be different between sexes as well as between species. For example, in the rat, steroid 16α-hydroxylase activity towards androst-4-ene-3,17-dione, develops in type B fashion in both males and females, but in females activity starts to disappear at 30 days of age and is undetectable by 40 days. It seems that the mono-oxygenase system develops largely as a unit, with the rate dependent on species and sex of the animal, and the particular substrate.

With some of the phase 2 metabolizing enzymes there may be strict **ontogenetic patterns of expression**. Sulphate conjugation ability occurs early in rats, whereas glucuronidation (of xenobiotics), and conjugation with glutathione and amino acids, only develop over about 30 days from birth.

These deficiencies in metabolic capacity in neonates may have significance for the toxicity of foreign compounds.

When the parent compound is responsible for the pharmacological or toxicological effect, reduced metabolic activity may lead to prolonged and exaggerated responses. For example, in mice there is very little oxidation of the side-chain of **hexobarbital,** and the sleeping time is excessively prolonged (by 70 times), as shown in table 5.18. Doses giving sleeping times of less than one hour in adult mice are fatally toxic to neonates. This activity correlates with the inability to metabolize the drug.

The converse is true of drugs requiring metabolic activation for toxicity. For

Table 5.18.   Effect of age on metabolism and duration of action of hexobarbital.

Age (days)	Percent hexobarbital metabolized *in vitro*† in 1 h (guinea-pigs)	Percent hexobarbital metabolized *in vivo* in 3 h (mice)	Sleeping time in mice (min)		
			(10 mg/kg)	(50 mg/kg)	(100 mg/kg)
1	0	0	>360	Died	Died
7	2·5–3·5	11–24	107 ± 26	243 ± 30	>360
21	13–21	21–33	27 ± 11	64 ± 17	94 ± 27
Adult	28–39		<5	17 ± 5	47 ± 11

†Guinea-pig liver microsomes.
Data of Jondorf *et al.* (1958) *Biochem. Pharmac.*, **1**, 352.

example, **paracetamol** is less hepatotoxic to newborn than to adult mice, as less is metabolically activated in the neonate. This is due to the lower levels of cytochromes P-450 in neonatal liver (figure 5.27). Also involved in this is the hepatic level of glutathione, which is required for detoxication. Although levels of this tripeptide are reduced at birth, development is sufficiently in advance of cytochromes P-450 levels to ensure adequate detoxication (figure 5.27). The same effect has been observed with the hepatotoxin bromobenzene. (For further details of paracetamol and bromobenzene see Chapter 7.) Similarly, **carbon tetrachloride** is not hepatotoxic in newborn rats as metabolic activation is required for this toxic effect, and the metabolic capability is low in the neonatal rat.

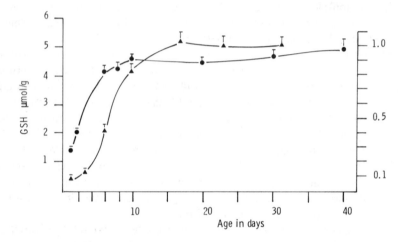

FIGURE 5.27. Development of hepatic reduced glutathione and cytochrome P-450 levels with age in mice. Glutathione (●), cytochrome P-450 (▲).
Data of Hart and Timbrell (1979) *Biochem. Pharmac.*, **28**, 3015.

The deficiency in phase 2 metabolizing enzymes may also influence the toxicity of foreign compounds. For example, glucuronosyl transferase activity in human neonates is often low and hence conjugation is impaired, which may be important for the detoxication of a drug. A relative deficiency of UDP-glucuronic acid may also contribute to the lower level of conjugation. Thus, with **chloramphenicol** which is 90% conjugated with glucuronc acid, neonatal deficiency of glucuronide conjugation can lead to severe cyanosis and death in some human infants due to the high plasma levels of unchanged drug. With paracetamol metabolism in newborn infants however, sulphate conjugation, which is normal in the neonate, will compensate for the reduction in glucuronidation. When conjugation of bilirubin with glucuronic acid is impaired **neonatal jaundice** results leading to brain damage due to the elevated levels of free bilirubin. This may be exacerbated by the administration of drugs which displace bilirubin from its binding sites on plasma albumin.

exposure to foreign compounds may result in pathological damage which may influence the disposition of the compound on subsequent exposure.

The absorption of foreign compounds from the gastrointestinal tract may be altered in certain malabsorption syndromes, and may be increased after subcutaneous or intramuscular injection if vasodilation accompanies the particular disease. The disposition of foreign compounds, once absorbed, can be influenced by changes in plasma proteins, which are sometimes reduced in certain disease states. Consequently, for foreign compounds which are highly protein-bound, the plasma concentration of the free compound may be significantly increased in such circumstances. This may alter renal excretion, increasing it in some cases, but could also increase the toxicity of a drug with a narrow therapeutic ratio if the compound dissociated from the protein was responsible for the toxicity. Thus, thiopental anaesthesia is prolonged when plasma albumin is reduced by chronic liver disease, and more unbound diphenylhydantoin and sulphonamides result from changes in plasma proteins in chronic liver disease.

Hepatic disease and damage clearly have the potential to be major factors in the metabolism of foreign compounds. Thus, in patients with liver necrosis due to paracetamol, the half-life of the latter was increased from 2 to 8 h and that of **antipyrine** from 12 to 24 h. Acute hepatic necrosis in animals caused by the administration of hepatotoxins resulted in the plasma half-lives of barbiturates, **diphenylhydantoin** and antipyrine being approximately doubled. However, there may be several factors operating such as *displacement* of a drug from plasma-protein binding sites by bilirubin. In liver damage, plasma-bilirubin levels may be high due to lack of conjugation with glucuronic acid. This may alter elimination of some drugs such as tolbutamide. The level of plasma albumin may often be reduced by liver disease as this is the site of synthesis of albumin, and hence binding to plasma albumin will tend to be reduced. However, the effects of liver disease can be somewhat unpredictable. For example, in patients with cirrhotic liver, the glucuronidation of **chloramphenicol** and the acetylation of **isoniazid** are both reduced and hence the half-lives are prolonged. However, chronic liver disease did not affect the hepatic clearance of **lorazepam** or **oxazepam** despite the fact that glucuronidation is the route of metabolism for these compounds. In general, the formation of glucuronic acid and sulphate conjugates tend to be *impaired* in liver diseases such as hepatitis, obstructive jaundice and cirrhosis.

Hepatic damage may also affect the disposition of some drugs by altering **hepatic blood flow**. For example, **Indocyanine Green** is not metabolized and the clearance is related to hepatic blood flow. In cholestatic liver damage clearance is reduced to 70% of the control. Both **antipyrine** and **aminopyrine** are cleared by metabolism in the liver, yet only the clearance of antipyrine is affected by the cholestatic damage. Although it seems that hepatocellular disease will generally affect drug clearance, it is still difficult to extrapolate from the known effects on the disposition of a particular drug to the likely effects on another. It is clearly important to know the relevant factors in the disposition of a particular drug, and also the severity and type of pathological damage. Thus, liver cirrhosis will alter blood flow but will not necessarily affect enzyme levels. Mild to moderate hepatitis in humans was found to have no influence on hepatic cytochromes P-450 content, yet severe hepatitis and cirrhosis reduced the content by 50%. The liver synthesizes cholinesterases, and therefore drugs hydrolysed by these enzymes, such as **aspirin, procaine** and **succinylcholine**, show reduced metabolism. Thus, liver damage and disease may have a number of effects on the disposition of a compound which may be due to:

(a) alteration in enzyme levels or activities
(b) alteration in blood flow
(c) alteration in plasma albumin and hence binding.

Decreased ability to form glucuronides may also occur in **Gilbert's disease** and the **Crigler–Najjer syndrome**. In these genetic disorders, glucuronyl transferase is reduced and consequently bilirubin conjugation may be affected when drugs which are also conjugated with glucuronic acid are administered.

Renal disease is another important factor, particularly if renal excretion is the major route of elimination for the pharmacologically or toxicologically active compound. This may be particularly important with drugs showing a low therapeutic ratio, such as digoxin and the aminoglycoside antibiotics. As already mentioned, plasma protein binding may be affected in renal disease, such binding being reduced particularly with regard to organic acids. Increased toxicity of drugs undergoing significant metabolism, such as chloramphenicol, has been found in uraemic patients. The half-lives of a number of drugs are prolonged in renal failure, although this effect is variable and by no means the general rule, different drugs being differently affected.

Chronic **renal disease** may also affect metabolism, not necessarily because of impaired metabolism in the kidney, but because of an indirect effect of renal failure on liver metabolism. For example, in animals with renal failure it was observed that there was a decrease in hepatic cytochromes P-450 content, and consequently zoxazolamine paralysis time and ketamine narcosis time were prolonged.

Cardiac failure may also affect metabolism by altering hepatic blood flow. However even after heart attack without hypotension or cardiac failure, metabolism may be affected. For example, the plasma clearance of lidocaine is reduced in this situation. Other diseases such as those which affect hormone levels: **hyper-** or **hypo-thyroidism**, lack of or excess **growth hormone**, and **diabetes** can alter the metabolism of foreign compounds.

Infectious diseases may also affect metabolism, partially by increasing levels of the endogenous compound interferon which will inhibit some metabolic pathways.

Tissue and organ specificity

The disposition or localization, and in some cases metabolism, of foreign compounds may be dependent upon the characteristics of a particular tissue or organ which may in turn affect the toxicity. There are many examples of organotropy in toxicology, but the mechanisms underlying such organ-specific toxic effects are often unknown.

It is clear that a foreign compound which is chemically similar to, or at least has certain structural similarities with, an endogenous compound may become localized in a particular tissue(s). An example of this is **6-hydroxydopamine**, a single dose of which is selectively toxic to sympathetic nerve endings. Because of its similarity to noradrenaline, 6-hydroxydopamine is distributed specifically to the sympathetic nerve endings. There it is oxidized to a quinone (figure 7.29), which binds covalently to the nerve endings and permanently inactivates them. In other cases, however, the mechanism of specificity is less clear. Certain carcinogens show organ or tissue specificity, such as *trans*-4-dimethylamino-stilbene, which causes earduct tumours after repeated administration, or hydrazine, which causes lung tumours (figure 5.28). This may reflect the ability of the tissue to carry out repair of damaged macromolecules, such as DNA. Thus, if DNA repair is poor in a particular tissue, such as nervous tissue, that tissue may be particularly susceptible to certain carcinogens. The carcinogen **diethylstilboestrol** induces tumours in those female organs particularly exposed to oestrogens, namely the mammary glands, uterus and vagina (figure 6.19). In male hamsters, however, this compound causes kidney tumours.

Although certain organs are prime targets for toxic effects because of their anatomical position and function, such as the gut, liver and kidney, they may not necessarily be the most susceptible. Such organs may have a particular ability to cope with a toxic insult, and this has been termed *reserve functional capacity*. Examples of organ-specific toxicity are the lung toxicity of the natural furan ipomeanol (figure 7.23), 3-methylfuran (figure 5.28), the herbicide paraquat

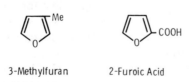

FIGURE 5.28. Structure of 4-dimethylaminostilbene, hydrazine, 2-methylfuran and 2-furoic acid.

(figure 3.16) and the kidney toxicity of chloroform (figure 7.17), 2-furoic acid (figure 5.28) and *p*-aminophenol (figure 5.7). However, some of these compounds, such as 3-methylfuran and paraquat, are rarely, if ever, hepatotoxic, even after oral administration. The lung toxin ipomeanol is considered in greater detail in the final chapter (see page 319).

The phospholipidoses caused by certain amphiphilic drugs typified by chlorphentermine (figure 3.19) tend to be organ- and tissue-specific, occurring particularly in the lungs and adrenals. This correlates with the accumulation of the drug, which is localized in the fatty tissue, particularly that associated with the adrenals and lungs (pages 52–3). Toxicity to target organs is discussed in greater detail in Chapter 6.

Dose

It is clear from Chapter 2 that the dose of a toxic compound is a major factor in its toxicity. This may be a simple relationship resulting from the increasing concentration of toxin at its site of action, with a proportional increase in response with increasing dose. However, the size of the dose may also influence the disposition or metabolism of the compound.

Thus, a large dose may be ineffectively distributed and remain at the site of administration as a depot. A large dose of a compound given orally, for instance, may not be all absorbed, depending on the rates of absorption and transit time within the gut. Saturable active absorption processes would be particularly prone to dose effects, which could result in unexpected dose–response relationships.

Once a toxic compound has been absorbed, the disposition of it *in vivo* may also be affected by the dose. Thus, saturation of plasma-protein binding sites may lead to a significant rise in the plasma concentration of free compound, with possible toxic effects. This, of course, depends on the fraction of the compound that is bound, and is only of significance with highly protein-bound substances. The result of this is a disproportionate rise in toxicity for a small increase in dose.

Saturation of the processes involved in the elimination of a foreign compound from the plasma, such as metabolism and excretion, may also have toxicological consequences. Thus, ethanol exhibits zero-order elimination kinetics at readily attainable plasma concentrations, because the metabolism is readily saturated. Therefore, once these plasma concentrations have been attained, the rate of elimination from the plasma is constant. Increasing the dosage of ethanol leads to accumulation and the well known toxic effects.

The biliary excretion of **furosemide** is also an active process and saturable, and after high doses of furosemide there is a disproportionate increase in the plasma level of the drug. This appears to be responsible for the toxic dose threshold, above which hepatic necrosis ensues (figures 3.33 and 3.34, pages 68–9). Similarly, saturation of the biliary excretion of proxicromil in dogs is believed to be the cause of the hepatotoxicity (figure 5.9).

The metabolism of toxic compounds may also be influenced by dose, with possible toxicological consequences. Saturation of metabolic routes may increase toxicity if the parent compound is more toxic than its metabolites, or if minor but toxic routes become more important. Conversely, toxicity may not increase proportionately with dose if the pathway responsible becomes saturated. For example, paracetamol hepatotoxicity is dose-dependent, but only above a threshold dose. This threshold is the result of saturation of the glutathione conjugation pathway. Another example is isoniazid hepatotoxicity, which results from an acetylated metabolite. Giving large doses of isoniazid to experimental animals does not cause hepatic necrosis, whereas giving several smaller doses does, probably because acetylation is saturated at high doses and the drug is metabolized by other routes. These and other examples such as salicylic acid and vinyl chloride are discussed in greater detail in the final chapter. The problem of **dose-dependent metabolism** is also important in considerations of the extrapolation from animals to man in the prediction of risk and safety assessment. For example, the food additive **estragole** undergoes side-chain, aliphatic oxidation followed by conjugation to form a carcinogenic metabolite (figure 5.29). Estragole is carcinogenic in mice at a dose of around 511 mg/kg, but humans are only exposed to around 1 μg/kg per day. When the metabolism of estragole in mice and rats is examined it is found that it is dose-dependent, with the proportion of the dose metabolized to the toxic metabolite increasing by ten-fold from about 1 to 10%. In humans the pathway only represents 0·3% of the dose at the normal daily intake. Thus, in this case the ratio of the amount of toxic metabolite required to produce tumours in mice to that which man is normally exposed to is about 15 million to one. These findings clearly have implications with regard to setting *safe* limits for daily intake of such substances. The data also underline the crucial importance of metabolic data gathered at different doses for the sensible interpretation and utilization of toxicity data. Another example of this is the case of **saccharin**, the artificial sweetener. This compound caused bladder cancer in experimental animals when these were exposed to levels of 5–7% of the diet but not if exposed to levels up to 5%. However, pharmacokinetic studies revealed that at these high exposure levels the plasma clearance of saccharin was saturated and therefore tissue levels would be higher than expected on the basis of a linear extrapolation from lower doses. Consequently, prediction of the incidence of bladder tumours at the lower exposure levels to which humans would be exposed could not reasonably be based on a simple linear extrapolation from the data obtained at the high exposure levels.

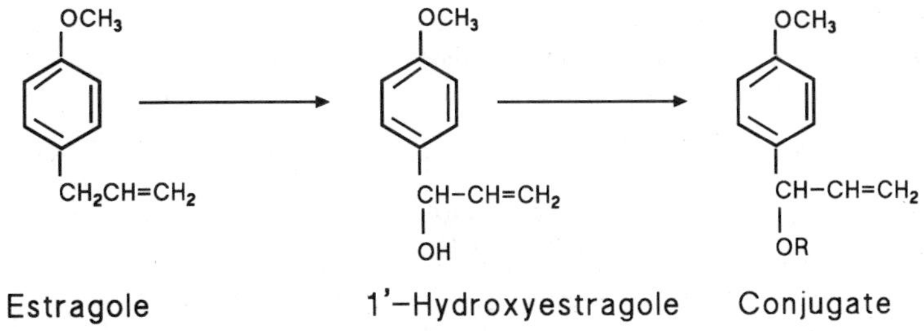

FIGURE 5.29. Metabolic activation of estragole.

Enzyme induction and inhibition

Most biological systems, and especially humans, are exposed to a large number of different chemicals in the environment. Thus, pesticides, natural contaminants in food, industrial chemicals and agricultural pollutants may all contaminate the environment and thereby affect various biological systems in that environment. Humans and certain other animals may also be exposed to drugs and food additives, and humans are exposed to substances in the workplace. These chemicals may modify their own disposition and that of other chemicals in several ways. One way this may occur is by an effect on the enzymes involved in the metabolism of foreign compounds; these enzymes may be induced or inhibited. By altering the routes or rates of metabolism of a foreign compound, either induction or inhibition clearly can have profound effects on the biological activity of the compound in question.

Enzyme induction

Although first reported with the microsomal mono-oxygenases, it is now known that a number of the enzymes involved in the metabolism of foreign compounds are inducible. Thus, as well as the cytochromes P-450, NADPH cytochrome P-450 reductase, cytochrome b_5, glucuronosyl transferases, epoxide hydrolases and glutathione transferases are also induced to various degrees. However, this discussion concentrates on the induction of the mono-oxygenases with mention of other enzymes where appropriate.

The induction of the microsomal enzymes has been demonstrated in many different species including humans, and in various different tissues as well as the liver. Induction usually results from repeated or chronic exposure although the extent of exposure is variable. The result of induction is an increase in the amount of an enzyme; induction requires *de novo* protein synthesis, and therefore an

Table 5.19.   Major types of cytochrome P-450 enzyme inducing agent.

Type	Examples	Isoenzymes induced
Barbiturate	Phenobarbital	CYP 2B1/2; 2C CYP 3A1/2
Polycyclic	3-Methylcholanthrene	CYP 1A1
Hydrocarbon	TCDD; benzo[a]pyrene Isosafrole	CYP 1A2
Alcohol/acetone	Acetone; isoniazid;	CYP 2E1
Steroid	Pregnenolone 16α-carbo nitrile;dexamethazone	CYP 3A1/2
Clofibrate	Diethylhexylphthalate; Ciprofibrate;Nafenopin	CYP 4A1/2

TCDD: 2,3,7,8-tetrachlordibenzo-*p*-dioxin.

increase in the apparent metabolic activity of a tissue *in vitro* or animal *in vivo*. Consequently, inhibitors of protein synthesis, such as cycloheximide, inhibit induction. It is a reversible, cellular response to exposure to a substance. Thus, it can be shown in isolated cells, such as hamster foetal cells in culture, that exposure to benzo[*a*]anthracene induces aryl hydrocarbon hydroxylase activity (AHH), one of the isoenzymes of cytochrome P-450.

A large variety of substances have been shown to be inducers, probably numbering several hundred. Apart from the fact that many are lipophilic and organic, there are no common factors and many chemical classes of compound are included (table 5.19). However, there are several types of induction which can be differentiated, and within some of these types inducers show certain structural similarities. For example, inducers of the **polycyclic hydrocarbon type** tend to be planar molecules.

Planar and non-planar **polychlorinated biphenyls** (figure 5.30) differ in the type of induction they will cause. Thus, 3,3′,4,4′,5,5′-hexachlorobiphenyl is a planar molecule which is an inducer of the polycyclic hydrocarbon type. 2,2′,4,4′, 6,6′-Hexachlorobiphenyl is a non-planar molecule due to the steric hindrance between the chlorine atoms in the 2- and 6-positions, and is a **phenobarbital type** of inducer. The variety and type of inducing agents are shown in table 5.19. Some compounds may indeed be **mixed types** of inducers, and thus mixtures of planar and non-planar polychlorinated biphenyls are found to act as inducers of both the polycyclic and phenobarbital type.

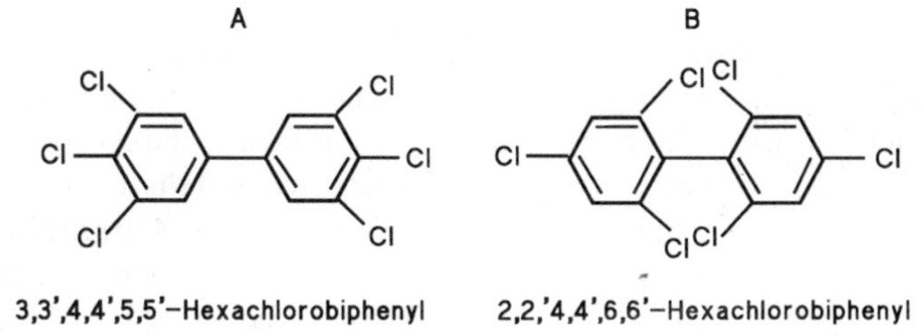

FIGURE 5.30. The structures of planar (A) and non-planar (B) hexachlorobiphenyls.

As already discussed in Chapter 4, cytochrome P-450 has many forms or isoenzymes which differ in their ability to catalyse particular reactions. Some of these forms of cytochrome P-450 are found in normal liver tissue and are 'constitutive', whereas others are only apparent after induction. Constitutive as well as non-constitutive forms of cytochrome P-450 are inducible. Some of the major forms of cytochrome P-450 which are induced are shown in table 5.19. It should be noted, however, that this is not an exhaustive list and there are species and tissue differences in the constitutive and induced forms of cytochrome P-450.

Thus, induction can change the *proportions* of isoenzymes in a particular tissue, and may increase the activity of a normally insignificant form by many times. Although phenobarbital induction increases the overall concentration of cytochromes P-450 in the liver by about three-fold, specific isoenzymes may be increased up to 70-fold. Treatment with **3-methylcholanthrene** can increase a specific form of the enzyme by a similar order.

Induction of the microsomal enzymes may also have other effects as well as the increased production of particular enzymes and isozymes. Here again the different types of inducer vary. Thus, the barbiturate type of inducer differs significantly from the polycyclic hydrocarbon type as can be seen from the list below.

Some of the characteristic changes caused by the **barbiturate type** of inducer are:

(1) increase in smooth endoplasmic reticulum
(2) increase in liver blood flow
(3) increase in bile flow
(4) increase in protein synthesis
(5) liver enlargement
(6) increase in phospholipid synthesis
(7) increase in cytochromes P-450 content ($3\times$)
(8) increase in NADPH cytochrome P-450 reductase ($3\times$)
(9) increase in glucuronosyl transferases
(10) increase in glutathione transferases
(11) increase in epoxide hydrolases
(12) induction of cytochrome P-450 mostly occurs in the centrilobular area of the liver.

The **polycyclic hydrocarbon type** of inducer does not have such major effects, only causing slight liver enlargement and having no effect on liver blood or bile flow. The increase in cytochromes P-450 is not confined to the centrilobular area of the liver, protein synthesis is only slightly increased and there is no increase in phospholipid synthesis. Other enzymes than cytochromes P-450 are also induced by polycyclic hydrocarbons, although generally to a lesser extent than with barbiturate induction. NADPH cytochrome P-450 reductase is not induced by polycyclic hydrocarbons however.

With the **clofibrate type** of inducer other changes are also apparent. Thus, there is a proliferation in the number of **peroxisomes** (an intracellular organelle), as well as induction of a particular form of cytochrome P-450 involved in fatty acid metabolism. A number of other enzymes associated with the role of this organelle in fatty acid metabolism are also increased, such as **carnitine acyltransferase** and **catalase**. This phenomenon is discussed in more detail in Chapter 6. The onset of the inductive response is in the order of a few hours (3–6h after polycyclic hydrocarbons, 8–12h after barbiturates), is maximal after 3–5 days with barbiturates (24–48h with polycyclic hydrocarbons) and lasts for at least 5 days (somewhat longer with polycyclic hydrocarbon induction). The magnitude of the inductive effect may depend on the size and duration of dosing with the inducer, and will also be influenced by the sex, species, strain of animal and the tissue exposed.

It is clear from these comments that the biochemical and toxicological effects

Table 5.20.   Effect of various enzyme inducers on metabolism of different compounds.

Compound	Inducer				
	Control	Pb	PCN	3MC	Aro
	nmol product/min/nmol cyt P-450				
Ethylmorphine	13·7	16·8	24·9	6·4	9·5
Aminopyrine	9·9	13·9	9·7	7·6	13·7
Benzphetamine	12·5	45·7	6·6	5·7	15·8
Caffeine	0·48	0·65	—	0·52	0·64
Benzo[a]pyrene	0·14	0·14	0·14	0·33	—

Pb: phenobarbital; PCN: pregnenolone-16α-carbonitrile; 3MC: 3-methylcholanthrene; Aro: arochlor 1254.
Data from Powis *et al.* (1977) In *Microsomes and Drug Oxidations*, edited by V. Ullrich, A. Roots, A. Hildebrandt, R. W. Estabrook and A. H. Conney, p. 137 (Oxford: Pergamon Press).

seen after various inducers may be markedly different. This is illustrated by the effects of different inducers on the metabolism of various substrates examined *in vitro* and shown in table 5.20. It can be seen that in some cases the inducers cause no change in the metabolism, whereas in other cases metabolism is increased or even decreased. These effects are compounded by species and tissue differences in response to inducers and the differences in isoenzymes present in these species and tissues.

Induction therefore may cause:

(a) increased rate of metabolism of a foreign compound through one pathway
(b) altered metabolite profiles if the foreign compound is metabolized by several routes and only one is induced
(c) no effect on metabolism if the particular isoenzymes induced are not involved in metabolism of the particular compound
(d) decreased metabolism if induction increases levels of certain isoenzymes at the expense of the one(s) metabolizing the compound in question.

Depending on the role of metabolism in the toxicity of a compound therefore, enzyme induction may increase, decrease or cause no change in the toxicity of a particular compound. The effects of induction need to be considered in the light of distribution and excretion as competing processes (figure 4.67 and 6.1). However, the consequences can be simply summarized and explained as follows:

(i) If a metabolite is responsible for the toxic effect of a compound, then induction of the enzyme responsible may increase that toxicity. However, if there is only one route of metabolism and elimination is dependent on this then only the rate of metabolism to the single metabolite will be increased rather than the total amount. This may not increase toxicity if no other factors are involved or are not time-dependent. However, as most toxic effects are multi-stage, involving repair and protection, this is unlikely.
(ii) If the parent compound is responsible for a toxic effect, then induction

Table 5.21.   Effect of pentobarbital pretreatment on the duration of the pharmacological effect and disposition of pentobarbital in rabbits.

Pretreatment	Sleeping time (min)	Plasma level of pentobarbital on awakening (μg/ml)	Pentobarbital half-life in plasma (min)
None	67 ± 4	$9 \cdot 9 \pm 1 \cdot 4$	79 ± 3
Pentobarbital	30 ± 7	$7 \cdot 9 \pm 0 \cdot 6$	26 ± 2

Rabbits were pretreated with three daily doses of pentobarbital (60 mg/kg; s.c.) then given a single challenge dose of pentobarbital (30 mg/kg; i.v.).
Data from Remmer (1962) In *Metabolic Factors Controlling Duration of Drug Action, Proceedings of First International Pharmacological Meeting*, Vol. 6, edited by B. B. Brodie and E. G. Erdos (New York: Macmillan).

of metabolism may decrease that toxicity. However, induction of metabolism may lead to a different toxic effect due to a metabolite.

(iii) If a foreign compound is metabolized by several routes, then induction may alter the balance of these routes. This may lead to either increases or decreases in toxicity.

(iv) Induction may change the stereochemistry of a reaction.

Although some of these principles will be illustrated in more detail by the examples in Chapter 7, it is worthwhile examining some briefly at this point. A simple example is the pharmacological effect of a barbiturate, measured as sleeping time, which is dramatically reduced by induction of the enzymes of metabolism (table 5.21). This effect correlates with the plasma half-life, whereas the plasma level of pentobarbital on awakening is similar in the control and the induced groups (table 5.21). A toxicological example is afforded by **paracetamol** which is metabolized by several routes. The hepatotoxicity of paracetamol in the rat is increased by induction with phenobarbital due to an increase in the cytochrome P-450 isoenzyme which activates the drug. However, in the hamster hepatotoxicity is decreased due to an increase in glucuronidation, a detoxication pathway induced by phenobarbital (see Chapter 7). With the hepatotoxin **bromobenzene**, 3-methylcholanthrene induction decreases the toxicity. This is due to an increase in an alternative, non-toxic pathway, 2,3-epoxidation and an increase in the detoxication pathway catalysed by epoxide hydrolase (Chapter 7, table 7.12). With the lung toxic compound **ipomeanol**, phenobarbital induction decreases the toxicity, metabolic activation and the LD_{50}, whereas pretreatment with 3-methylcholanthrene changes the target organ from the lung to the liver (see Chapter 7).

The pharmacological action of **codeine** is increased by induction as this increases demethylation to morphine. Induction by phenobarbital decreases the toxicity of **organophosphates**, but increases that of **phosphorothionates**. Studies with the drug **warfarin** have shown that induction by both phenobarbital and 3-methylcholanthrene will change the stereochemistry of the product, as can be seen in table 5.22. Thus, hydroxylation in the 8-position in the R-isomer is increased 12 times compared with only four times with the S-isomer following 3-methylcholanthrene induction.

Table 5.22. Influence of cytochrome P-450 induction on the *in vitro* metabolism of R- and S-warfarin.

Inducer	Hydroxylated warfarin metabolites			
	R-isomer		S-isomer	
	7-OH	8-OH	7-OH	8-OH
	(nmol warfarin metabolite/nmol P-450)			
Uninduced	0·22	0·04	0·04	0·01
Phenobarbitone	0·36	0·07	0·09	0·02
3-Methylcholanthrene	0·08	0·50	0·04	0·04

Data from Gibson, G.G. and Skett, P. (1986) *Introduction to Drug Metabolism* (London: Chapman & Hall).

Thus, the importance of enzyme induction is that it may alter the toxicity of a foreign compound. This can have important clinical consequences and underlie drug interactions. Thus, the antitubercular drug rifampicin is thought to increase the hepatotoxicity of the drug isoniazid, and **alcohol** may increase susceptibility to the hepatotoxicity of paracetamol. However, it should also be noted that induction can alter the metabolism of endogenous compounds. For example, the antitubercular drug, **rifampicin**, is a microsomal enzyme inducer in human subjects. As well as increasing the toxicity of drugs such as isoniazid this compound also alters steroid metabolism, and may lead to reduced efficacy of the contraceptive pill. Similarly, in birds exposed to chlorinated hydrocarbons induction of the enzymes involved in steroid metabolism is believed to lead to alterations in the production of eggs resulting in greater fragility and therefore increased breakage.

That the influence of environmental agents is sufficient to cause significant changes in xenobiotic metabolism in humans has been shown in a number of studies. This is illustrated by studies of the effects of cigarette smoking and cooking meat over charcoal on the metabolism of the drug **phenacetin** in human volunteers. Both these activities produce polycyclic hydrocarbons such as benzo[a]pyrene which is a potent microsomal enzyme inducer. Eating charcoal grilled steak was shown to cause a significant increase in the rate of metabolism of phenacetin by de-ethylation to paracetamol (figure 5.20). This was indicated by the plasma level of phenacetin, which was significantly lower (20–25%) in human volunteers after eating meat exposed to charcoal compared with foil-wrapped meat. There was no decrease in half-life as phenacetin undergoes a significant first-pass effect, and enzyme induction in the gastrointestinal tract may have been a factor in this study responsible for a significant proportion of the increased metabolism. Cigarette smoking similarly increased the rate of metabolism of phenacetin. A study comparing **antipyrine** metabolism in Caucasians with that in Asians revealed that there were significant differences in the rate of metabolism between the two ethnic groups. The greater rate of metabolism (shorter half-life, increased clearance) in the Caucasians was ascribed at least in part to the influence of dietary factors such as eating meat and exposure to coffee, cigarette smoke and alcohol in the Caucasians.

MECHANISMS OF INDUCTION AND GENE REGULATION AND EXPRESSION

The mechanisms of induction of the cytochromes P-450 will be considered in terms of the different types of inducer. However, some general principles can be considered first.

Induction involves increased synthesis of enzyme protein which may be detected as an increase in total enzyme level as with phenobarbital induction or increase in a particular isoenzyme. Protein synthesis is increased and this usually seems to be necessary as inhibition of protein synthesis results in inhibition of induction. The increased protein synthesis may involve increased mRNA synthesis and inhibitors of this, such as **actinomycin D**, block induction. For a simple diagram explaining the relationship of protein synthesis to DNA see figure 6.22.

Some of the possible mechanisms of induction are:

(1) increased synthesis of mRNA or precursor rRNA
(2) increased stability of mRNA or rRNA
(3) decreased haem degradation
(4) decreased apoprotein degradation
(5) increased transport of RNA
(6) effects on DNA-dependent RNA polymerase.

The polycyclic type of inducer has been the most successfully studied and this will be described first. The discovery of strains of mice which are non-responsive to induction by polycyclic hydrocarbons (see below) has been a major factor in the study of mechanisms of induction. Thus, strains of mice (DBA/2) were found in which aryl hydrocarbon hydroxylase activity (AHH) could not be induced by 3-methylcholanthrene in comparison with responsive mice (C57BL/6). It was found that this non-responsiveness was dependent on a recessive gene at a single locus. The alleles were known as Ah^b for the responsive trait and Ah^d for the non-responsive, recessive trait. It is significant that the responsive mouse strain is often more susceptible to the various carcinogenic effects of polycyclic hydrocarbons. Other toxic effects are also associated with the possession of responsiveness. For example, mice responsive to AHH induction are susceptible to corneal damage caused by paracetamol, whereas non-responsive mice are not. Variations in responsiveness to induction have also been observed in strains of rats, and there may be a genetic factor operating in humans. Polycyclic hydrocarbons such as 3-methylcholanthrene and benz[a]anthracene induce aryl hydrocarbon hydroxylase, which is known to be cytochrome CYP 1A1. However, by far the most potent inducer was found to be **tetrachloro-dibenz-dioxin (TCDD)** which could in fact cause induction in the non-responsive mice. TCDD will cause an increase in the rate of transcription of the CYP 1A1 gene in cells in culture within minutes.

It was subsequently discovered that the non-responsive mice (Ah^d) possessed a defective cytosolic receptor rather than being deficient for the cytochrome P-450 gene. Thus, this receptor had reduced binding affinity for polycyclic hydrocarbons which could be overcome with larger amounts of a potent inducer such as TCDD. The non-responsive trait was due to a defective regulatory gene. Thus, a cytosolic receptor is involved in the induction by polycyclic hydrocarbons. This receptor binds the ligand (hydrocarbon), and the *liganded receptor* (hydrocarbon–receptor complex) is translocated to the nucleus (figure 5.31). The liganded receptor undergoes a conformational change which allows it to bind to DNA. This binding is to a **DNA recognition sequence**. This sequence is a DNA domain upstream of the CYP 1A1 gene, which is part of a polycyclic hydrocarbon responsive **transcriptional enhancer**. An enhancer is a segment of DNA which can increase the transcription rate. This causes an alteration in *chromatin structure* leading to increased accessibility of regulatory DNA and increased transcription of the CYP 1A1 gene. There is thus an increase in mRNA which results in synthesis of the protein portion of cytochrome P-450. The receptor is a protein with a molecular weight in the region of 95–126 kDa and it controls other enzyme activities as well as cytochrome CYP 1A1, including CYP 1A2. It may have at least three functional domains: ligand binding, DNA binding, a domain for binding a transcription factor(s). Regulation of the cytochrome P-450 genes by inducers such as TCDD is clearly very complex and may also involve *post-transcriptional stabilization of mRNA*, especially for CYP 1A2.

Pretreatment with phenobarbital leads to an increase in mRNA in the liver within 4 h. The mRNA, which is translatable, polysomal and is coded for a cytochrome P-450, is increased 14-fold. The major effect seems to be an increase

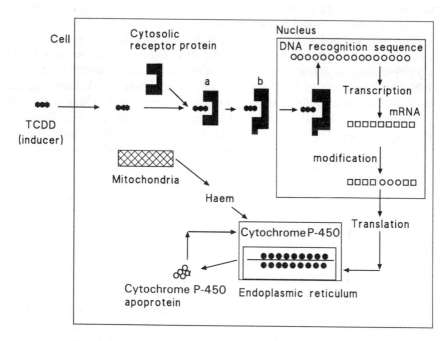

FIGURE 5.31. A mechanism for the receptor mediated induction of cytochrome P-450 by a polycyclic hydrocarbon such as TCDD. The inducer-receptor complex, a, undergoes a transformation to yield a complex, b, which binds to the DNA.

in *gene transcription rate*. The result is an increase in CYP 2B1 and CYP 2B2, the latter being a *constitutive form* of cytochrome P-450. CYP 2B2 is inactive in untreated rat liver, but is a constitutive form in lung and testis. Phenobarbital pretreatment also induces CYP 2C and CYP 3A1 and 3A2. However, no receptor has yet been isolated for phenobarbital, although the potency of this inducer is very much less than that of TCDD. Therefore the overall mechanism for this type of induction is still unknown. One or more of the mechanisms indicated above might be operable. It is also possible that there is an endogenous inducing agent which is also a substrate for cytochrome P-450. If levels of this agent were to rise, for instance if a cytochrome P-450 were inhibited by an exogenous substance, this might then activate RNA transcription and produce cytochromes P-450 protein via a cytosolic receptor. It is noteworthy that many inducers are also acute inhibitors of cytochromes P-450. Alternatively, there may be an *endogenous derepressor* molecule which is normally metabolized by cytochromes P-450. Again if a cytochrome P-450 is inhibited the resulting accumulation of derepressor substance could lead to derepression and activation of transcription. This would be similar to other methods of feedback control seen in biochemistry. Such a mechanism might also account for the enormous variety of chemical structures which will act as phenobarbital type inducing agents, and for which it would be difficult to envisage a single receptor.

Cytochrome P-450 involved in steroid metabolism (CYP 3) is induced by compounds such as **pregnenolone 16α-carbonitrile** and **dexamethazone**. This also involves *transcriptional activation* and a rise in mRNA levels and possibly also *post-transcriptional regulation* and *stabilization*. The glucocorticoid receptor may also be involved with some of the isoenzymes of this group, but not with CYP 3A1. With the hypolipidaemic drug-inducible cytochromes P-450 (CYP 4A) there is transcriptional activation within 1 h of administration of **clofibrate**. The levels of CYP 4A1 are very low in liver and kidney, but are markedly increased by the inducers. As well as cytochromes P-450, a number of other enzymes involved in peroxisomal β-oxidation reactions are also induced. A receptor for clofibrate has been isolated, but it has much less affinity for the receptor than TCDD. In contrast to the other cytochromes P-450 discussed, with CYP 2E1, which is induced with **ethanol** or **acetone** exposure, there is no increase in mRNA. Here there seems to be post-transcriptional regulation such as protection of the enzyme protein against degradation by binding of the inducer. In diabetes, which also leads to induction of this form of cytochrome P-450, stabilization of mRNA leads to an increase in the amount of this molecule rather than an increase in synthesis.

As well as induction of the synthesis of the apoprotein portion of cytochrome P-450, there is also induction of the synthesis of the haem portion. Clearly it is also necessary to have an increased amount of haem if there is an increase in the amount of the enzyme apoprotein being synthesized. Thus the rate-limiting step in haem synthesis, the enzyme δ-**aminolaevulinate synthetase** is inducible by both phenobarbital and TCDD. This is the result of transcriptional activation of the gene which codes for the δ-aminolaevulinate synthetase. It may be that the decrease in the haem pool, which results from incorporation of haem into the newly synthesized apoprotein, leads to derepression of the gene and hence increased mRNA synthesis. The gene repression could be haem-mediated or haem may modulate P-450 genes.

INDUCTION OF NON-CYTOCHROME P-450 ENZYMES

As already indicated above, other enzymes as well as cytochromes P-450 may be induced by a variety of compounds. Some of the important enzymes known to be induced and their inducers are shown in table 5.23. These other enzymes may have a crucial bearing on the overall toxicity of a compound as seen in examples in Chapter 7.

Table 5.23.   Induction of enzymes other than cytochrome P-450.

Enzyme	Inducing agent*
Glucuronosyl transferases	Pb, 3-MC, TCDD, PCBs
NADPH Cytochrome P-450 reductase	Pb, PCBs, isosafrole
Epoxide hydrolases	Pb, 3-MC, Arochlor, isosafrole
Glutathione transferases	Pb, 3-MC, TCDD
Cytochrome b_5	2-acetylaminofluorene, BHT
Catalase	Clofibrate, phthalates
Carnitine acyl transferase	Clofibrate

*Pb, phenobarbital; BHT, butylated hydroxy toluene; 3-MC, 3-methylcholanthrene; TCDD, 2,3,7,8-tetrachlorodibenzdioxin; PCBs, polychlorinated biphenyls.

DETECTION OF ENZYME INDUCTION

Enzyme induction may be detected in a variety of ways:

(1) a decrease in half-life or plasma level
(2) a change in proportions of metabolites
(3) a change in a pharmacological effect such as sleeping time after a barbiturate or paralysis time after zoxazolamine
(4) a change in a toxic effect
(5) an increase in plasma γ-glutamyl transferase
(6) an increase in the urinary excretion of D-glucaric acid
(7) an increase in urinary 6-β-hydroxycortisol excretion in humans.

Enzyme inhibition

In contrast to enzyme induction, inhibition usually only requires a *single* dose or exposure. Although environmentally it may be of less consequence than induction, with drug interactions it is probably of greater importance. Many of the different enzymes involved in the metabolism of foreign compounds may be inhibited with consequences for the biological activity of those compounds. As with enzyme induction, the consequences of inhibition will depend on the role of metabolism in the mechanism of toxicity or other biological activity. Thus, enzyme inhibition may increase or decrease toxicity or cause no change.

Enzyme inhibition will be considered in the broadest sense as a decrease in the activity of an enzyme *in vivo* or *in vitro* which is caused by a foreign compound. Inhibition may be divided into several types involving:

(1) interaction of the compound with the active site of the enzyme to give a *complex*, this may involve *reversible* or *irreversible binding*
(2) competitive inhibition by two different compounds at the same enzyme active site
(3) destruction of the enzyme
(4) reduced synthesis
(5) allosteric effects
(6) lack of cofactors.

Each of these types are examined in the following sections.

1. COMPLEXES

There are a number of well known inhibitors of cytochromes P-450 which form *stable complexes* with the enzyme. Thus SKF 525A, piperonyl butoxide, triacetyloleandomycin, **amphetamine, isoniazid** and **cimetidine** are a few of those known to form complexes with cytochromes P-450. In the case of several of these inhibitors, metabolism catalysed by cytochromes P-450 takes place and the metabolite so produced inhibits the enzyme. Thus, amphetamine is believed to be metabolized to a nitroso intermediate which complexes with the enzyme. **Piperonyl butoxide** is a commonly used microsomal enzyme inhibitor. It is active both *in vivo* and *in vitro*, and is metabolized to a reactive carbene which binds to the enzyme. It may also inhibit δ-aminolaevulinic acid synthetase, the rate-limiting step in haem biosynthesis. It is used as an insecticide **synergist** (see Chapter 2) to block the microsomal enzyme-mediated metabolism of insecticides. For example, when a structurally similar synergist, 2,3-methylenedioxy-naphthalene, is mixed with the insecticide **carbaryl**, the toxicity of carbaryl to house-flies occurs at around $5\,\mu g/g$, whereas no toxicity is observed in mice exposed to a dose of carbaryl of $750\,mg/kg$. This impressive and selective difference in toxicity is partly due to the fact that mice are able to metabolize and therefore remove the synergist, 2,3-methylenedioxynaphthalene, and so metabolism of the carbaryl is not inhibited. In contrast, the house-fly does not metabolize the synergist, and so microsomal metabolism of carbaryl is inhibited giving rise to greater toxicity. This is an example of the creative use of inhibition in selective toxicity.

Triacetyloleandomycin, an antibiotic, is associated with a number of drug interactions which may be the result of inhibition of cytochromes P-450. Thus, when co-administered with oral contraceptives, triacetyloleandomycin may cause problems of cholestasis, and with carbamazepine and theophylline signs of

Table 5.24. Effect of the enzyme inhibitor triacetyloleandomycin on metabolism of hexobarbital.

Treatment	Hexobarbital metabolism (nmol/min/mg)	Sleeping time (min)
Control	$1\cdot8\pm0\cdot7$	22 ± 8
TAO (1 h)	$1\cdot7\pm0\cdot7$	27 ± 9
TAO (24 h)	$1\cdot2\pm0\cdot7$	40 ± 18
TAO×4	$0\cdot3\pm0\cdot1$	168 ± 58

*TAO, triacetyloleandomycin: 1 h, 1 h after the dose; 24h, 24 h after the dose; × 4, 4 daily doses. All the doses were 1 mmol/kg.

Data from Pessayre *et al.* (1981) *Biochem. Pharmacol.*, **30**, 559.

neurologic intoxication can occur. Again it is metabolized by cytochromes P-450, first by demethylation and then by oxidation, and the resulting metabolite forms a stable complex with the enzyme. The inhibition has a marked effect on metabolism and biological activity of compounds such as hexobarbital (table 5.24). Although one dose is sufficient, the inhibition increases with time and clearly repeated dosing markedly increases the inhibition.

The type of inhibition in which a complex is formed will tend to be *non-competitive*, although if initial metabolism is required then there may be competition at this stage. Thus, piperonyl butoxide itself is a competitive inhibitor, but its metabolite is a non-competitive inhibitor.

Metyrapone, another compound widely used as an inhibitor of cytochromes P-450, binds to the reduced form of the enzyme and acts as a non-competitive inhibitor.

A *clinically* important example of microsomal enzyme inhibition is the interaction between the drugs **isoniazid** and **diphenylhydantoin**. Termination of the pharmacological activity of diphenylhydantoin depends on microsomal metabolism (figure 7.35). This may be inhibited by isoniazid when the two drugs are administered together. When this occurs the plasma half-life of diphenylhydantoin is significantly increased, and signs of toxic effects to the central nervous system are apparent. This interaction is seen to a greater extent in slow acetylators which have higher plasma levels of isoniazid. **SKF 525A** is another inhibitor of this type which is active both *in vivo* and *in vitro*, but seemingly may sometimes act as a competitive inhibitor and at other times as a non-competitive inhibitor. This compound not only inhibits cytochromes P-450-mediated reactions, but also inhibits glucuronosyl transferases and esterases.

Inhibitors of cytochromes P-450 may be specific for only one form and SKF 525A, for example, is specific for the phenobarbital inducible form(s), whereas the inhibitor **7,8-benzoflavone** is specific for the polycyclic hydrocarbon inducible form.

Other important enzyme inhibitors of this type are the organophosphorus compounds. Thus, after metabolism to the oxygen analogues, the insecticides parathion and malathion (figures 4.25 and 5.10) form complexes with the enzyme acetylcholinesterase as described in more detail in Chapter 7.

Carbamates such as carbaryl (figure 4.46) will also inhibit esterases, although usually less potently than the organophosphorus compounds. The toxicity of compounds metabolized by hydrolysis may thus be altered by such inhibitors. For example, *bis-p-nitrophenylphosphate* (BNPP) is an organophosphate type of inhibitor (figure 5.32) which blocks the hydrolysis of compounds such as phenacetin and acetylisoniazid (figures 5.20 and 4.45). Thus, **phenacetin** will cause methaemoglobinaemia in both experimental animals and man due to the metabolite 2-hydroxyphenetidine (figure 5.20). This metabolite is a product of deacetylation and hydroxylation, and the deacetylation step, catalysed by an acyl amidase, is inhibited by BNPP. Consequently methaemoglobinaemia can be prevented in experimental animals by treatment with the organophosphate. BNPP has also been used as an experimental tool to study **isoniazid** toxicity as it

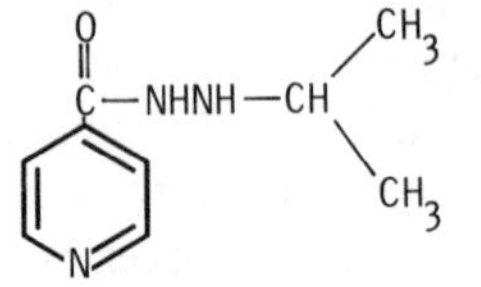

Iproniazid

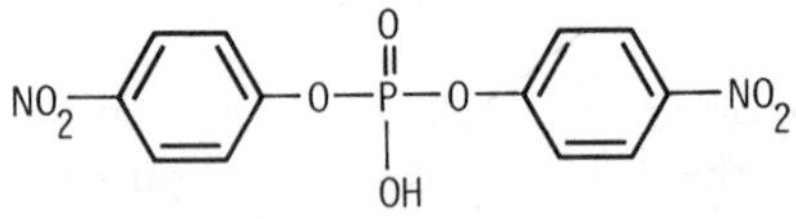

bis-p-Nitrophenyl phosphate

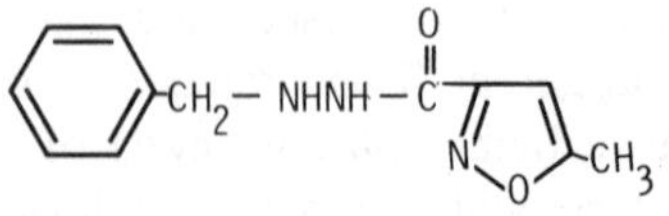

Isocarboxazid

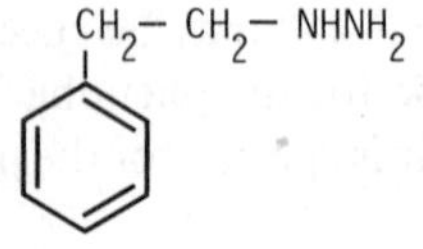

Phenelzine

FIGURE 5.32. Structures of the enzyme inhibitors bis-*p*-nitrophenyl phosphate (BNPP), iproniazid, phenelzine and isocarboxazid.

inhibits the hydrolysis of acetylisoniazid (see figure 7.15). Organophosphorus compounds may cause prolonged and cumulative inhibition of cholinesterases in humans occupationally exposed, which can have important toxicological consequences as discussed in Chapter 7.

Another enzyme involved in the metabolism of foreign compounds which may be purposefully inhibited is aldehyde dehydrogenase. This is irreversibly and non-competitively inhibited by the drug **disulphiram** (Antabuse). This inhibition, which lasts for about 24 h, is used in the treatment of alcoholism. Intake of alcohol after disulphiram leads to unpleasant effects due to the accumulation of acetaldehyde. This is not removed by metabolism as the enzyme aldehyde dehydrogenase is inhibited by the disulphiram. Another example of the clinical use of enzyme inhibition is the use of **allopurinol** which inhibits xanthine oxidase *in vivo* and consequently inhibits the metabolism of compounds such as the anti-cancer drug, 6-mercaptopurine (figure 4.18), thereby increasing its efficacy. Disulphiram also inhibits cytochromes P-450, and consequently other drug interactions may occur, such as with diphenylhydantoin, in a similar manner to that described for isoniazid. Disulphiram reduces the toxicity of **1,2-dimethyl-hydrazine**, a colon carcinogen (figure 5.33). The *N*-oxidation of the intermediate metabolite, azomethane, essential for the carcinogenicity, was found to be inhibited by disulphiram.

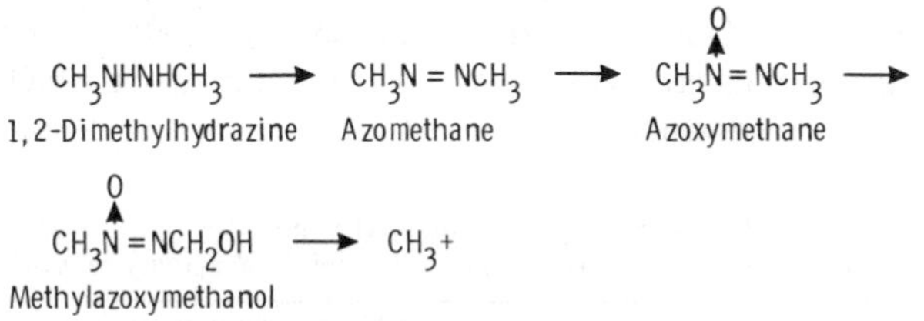

FIGURE 5.33. Metabolism of 1,2-dimethylhydrazine.

Monoamine oxidase inhibitors such as **phenelzine** and **isocarboxazid** (figure 5.32) are used clinically for the treatment of depression. These compounds also form complexes with the enzyme and irreversibly block its action. With iproniazid, complete inhibition of the enzyme lasts for 24 h and normal activity is not regained for 5 days. This inhibition may have important toxicological consequences because endogenous compounds and amines in food are also metabolized by monoamine oxidase. Consequently, intake of significant quantities of such amines as tyramine found in cheese, may lead to life threatening hypertension as the amine accumulates in patients who have taken monoamine inhibiting drugs.

2. COMPETITIVE INHIBITION

Any two compounds which are metabolized by the same enzyme may competitively inhibit the metabolism of the other. The extent of this will depend on the affinity each compound has for the enzyme. One example where this is important toxicologically is in the treatment of **ethylene glycol** and **methanol** poisoning. Both of these compounds are toxic as a result of metabolism by the enzyme alcohol dehydrogenase (see Chapter 7). Consequently one method of treatment is to reduce this by administration of ethanol, which has a greater affinity for the enzyme and so reduces metabolism and toxicity.

With cytochromes P-450 Type I ligands which bind to the enzyme as substrates, but not to the iron, act as competitive inhibitors. Thus, dichlorobiphenyl, a high-affinity Type I ligand for P-450, is a competitive inhibitor of the O-demethylation of *p*-nitroanisole.

With competitive inhibition the K_m of the enzyme is found to change but the V_{max} remains the same, whereas with non-competitive types of inhibition the reverse is the case. With uncompetitive inhibition, where the inhibitor interacts with the enzyme substrate complex, both V_{max} and K_m change.

3. DESTRUCTION

A number of foreign compounds will destroy enzymes such as cytochromes P-450. Thus, halogenated alkanes, hydrazines, compounds containing carbon–carbon double and triple bonds may all interact with and destroy cytochrome P-450. In many cases this is **suicide inhibition**, whereby the substrate is metabolized by the enzyme but the product reacts with and destroys the enzyme. Thus drugs containing the alkene group such as **secobarbital, allobarbital, fluoroxene** and compounds such as **allylisopropylacetamide** and **vinyl chloride** (figure 7.4 and see Chapter 7) all destroy hepatic cytochrome P-450. Similarly, drugs such as ethinyloestradiol and norethindrone which contain the alkyne group will also destroy the enzyme. Carbon tetrachloride is an example of

Table 5.25. Effect of allylisopropylacetamide (AIA) on cytochrome P-450 and drug metabolizing activity *in vitro* and pharmacological activity *in vivo*.

Parameter	% Non-AIA treated animal		
	Control* (non-induced)	Pb induced	3-MC induced
Cyt P-450	84	33	74
Hexobarbital hydroxylase	80	22	62
Ethylmorphine *N*-demethylase	62	8	35
p-Chloro-*N*-methylaniline *N*-demethylase	75	51	75
	Control animal	AIA treated	
Hexobarbital sleeping time (mins)	38	236	
Zoxazolamine paralysis time (mins)	258	478	

*Results for *in vitro* enzyme activity are expressed as % of the corresponding non-AIA treated control. Sleeping and paralysis times are expressed as absolute values in minutes.
Data from Farrell, H. and Correia, M. A. (1980) *J. Biol. Chem.*, **255**, 10 128, and Unseld, B. and de Matteis, F. (1978) *Int. J. Biochem.*, **9**, 865.

a halogenated alkane which destroys cytochrome P-450, as will several aliphatic and aromatic hydrazines such as ethylhydrazine and phenylhydrazine.

The cytochrome P-450 destroyed may be a specific isoenzyme, as is the case with **carbon tetrachloride** and allylisopropylacetamide (table 5.25). Indeed, with carbon tetrachloride the isoenzyme destroyed is the one which is responsible for the metabolic activation (CYP 1A2). With allylisopropylacetamide it is the phenobarbital-inducible form of the enzyme which is preferentially destroyed as can be seen from table 5.25. It seems that it is the haem moiety which is destroyed by the formation of covalent adducts between the reactive metabolite, such as the trichloromethyl radical formed from carbon tetrachloride (see Chapter 7), and the porphyrin ring.

4. REDUCED SYNTHESIS

The synthesis of enzymes may be decreased resulting in a decrease in the *in vivo* activity. With cytochrome P-450 there are a number of ways in which this occurs. Thus, administration of the metal **cobalt** to animals will decrease levels of cytochromes P-450 by inhibiting both the synthesis and increasing the degradation of the enzyme. Thus, cobalt inhibits δ-aminolaevulinic acid synthetase, the enzyme involved in haem synthesis. Cobalt will also increase the activity of haem oxygenase which breaks down the haem portion to biliverdin. The compound **3-amino,1,2,3-triazole** decreases cytochromes P-450 levels by inhibiting porphyrin synthesis.

Clearly any agent which inhibits protein synthesis will lead to a general decrease in the levels of enzymes.

5. ALLOSTERIC EFFECTS

When **carbon monoxide** interacts with haemoglobin as well as competing with oxygen, it also causes an allosteric change which affects the binding of oxygen. Although haemoglobin is strictly speaking not an enzyme, this allosteric effect has important toxicological implications as discussed in more detail in Chapter 7.

6. LACK OF COFACTORS

Clearly, where cofactors are involved in a metabolic pathway, a lack of these will result in a decrease in metabolic activity. Thus depletion of hepatic glutathione by compounds such as **diethyl maleate** or inhibition of its synthesis by compounds such as **buthionine sulphoximine** will decrease the ability of the animal to conjugate foreign compounds with glutathione. **Salicylamide** and **borneol** will deplete animals of UDP-glucuronic acid by forming glucuronide conjugates. Galactosamine (figure 7.40) will inhibit the synthesis of UDP-glucuronic acid. Sulphate required for conjugation is easily depleted by giving large doses of compounds which are conjugated with it, such as paracetamol. Salicylamide also inhibits sulphate conjugation and the combined effect of inhibition of glucuronidation and sulphate conjugation, the two major pathways for elimination of paracetamol, markedly increases the hepatotoxicity of the latter.

Bibliography

General

ALVARES, A. P. and PRATT, W. B. (1990) Pathways of drug metabolism. In *Principles of Drug Action, The Basis of Pharmacology*, edited by W. B. Pratt and P. Taylor (New York: Churchill Livingstone).

CALDWELL, J. and JAKOBY, W. B. (editors) (1983) *Biological Basis of Detoxication* (New York: Academic Press).

GIBSON, G. G. and SKETT, P. (1986) *Introduction to Drug Metabolism* (London: Chapman & Hall).

HODGSON, E. and GUTHRIE, F. E. (editors) (1980) *Introduction to Biochemical Toxicology*, Chapters 6, 7 and 8 (New York: Elsevier-North Holland).

JAKOBY, W. E. (editor) (1980) *Enzymatic Basis of Detoxication* (New York: Academic Press).

PARKE, D. V. (1968) *The Biochemistry of Foreign Compounds*, Chapter 6 (Oxford: Pergamon).

Chemical Factors

ARIENS, E. J., WUIS, E. W. and VERINGA, E. J. (1988) Stereoselectivity of bioactive xenobiotics. *Biochem. Pharmacol.*, **37**, 9.

LEWIS, D. F. V. (1990) MO-QSARs: a review of molecular orbital-generated quantitative structure–activity relationships. In *Progress in Drug Metabolism*, edited by G. G. Gibson (London: Taylor & Francis).

HANSCH, C. (1972) Quantitative relationships between lipophilic character and drug metabolism. *Drug Metab. Rev.*, **1**, 1.

TRAGER, W. F. and JONES, J. P. (1987) Stereochemical considerations in drug metabolism. In *Progress in Drug Metabolism*, Vol. 10, edited by J. W. Bridges, L. F. Chasseaud and G. G. Gibson (London: Taylor & Francis).

TUCKER, G. T. and LENNARD, M. S. (1990) Enantiomer specific pharmacokinetics. *Pharmac. Ther.*, **45**, 309.

Species, strain and sex effects

ADAMSON, R. H. and DAVIES, D. S. (1973) Comparative aspects of absorption, distribution, metabolism and excretion of drugs. In *International Encyclopaedia of Pharmacology and Therapeutics*, Section 85, Chapter 9 (Oxford: Pergamon).

CALABRESE, E. J. (1985) *Toxic Susceptibility: Male/Female Differences* (New York: John Wiley).

CALDWELL, J. (1981) The current status of attempts to predict species differences in drug metabolism. *Drug Metab. Rev.*, **12**, 221.

CALDWELL, J. (1982) Conjugation reactions in foreign compound metabolism: definition, consequences and species variations. *Drug Metab. Rev.*, **13**, 745.

CALDWELL, J. (1986) Conjugation mechanisms of xenobiotic metabolism: Mammalian aspects. In *Xenobiotic Conjugation Chemistry*, edited by G. D. Paulson, J. Caldwell, D. H. Hutton and J. M. Menn (Washington: American Chemical Society).

GUSTAFSSON, J. Å. (1983) Sex steroid induced changes in hepatic enzymes. *Annu. Rev. Physiol.*, **45**, 51.

HATHWAY, D. E., BROWN, S. S., CHASSEAUD, L. F. and HUTSON, D. H. (reporters) (1970–1981) *Foreign Compound Metabolism in Mammals*, Vols. 1–6 (London: The Chemical Society). (This series contains chapters on species, strain and sex differences in metabolism.)

HUCKER, H. B. (1970) Species differences in drug metabolism. *Annu. Rev. Pharmac.*, **10**, 99.

JONDORF, W. R. (1981) Drug metabolism as evolutionary probes. *Drug Metab. Rev.*, **12**, 379.

KATO, R. and YAMOZOE, Y. (1990) Sex-dependent regulation of cytochrome P450 expression. In *Frontiers of Biotransformation*, Vol. II, *Principles, Mechanisms and Biological Consequences of Induction*, edited by K. Ruckpaul and H. Rein (London: Taylor & Francis).

SMITH, R. L. (1970) Species differences in the biliary excretion of drugs. In *Problems of Species Difference and Statistics in Toxicology, Proc. Eur. Soc. Study Drug Tox.*, Vol. XI, edited by S. B. Baker, J. Tripod and J. Jacob.

WALKER, C. H. (1980) Species variations in some hepatic microsomal enzymes that metabolise xenobiotics. *Prog. Drug Metab.*, **5**, 113.

WILLIAMS, R. T. (1974) Interspecies variations in the metabolism of xenobiotics. *Biochem. Soc. Trans.*, **2**, 359.

Genetic factors

IDLE, J. (1988) Pharmacogenetics. Enigmatic variations. *Nature*, **331**, 391.

IDLE, J. R. and SMITH, R. L. (1979) Polymorphisms of oxidation at carbon centers of drugs and their clinical significance. *Drug. Metab. Rev.*, **9**, 301.

NEBERT, D. W. (1986) The genetic regulation of drug-metabolising enzymes. *Drug Metab. Disp.*, **16**, 1.

NEBERT, D. W. and WEBER, W. W. (1990) Pharmacogenetics. In *Principles of Drug Action. The Basis of Pharmacology*, edited by W. B. Pratt and P. Taylor (New York: Churchill Livingstone).

OMENN, G. S. and GELBOIN, H. V. (1984) *Genetic Variability in Responses to Chemical Exposure, Banbury Report* **16**, Cold Spring Harbor Laboratory (New York: Cold Spring Harbor).

PRICE EVANS, D. A. (1989) *N*-Acetyltransferase. *Pharmac. Ther.*, **42**, 157.

VESELL, E. S. (1975) Genetically determined variations in drug disposition and response in man. In *Handbook of Experimental Pharmacology*, Vol. 28, Part 3, *Concepts in Biochemical Pharmacology*, edited by J. R. Gillette and J. R. Mitchell (Berlin: Springer).

WEBER, W. W. (1987) *The Acetylator Genes and Drug Response* (New York: Oxford University Press).

WEBER, W. W. and HEIN, D. W. (1985) *N*-Acetylation pharmacogenetics. *Pharmacol. Rev.*, **37**, 25.

Diet

ANGELI-GREAVES, M. and MCLEAN, A. E. M. (1979) The effect of diet on the toxicity of drugs. In *Drug Toxicity*, edited by J. W. Gorrod (London: Taylor & Francis).

CAMBELL, T. C., HAYES, J. R., MERRILL, A. H., MASO, M. and GOETCHIUS, M. (1979) The influence of dietary factors on drug metabolism in animals. *Drug Metab. Rev.*, **9**, 173.

DOLLERY, C. T., FRASER, H. S., MUCKLOW, J. C. and BULPITT, C. J. (1979) Contribution of environmental factors to variability in human drug metabolism. *Drug Metab. Rev.*, **9**, 207.

KRISHNASWAMY, K. (1983) Drug metabolism and pharmacokinetics in malnutrition. *TIPS*, **4**, 295.

PARKE, D. V. and IAONNIDES, C. (1981) The role of nutrition in toxicology. *Annu. Rev. Nutr.*, **1**, 207.

Age

GILLETTE, J.R. and STRIPP, B. (1975) Pre- and postnatal enzyme capacity for drug metabolite production. *Fed. Proc.*, **34**, 172.

KLINGER, W. (1990) Biotransformation of xenobiotics during ontogenetic development. In *Frontiers of Biotransformation*, Vol. II, *Principles, Mechanisms and Biological Consequences of Induction*, edited by K. Ruckpaul and H. Rein (London: Taylor & Francis).

NEIMS, A.H., WARNER, M., LOUGHNAN, P.M. and ARANDA, J.V. (1976) Developmental aspects of the hepatic cytochrome P-450 mono-oxygenase system. *Annu. Rev. Pharmac. Toxicol.*, **16**, 427.

SCHMUCKER, D.L. (1979) Age-related changes in drug disposition. *Pharmac. Rev.*, **30**, 445.

SKETT, P. and GUSTAFSSON, J.-Å (1979) Imprinting of enzyme systems of xenobiotic and steroid metabolism. *Rev. Biochem. Tox.*, **1**, 27.

Pathological conditions

HOYUMPA, A.M. and SCHENKER, S. (1982) Major drug interactions: effect of liver disease, alcohol and malnutrition. *Annu. Rev. Med.*, **33**, 113.

KATO, R. (1977) Drug metabolism under pathological and abnormal physiological states in animals and man. *Xenobiotica*, **7**, 25.

PRESCOTT, L.F. (1975) Pathological and physiological factors affecting drug absorption, distribution, elimination and response in man. In *Handbook of Experimental Pharmacology*, Vol. 28, Part 3, *Concepts in Biochemical Pharmacology*, edited by J.R. Gillette and J.R. Mitchell (Berlin: Springer).

WILKINSON, G.R. and SCHENKER, S. (1975) Drug disposition and liver disease. *Drug Metab. Rev.*, **4**, 139.

Tissue and organ specificity

CHAMBERS, P.L. and GUNSEL, P. (editors) (1979) Mechanism of toxic action on some target organs. *Archs Toxicol.*, Suppl. 2.

GRAM, T.E. (editor) (1980) *Extrahepatic Metabolism of Drugs and Other Foreign Compounds* (Jamaica, New York: Spectrum).

Enzyme induction

ALVARES, A.P., PANTUCK, E.J., ANDERSON, K.E., KAPPAS, A. and CONNEY, A.H. (1979) Regulation of drug metabolism in man by environmental factors. *Drug Metab. Rev.*, **9**, 185.

BRESNICK, E. and HOUSER, W.H. (1990) The induction of cytochrome P-450c by cytosolic hydrocarbons proceeds through the interaction of a 4S cytosolic binding protein. In *Frontiers of Biotransformation*, Vol. II, *Principles, Mechanisms and Biological Consequences of Induction*, edited by K. Ruckpaul and H. Rein (London: Taylor & Francis).

CONNEY, A.H. (1967) Pharmacological implications of microsomal enzyme induction. *Pharmac. Rev.*, **19**, 317.

PARK, B.K. (1982) Assessment of the drug metabolism capacity of the liver. *Brit. J. Clin. Pharmacol.*, **14**, 631.

PARKE, D.V. (1990) Induction of cytochromes P450—General principles and biological consequences. In *Frontiers of Biotransformation*, Vol. II, *Principles, Mechanisms and Biological Consequences of Induction*, edited by K. Ruckpaul and H. Rein (London: Taylor & Francis).

SNYDER, R. and REMMER, H. (1979) Classes of hepatic microsomal mixed function oxidase inducers. *Pharmac. Ther.*, **7**, 203.

TUKEY, R. H. and JOHNSON, E. F. (1990) Molecular aspects of regulation and structure of the drug-metabolizing enzymes. In *Principles of Drug Action*, edited by W. B. Pratt and P. Taylor (New York: Churchill Livingstone).

WHITLOCK, J. P. (1989) The control of cytochrome P-450 gene expression by dioxin. *TIPS*, **10**, 285.

Enzyme inhibition

ANDERS, M. W. (1971) Enhancement and inhibition of drug metabolism. *Annu. Rev. Pharmac.*, **11**, 37.

ORTIZ DE MONTELLANO, P. R. and CORREIA, M. A. (1983) Suicidal destruction of cytochrome P-450 during oxidative metabolism, *Annu. Rev. Pharmac. Toxicol.*, **23**, 481.

TESTA, B. and JENNER, P. (1981) Inhibitors of cytochrome P-450s and their mechanisms of action. *Drug Metab. Rev.*, **12**, 1.

Chapter 6

Toxic responses to foreign compounds

Introduction

There are many ways in which an organism may respond to a toxic compound, and the type of response depends upon numerous factors. Although many of the toxic effects of foreign compounds have a biochemical basis, the expression of the effects may be very different. Thus, the development of tumours may be one result of an attack on nucleic acids, another might be the birth of an abnormal offspring. The interaction of a toxic compound with normal metabolic processes may cause a physiological response such as muscle paralysis, or a fall in blood pressure, or it may cause a tissue lesion in one organ. The covalent interaction between a toxic foreign compound and a normal body protein may in some circumstances cause an immunological response, in others a tissue lesion.

Thus, although all these toxic responses may have a biochemical basis, they have been categorized according to the manifestation of the toxic effect. Therefore although there will be overlap between some of the types of toxic response, for the purposes of this discussion it is convenient to divide them into the following:

 (i) direct toxic action: tissue lesions
 (ii) pharmacological, physiological and biochemical effects
 (iii) teratogenesis
 (iv) immunotoxicity
 (v) mutagenesis
 (vi) carcinogenesis.

Toxic responses may be detected in a variety of ways in animals and some of these have already been alluded to in previous chapters. Toxic responses may be the **all-or-none type** such as the death of the organism or they may be **graded responses**. Thus the main means of detection are:

 (a) Death: the LD_{50} assay has been utilized as an indicator of toxicity although it will be increasingly superseded by other assays;
 (b) Pathological change: this could be the development of a tumour or

destruction of tissue but it would be detectable by observation either macroscopically or microscopically;

(c) Biochemical change: this might involve an effect on an enzyme such as inhibition or alteration in a particular metabolic pathway. Alternatively the appearance of an enzyme or other substance in body fluids may indicate leakage from tissue due to damage and be indicative of pathological change;

(d) Physiological change: this could be measured in the whole conscious animal as for example a change in blood pressure, in temperature or in a response to a particular stimulus;

(e) Changes in normal status: there are a number of *markers of toxicity* which are simple to determine yet indicate a toxic response. Thus changes in body weight, food and water intake, urine output and organ weight may all be sensitive indicators of either general or specific toxicity. Thus animals often consume less food and lose weight after exposure to toxic compounds and increased organ weight may be due to a tumour, fluid or triglyceride accumulation, hypertrophy or enzyme induction. These changes may of course be confirmed by chemical, biochemical or histopathological measurements.

Examples of some of these manifestations of toxicity will be covered in the various sections of this chapter.

Direct toxic action: tissue lesions

Introduction

Some toxic compounds cause *direct* damage to tissues leading to the death of some or all of the cells in an organ for instance. The damage may be *reversible* or *irreversible* and the overall toxic response will depend on many factors including the *importance* of the particular tissue to the animal, the degree of *specialization* and its *reserve* functional capacity and ability to *repair* the damage. It is beyond the scope of this book to examine all the types of tissue lesion, although the major types of toxic damage are discussed. However, it is instructive to discuss first the principles underlying this type of toxicity, the factors which affect it and the general mechanisms of cellular damage.

Target organ toxicity

Although any organ or tissue may be a target for a toxic compound, such compounds often damage specific organs. Therefore it is instructive first to examine the principles underlying the susceptibility of certain organs to damage by toxic substances.

Thus, there are a number of different reasons why an organ might be a target:

(a) its blood supply
(b) the presence of a particular enzyme or biochemical pathway
(c) the function or position of the organ
(d) the vulnerability to disruption or degree of specialization
(e) the ability to repair damage
(f) the presence of particular uptake systems
(g) the ability to metabolize the compound and the balance of toxica-
     tion/detoxication systems
(h) binding to particular macromolecules.

Thus, those organs well supplied with blood such as the kidneys will be exposed to foreign compounds to a greater extent than poorly perfused tissue such as the bone. Therefore, factors which affect the absorption, distribution and excretion of foreign compounds and the physicochemical characteristics of those compounds can all affect the toxicity to particular target organs. Those organs which are metabolically active such as the liver may be more vulnerable than metabolically less active tissues such as the skin. This is due to the fact that many compounds require metabolic activation in order to be toxic (see Chapter 4, figure 4.67 and Chapter 7). However the toxicity to the organ will depend on the balance of toxication and detoxication pathways and other factors such as the ability to repair toxic damage. Thus, although the liver may have the greatest metabolic activity if this extends to detoxication pathways as well as toxication pathways, overall the toxicity may be less than in an organ which has lower activity for the toxication pathway, but lacks the detoxication pathway. Organs or tissues which are at sites of entry such as the gastrointestinal tract may be exposed to higher concentrations of foreign compounds prior to dilution by blood and other fluids, and hence may be more susceptible than deep tissues such as muscle. Such tissues as the skin, gastrointestinal tract and lungs may suffer local irritation as they tend to be exposed to substances as a result of their position and function. Highly specialized and vital organs such as the central nervous system (CNS) are susceptible to disruption and are not easily repaired in comparison with adipose tissue, for example, which is less vital and less specialized. Uptake into and concentration of foreign compounds by some organs, in some cases such as the kidney and liver as a result of excretory function, may impart vulnerability. Binding to specific macromolecules such as **melanin** in the eye, for example, may lead to target organ toxicity. Some of the interrelationships between metabolism, distribution and excretion which may affect toxicity are shown in figure 6.1. Thus, the distribution of the parent compound or metabolite(s) into the target tissue(s), metabolism and excretion in such tissues and the interaction with receptors or other critical cellular macromolecules are all dynamic processes occurring at particular rates. Factors which affect these processes therefore will influence toxicity and the particular target organ, and may even change the target organ.

Thus, it is clear from figure 6.1. that if the parent compound is toxic, factors such as enzyme induction or inhibition which change metabolism (m) or change distribution to tissues (d) or excretion (e) will tend to change the toxicity. This is provided that the toxic response is dose-related. Specific uptake systems will

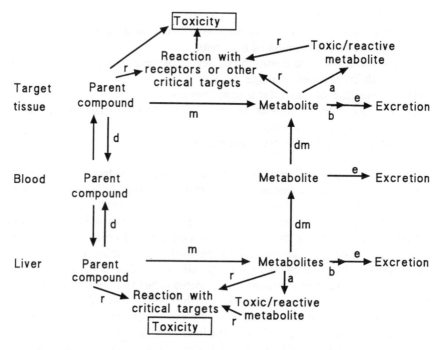

FIGURE 6.1. The interrelationships between the disposition and toxicity of a foreign compound. The parent compound may undergo distribution out of the blood (d) to other tissues and there cause toxicity by reaction with receptors (r). Alternatively metabolism (m) may also occur and give rise to toxic metabolites (a) which react with critical targets (r). Metabolites may also distribute back into the blood (b) and be excreted (e).

influence the distribution (d) and specific excretory routes (e) may be saturated. Metabolism may cause the appearance of another, different, type of toxicity. If a metabolite is toxic then factors which increase metabolism may increase the toxicity, provided that detoxication pathways (b) are not also increased and therefore compensate. However, the presence or absence of such pathways (b) may determine whether a tissue becomes the target. Toxic metabolites or proximate toxic metabolites may be produced in one organ such as the liver and transported to another (dm). The interaction of toxic or reactive metabolites with receptors (r) may be tissue specific and there may be protective agents such as glutathione in some tissues. These principles are further exemplified in this section with regard to the various target organs considered, and also in the final chapter with specific examples of toxic compounds.

Liver

As it is one of the portals of entry to the tissues of the body, the liver is exposed to many potentially toxic substances via the gastrointestinal tract from the diet, food additives and contaminants, and drugs and is frequently a target in experimental animals. In man, liver damage is less common and only around 9%

of adverse drug reactions affect the liver. By virtue of its position, structure, function and biochemistry the liver is especially vulnerable to damage from toxic compounds. Substances taken into the body from the gastrointestinal tract are absorbed into the hepatic-portal blood system and pass via the portal vein to the liver. Thus, after the gastrointestinal mucosa and blood, the liver is the next tissue to be exposed to a compound, and as it is prior to dilution in the systemic circulation, this exposure will often be at a higher concentration than that of other tissues. The liver, the largest gland in the body, represents around 2–3% of the body weight in man and other mammals such as the rat. It is served by two blood supplies, the portal vein which accounts for 75% of the hepatic blood supply, and the hepatic artery. The portal vein drains the gastrointestinal tract, spleen and pancreas and therefore supplies nutrients, and the hepatic artery supplies oxygenated blood (figure 6.2). The liver receives around 25% of the cardiac output, which flows through the organ at around $1-1\cdot3$ ml/min/g and

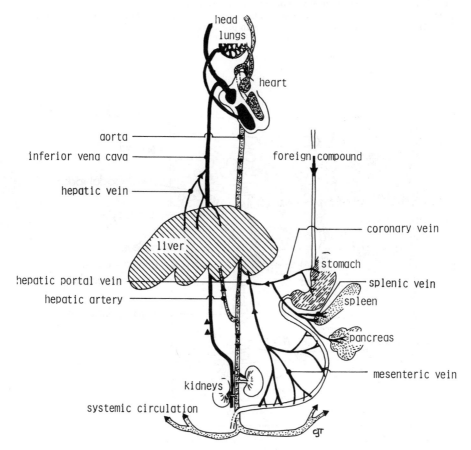

FIGURE 6.2. The vasculature suppling and draining the liver and its relationship to the systemic circulation.
From Timbrell, J. A., The liver as a target organ for toxicity, in *Target Organ Toxicity*, edited G. M. Cohen (Boca Raton, Fl.: CRC Press), 1989, with permission.

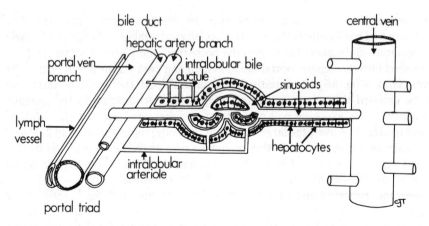

FIGURE 6.3. Schematic representation of the arrangement and relationship of vessels and sinusoids in the liver. The central vein drains into the hepatic vein.
From Timbrell, J. A., The liver as a target organ for toxicity, in *Target Organ Toxicity*, edited G. M. Cohen (Boca Raton, Fl.: CRC Press), 1989, with permission.

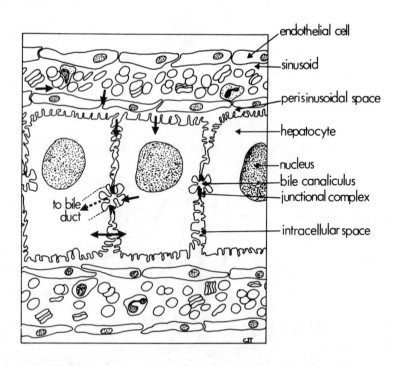

FIGURE 6.4. Diagrammatic representation of the arrangement of hepatocytes within the liver and the relationship to the sinusoids.
From Timbrell, J. A., The liver as a target organ for toxicity, in *Target Organ Toxicity*, edited G. M. Cohen (Boca Raton, Fl.: CRC Press), 1989, with permission.

drains via the hepatic vein into the inferior vena cava. In between the blood entering the liver via hepatic artery and portal vein and leaving via the hepatic vein, the blood flows through sinusoids (figure 6.3). Sinusoids are specialized capillaries with discontinuous basement membranes which are lined with Kupffer cells and endothelial cells. There are large fenestrations in the sinusoids which allow large molecules to pass through into the interstitial space and into close contact with the hepatocytes (figure 6.4). The liver is mainly composed of hepatocytes arranged as plates approximately two-cells thick, each plate bounded by a sinusoid (figure 6.4). The membranes of adjacent hepatocytes form the bile canaliculi into which **bile** is secreted. The bile canaliculi form a network which feed bile into ductules which become bile ducts (figure 6.3).

The structural and functional unit of the liver is the lobule, which is usually described in terms of the hepatic acinus (figure 6.5), based on the microcirculation in the lobule. When the lobule is considered in structural terms it may be described as either a classical or a portal lobule (see glossary). The **acinus** comprises a unit bounded by two portal tracts and terminal hepatic or central venules, where a portal tract is composed of a portal venule, bile ductule and hepatic arteriole (figure 6.5). Blood flows from the portal tract towards the central venules, whereas bile flows in the opposite direction. There are three circulatory zones in the acinus, with zone 1 receiving blood from the afferent venules and arterioles first, followed by zone 2 and finally zone 3. Thus, there will be metabolic differences between the zones because of the blood flow. Zone 1 will receive blood which is still rich in oxygen and nutrients, such as fats and other

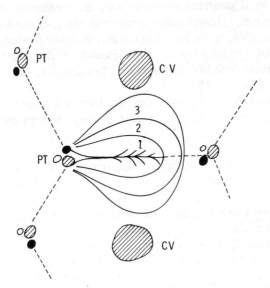

FIGURE 6.5. Schematic representation of a hepatic acinus. PT represents the portal tract, consisting of branches of the portal vein and hepatic artery and a bile duct. CV represents a branch of the central vein. The areas 1, 2 and 3 represent the various zones draining the terminal afferent vessel.

Adapted from Rappaport, A. M. (1969) In *Diseases of the Liver*, edited by L. Schiff (Philadelphia: J. B. Lippincott).

constituents. The hepatocytes in zone 3, however, will receive blood which has lost much of the nutrients and oxygen. Zone 1 approximates to the periportal region of the classical lobule and zone 3 to the **centrilobular** region. Zone 3, particularly where several acini meet, is particularly sensitive to damage from toxic compounds. The acinus is also a secretory unit, the bile it produces flowing into the terminal bile ductules in the portal tract.

The close proximity of the blood in the sinusoids with the hepatocytes allows efficient exchange of compounds, both endogenous and exogenous and consequently foreign compounds are taken up very readily into hepatocytes. For example, the drug **propranolol** is extensively extracted in the 'first pass' through the liver.

The liver is a target organ for toxic substances for four main reasons:

(a) The large and diverse metabolic capabilities of the liver enable it to metabolize many foreign compounds, but as metabolism does not always result in detoxication this may make it a target (see Chapter 7, carbon tetrachloride and paracetamol).

(b) The liver also has an extensive role in intermediary metabolism and synthesis and consequently interference with endogenous metabolic pathways may lead to toxic effects, as discussed in Chapter 7 (see galactosamine and ethionine).

(c) The secretion of bile by the liver may also be a factor. This may be due to the biliary excretion of foreign compounds leading to high concentrations, especially if saturated as occurs with the hepatotoxic drug **furosemide**. Alternatively, enterohepatic circulation can give rise to prolonged high concentrations in the liver. Interference with bile production and flow as a result of precipitation of a compound in the canalicular lumen or interference with bile flow may lead to damage to the biliary system and surrounding hepatocytes.

(d) The blood supply ensures that the liver is exposed to relatively high concentrations of toxic substances absorbed from the gastrointestinal tract.

The hepatocytes, or parenchymal cells, represent about 80% of the liver by volume, and are the major source of metabolic activity. However, this metabolic activity varies depending on the location of the hepatocyte. Thus zone 1 hepatocytes are more aerobic and therefore are particularly equipped for pathways such as the β-oxidation of fats, and they also have more **glutathione** and **glutathione peroxidase**. These hepatocytes also contain alcohol dehydrogenase and are able to metabolize **allyl alcohol** to the toxic metabolite acrolein which causes necrosis in zone 1. Conversely, zone 3 hepatocytes have a higher level of **cytochromes P-450** and **NADPH cytochrome P-450 reductase**, and **lipid synthesis** is higher in this area. This may explain why zone 3 is the most often damaged and lipid accumulation is a common response (see carbon tetrachloride for instance; Chapter 7).

The Kupffer cells are known to contain significant **peroxidase** activity and also

acetyltransferase. The differential distribution of isoenzymes may also be a factor in the localisation of damage.

There are various types of toxic response which the liver sustains which reflect its structure and function. Viewed simply, liver injury is usually due either to the metabolic capabilities of the hepatocyte or involves the secretion of bile.

The various types of liver damage which may be caused by toxic compounds are discussed in the following sections.

Fatty liver (steatosis)

This is the accumulation of **triglycerides** in hepatocytes and there are a number of mechanisms underlying this response as is discussed below (see 'Mechanisms of toxicity'). The liver has an important role in lipid metabolism, and triglyceride synthesis occurs particularly in zone 3. Consequently, fatty liver is a common response to toxicity, often the result of interference with protein synthesis, and may be the only response as after exposure to **hydrazine**, **ethionine** and **tetracycline**, or it may occur in combination with necrosis as with **carbon tetrachloride**. It is normally a reversible response which does not usually lead to cell death, although it can be very serious as is the case with tetracycline-induced fatty liver in humans. Repeated exposure to compounds which cause fatty liver, such as alcohol, may lead to cirrhosis.

Cytotoxic damage

Many toxic compounds cause direct damage to the hepatocytes, which leads to cell death and necrosis. This is a general toxic response, not specific for the liver, and there are undoubtedly many mechanisms which underlie cytotoxicity but most are still poorly understood. The mechanisms underlying cytotoxicity in general are discussed below and several examples of hepatotoxins are discussed in more detail in Chapter 7.

The zone of the liver damaged may depend on the mechanism, but may also be the result of the microcirculation. Damage may be zonal, diffuse or massive. For example, **cocaine** and **allyl alcohol** cause zone 1 (periportal) necrosis. With allyl alcohol this is partly as a result of the presence of alcohol dehydrogenase and partly because this is the first area exposed to the compound in the blood. Conversely, carbon tetrachloride, bromobenzene and paracetamol cause zone 3 (centrilobular) necrosis as a result of metabolic activation occurring primarily in that region (see Chapter 7). Midzonal, zone 2 necrosis is less common, but has been described for the natural product **ngaione** and **beryllium**. Galactosamine causes diffuse hepatic necrosis (see Chapter 7), presumably because it interferes with a metabolic pathway which occurs in all regions of the liver lobule. The explosive **trinitrotoluene** (TNT) can cause massive liver necrosis.

Ischaemia may also be a component of cytotoxic damage, and consequently interference with liver blood flow by toxic compounds such as **phalloidin** which causes swelling of the endothelial cells lining the sinusoids, may cause or contribute towards cytotoxicity.

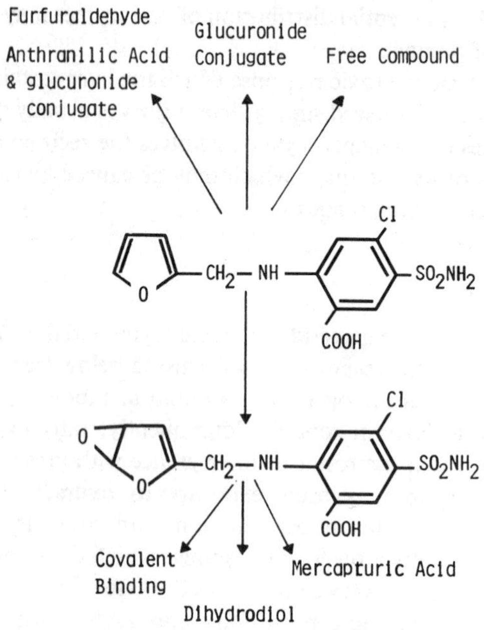

Furfuraldehyde

Anthranilic Acid
& glucuronide
conjugate

Glucuronide
Conjugate

Free Compound

Cl

$-CH_2-NH-$ $-SO_2NH_2$

COOH

Cl

$-CH_2-NH-$ $-SO_2NH_2$

COOH

Covalent
Binding

Mercapturic Acid

Dihydrodiol

FIGURE 6.6. The metabolism of furosemide. The epoxide intermediate shown in brackets is the postulated reactive intermediate.
From Timbrell, J.A., The liver as a target organ for toxicity, in *Target Organ Toxicity*, edited G.M. Cohen (Boca Raton, Fl.: CRC Press), 1989, with permission.

Other compounds cause liver necrosis because of biliary excretion. Thus the drug **furosemide** causes a dose-dependent centrilobular necrosis in mice. The liver is a target as a result of its capacity for metabolic activation *and* because furosemide is excreted into the bile by an active process which is saturated after high doses. The liver concentration of furosemide therefore rises disproportionately (figure 3.34), and metabolic activation allows the production of a toxic metabolite (figure 6.6). The drug **proxicromil** (figure 5.9) caused hepatic damage in dogs as a result of saturation of biliary excretion and a consequent increase in hepatic exposure.

Cholestatic damage

There are various types of interference with the biliary system, and this can lead to bile stasis or damage to the bile ducts, ductules or canaliculi. In some cases such as with **chlorpromazine**, damage to the hepatocytes may ensue. Thus, some foreign compounds such as the antibiotic **rifampicin**, interfere with bilirubin transport and conjugation giving rise to **hyperbilirubinaemia**. Other compounds, **icterogenin** for example, cause bile stasis and bilirubin deposits in the canaliculi. This canalicular damage may also be accompanied by damage to hepatocytes, such as caused by chlorpromazine. This drug is a surface active agent which can reach high concentrations in the bile and so directly damage the lining cells. It also can cause precipitation of insoluble substances in the lumen of the canaliculi. Accumulation of bile and its constituent bile salts may indeed be the cause of damage and some are surface active agents. Consequently, if high concentrations are reached, the cells of the biliary system and hepatocytes exposed can be damaged. The secondary bile acid **lithocholate** will cause direct damage to the canalicular membrane for example. Some compounds damage the bile ducts and ductules directly such as α-**naphthylisothiocyanate**. The result of the destruction of bile-duct lining cells will be cholestasis as debris from the necrotic cells will block the ductules.

Cirrhosis

This is a chronic lesion resulting from repeated injury and subsequent repair. It may result from either hepatocyte damage or cholestatic damage, each giving rise to a different kind of cirrhosis. Thus, carbon tetrachloride will cause liver cirrhosis after repeated exposure, but also compounds which do not cause acute necrosis, such as ethionine and alcohol may cause cirrhosis after chronic exposure.

Vascular lesions

Occasionally toxic compounds can directly damage the hepatic sinusoids and capillaries. One such toxic compounds is **monocrotaline**, a naturally occurring pyrrolozidine alkaloid, found in certain plants (*Heliotropium*, *Senecio* and *Crotolaria* species). Monocrotaline (figure 6.7) is metabolized to a reactive metabolite which is directly cytotoxic to the sinusoidal and endothelial cells causing damage and occlusion of the lumen. The blood flow in the liver is therefore reduced and ischaemic damage to the hepatocytes ensues. Centrilobular necrosis results and the venous return to the liver is blocked. Hence, this is known as **veno-occlusive disease** and results in extensive alteration in hepatic vasculature and function. Chronic exposure causes cirrhosis.

Liver tumours

Both benign and malignant liver tumours may arise from exposure to hepatotoxins and can be derived from various cell types. Thus, **adenomas** have been associated with the use of **contraceptive steroids** and exposure to **aflatoxin B_1**, and dimethylnitrosamine can produce hepatocellular carcinomas whereas vinyl chloride causes haemangiosarcomas derived from the vasculature (see Chapter 7).

Proliferation of peroxisomes

A response to exposure to certain foreign compounds which occurs predominantly in the liver and has been described only relatively recently, is the phenomenon of peroxisomal proliferation. Peroxisomes (microbodies) are

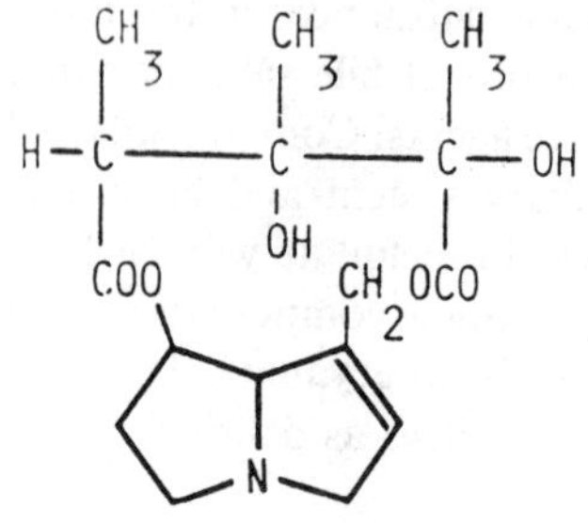

Monocrotaline

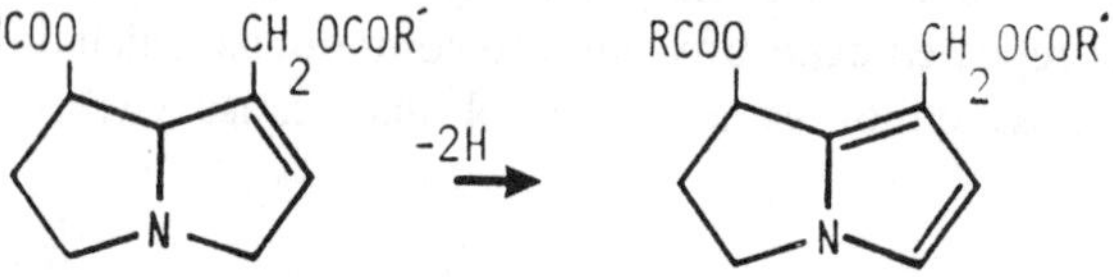

FIGURE 6.7. The structure of the pyrrolizidine alkaloid monocrotaline and the microsomal enzyme mediated metabolic activation of the pyrrolizidine alkaloid nucleus.
From Timbrell, J. A., The liver as a target organ for toxicity, in *Target Organ Toxicity*, edited G. M. Cohen (Boca Raton, Fl.: CRC Press), 1989, with permission.

organelles found in many cell types, but especially hepatocytes. Exposure of rodents to certain types of foreign compound leads to an increase in the *number* of these organelles and an increase in the *activities* of various enzymes. The function of the organelle is mainly the oxidative metabolism of lipids and certain other oxidative metabolic pathways. Thus, the enzymes for the β-oxidation of fatty acids are found. The importance of this phenomenon in toxicity and especially carcinogenicity is discussed later in this chapter. The types of compounds which cause the proliferation are generally acids or compounds which can be metabolized to acids. Thus, the hypolipidaemic drug **clofibrate** and a number of similarly acting drugs, will cause the proliferation. **Phthalate esters** which are commonly used as plasticizers are another group of compounds which have been shown to be active. The result of exposure is a large increase in the numbers of peroxisomes visible by electron microscopy and a large increase in some of the enzymes located in the organelle. Thus, in rats after 28 days exposure to clofibrate, peroxisomal volume was increased almost five-fold, liver weight was increased by 36% and the enzyme **carnitine acetyl CoA transferase** was increased six-fold. Other hypolipidaemic drugs have been reported to cause greater effects than illustrated here. Clearly there is a considerable increase in protein synthesis occurring. Such compounds also cause an increase in a cytochrome P-450 isoenzyme which metabolizes lauric acid, CYP 4.

The mechanisms underlying some of these types of injury will be discussed in general terms below and in Chapter 7.

Detection of hepatic damage

Simple quantitative tests can be used such as measurement of the liver weight/body weight ratio. Overt damage to the liver can be detected by light and electron microscopy of liver sections. However, damage can be detected by other non-invasive means such as the urinary excretion of conjugated bilirubin or the amino acid taurine.

Various parameters may be measured in plasma. Thus determination of the enzymes aspartate transaminase (**AST**) and alanine transaminase (**ALT**) is the most common means of detecting liver damage, the enzymes being raised several-fold in the first 24 h after damage. However, there are a number of other enzymes which may be used as markers. Plasma bilirubin can also be measured, being increased in liver damage, and plasma albumin is decreased by liver damage (although also by renal damage). Liver function may be determined using the hepatic clearance of a dye such as sulphobromophthalein.

Kidney

The function of the kidney is to filter waste products and toxins out of the blood while conserving essential substances such as glucose, amino acids and ions such as sodium. Anatomically the kidney is a complex arrangement of vascular endothelial cells and tubular epithelial cells, the blood vessels and tubules being intertwined. The functional unit of the kidney is the **nephron** (figure 6.8). Although all nephrons have their glomeruli and primary vascular elements in the cortex, the position of the nephron within the kidney does vary, in that it may lie completely within the cortex (*cortical* nephron) or the glomerulus may be located at the cortex-medulla boundary (*juxtamedullary* nephron). The environment both inside and outside the nephron varies along its length, and this influences the types of toxic effect produced by nephrotoxic agents.

The kidney is a target organ for toxicity for the following reasons:

1. Renal blood flow. The blood passing through the mammalian kidney represents 25% of the cardiac output despite the fact that the kidney only represents about 1% of the body mass. Therefore the exposure of kidney tissue to foreign compounds in the bloodstream, especially the cortex which receives more blood than the medulla, is relatively high.

2. The concentrating ability of the kidney. After glomerular filtration many substances are reabsorbed from the tubular fluid. Thus 98–99% of the water and sodium are reabsorbed. Hence, the concentration of foreign substances in the tubular lumen is considerably higher than that in the blood, and the tubular fluid : blood ratio may reach values of 500 : 1. For example, some **sulphonamides** when given in high doses may cause renal tubular necrosis as a result of the crystallization of less soluble, acetylated metabolites in the lumen of the tubule, due to the increased concentration in the tubular fluid and pH-dependent solubility (table 4.1). Similarly, **oxalate** crystals may occur in the renal tubules after ingestion of toxic amounts of ethylene glycol (see Chapter 7). Foreign substances may also be reabsorbed from the

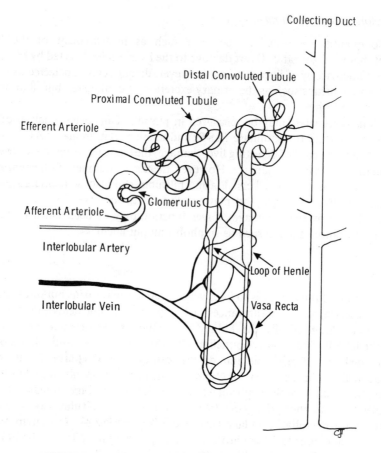

FIGURE 6.8. Schematic representation of a mammalian nephron.

tubular fluid along with water and endogenous compounds. In this case the tubular cells may contain relatively high concentrations of the foreign compound. The countercurrent exchange of small molecules may lead to very high concentrations of compounds in the interstitial fluid of the renal medulla. An example of a compound which is nephrotoxic as a result of uptake from the tubular fluid is the aminoglycoside antibiotic **gentamycin**. This drug is excreted into the tubular fluid by glomerular filtration, but binds to anionic phospholipids on the brush border of the proximal tubular cells. The resulting complex is absorbed into the cell by phagocytosis, is stored in secondary lysosomes and hence accumulates in the proximal tubular cells. Gentamycin causes a variety of biochemical derangements in the cell, and the lysosomes are destabilized such that their enzymes are released with resulting degradation of cellular components.

3. Active transport of compounds by the tubular cells. Compounds which are actively transported from the blood into the tubular fluid may accumulate in the proximal tubular cells, especially at concentrations where saturation

of the transport system occurs. Again concentrations to which tubular cells are exposed may be very much higher than in the bloodstream. An example of this is the drug **cephaloridine** which causes proximal tubular damage as discussed in more detail in Chapter 7.

4. Metabolic activation. Although the kidney does not contain as much cytochromes P-450 as the liver, there is sufficient activity to be responsible for metabolic activation, and other oxidative enzymes such as those of the prostaglandin synthetase system are also present. Such metabolic activation may underlie the renal toxicity of chloroform and paracetamol (see below Chapter 7). Other enzymes such as C–S lyase and glutathione transferase may also be involved in the activation of compounds such as hexachlorobutadiene (see Chapter 7). In some cases hepatic metabolism may be involved followed by transport to the kidney and subsequent toxicity.

It is clear that the tissues of the kidney are often exposed to higher concentrations of potentially toxic compounds than most other tissues. The toxic effects caused may be due to a variety of mechanisms ranging from the simple irritant effects of sulphonamides and oxalate crystals to the enzyme inhibition and tubular damage proposed to underlie aspirin-induced medullary lesions. It has been suggested that the latter is due to inhibition of prostaglandin synthesis by aspirin, giving rise to vasoconstriction of the vasa recta (figure 6.8), and hence reduction of blood flow and ischaemic damage to the kidney. The proximal tubule and hence the cortex, is the most common site of damage from foreign compounds, whereas the medulla is less commonly damaged and the collecting ducts only rarely. The glomerulus may be damaged by nephrotoxins such as the **aminoglycosides**, although these compounds also damage the proximal tubule. The proximal convoluted tubule is damaged by **chromium** whereas other metals such as mercury (see Chapter 7) damage the straight portion (pars recta). The pars recta section of the proximal tubule contains the highest concentration of cytochromes P-450, and this is one reason for its particular susceptibility to toxic injury. Another is the particular capacity for secretion of organic compounds. Hence, cephaloridine damages this section (see Chapter 7). The loop of Henle is particularly susceptible to damage from analgesics such as **aspirin** and **phenacetin, fluoride** and **2-bromoethylamine** which damages the thin limb and collecting ducts. This is manifested as papillary necrosis, following damage to the renal papillae. The distal convoluted tubule is less commonly a target, although **cisplatin,** the anti-cancer drug, damages this portion of the kidney. Interestingly the *trans* isomer does not cause kidney damage at the same doses as the *cis* isomer. **Amphoterecin** causes damage to both proximal and distal tubules. It is a surface-active compound which is believed to act by binding to membrane phospholipids, causing the cells to become leaky.

The kidney has a marked ability to compensate for tissue damage and loss, and consequently, unless it is evaluated immediately, nephrotoxic effects may not be recognized. Similarly, chronic toxicity may not be detected because of this compensatory ability.

Detection of kidney damage

There are a variety of ways in which kidney damage can be detected ranging from simple qualitative tests to more complex biochemical assays.

Simple tests include urine volume, pH and specific gravity measurement; kidney weight/body weight ratio; and detection of the presence of protein or cells in the urine.

Clearly overt pathological damage can be detected by light or electron microscopy, but this requires sacrifice of the animal. Damage can also be detected by measurement of a variety of urinary constituents. Thus, damage to tubular cells results in the leakage of enzymes such as γ-**glutamyltransferase** and **N-acetylglucosaminidase** into the tubular fluid, and therefore these can be detected and quantitated in the urine. Also, such damage will be reflected in altered renal tubular function leading to the excretion of glucose and amino acids. Measurement of urea or creatinine in the plasma will also indicate renal dysfunction if these are raised. The more common measurement is urea or **blood urea nitrogen** (BUN). Measurement of urea or creatinine in the urine and plasma allows the renal clearance of these endogenous compounds to be determined, and this will also indicate renal dysfunction. However, clearance of the polysaccharide inulin is a better indicator of glomerular filtration as it is less affected by other factors such as protein metabolism.

Different types of damage will cause different urinary profiles for endogenous compounds as has been investigated using high resolution proton nuclear magnetic resonance (NMR) of urine to generate urinary metabolite data patterns (see Bibliography).

Lung

The lung is a particularly vulnerable organ as regards toxic substances since it can be exposed to foreign compounds both in the external environment, and also internally from the bloodstream. The lungs receive 100% of the cardiac output and therefore are extensively exposed to substances in the blood. Also, the air breathed in may contain many potentially toxic substances: irritant gases such as **sulphur dioxide**, particles of dust, of metals and of other substances such as **asbestos,** solvents and other chemicals which may be local irritants. The lung is well adapted for the efficient exchange of gases between the ambient air and the blood. This also means it is efficient at absorbing potentially toxic substances from the air. The surface area is large, approximately 70 m^2 in humans, and the arrangement of air spaces, the alveoli and blood capillaries is such that the distance and cellular barrier between air and blood is minimal (see figure 3.10). Consequently, foreign compounds may be absorbed very rapidly if they are small enough to pass through pores in the alveolar membrane (8–10 Å radius), or are lipid soluble and able to cross the membrane by diffusion. Solid particles may be taken up by phagocytosis in some cases. The basic functional unit of the lung is the acinus, consisting of a respiratory bronchiole, alveolar ducts and sacs, alveoli and blood vessels (figure 3.10). There are various cell types within the lung and these exhibit different susceptibility to toxic damage.

Thus the lung is a target organ for the following reasons:

(1) it is perfused with the whole of the cardiac output of blood and so is exposed to substances in the blood
(2) it is exposed to substances in the air
(3) it has a high oxygen concentration and consequently may be damaged by reactive oxygen species.

However, the lungs are equipped with *defence mechanisms* especially with regard to the intake of foreign substances from the air. Thus, the upper airways of the respiratory system are lined with ciliated cells, and mucus is secreted which also lines the airways. Solid particles are therefore trapped by the mucus and cilia and are transported out of the respiratory system. Other substances may be removed after dissolving in the mucus and then being transported out by the *ciliary escalator*.

For substances entering the lung tissue from either the blood side or the air, metabolic capability and metabolic activation may be an important consideration in the toxicity. The lung is capable of metabolizing many foreign compounds, and the enzymes are present which catalyse both phase 1 and phase 2 reactions. However, the activity of the mono-oxygenase enzymes relative to the liver would depend on the particular cell type and species. Although overall the activity is generally less, for particular isoenzymes it may indeed be higher. For instance biphenyl hydroxylase activity is higher in rabbit lung, but lower in rat lung than the corresponding liver. Aniline hydroxylase activity, however, is lower in rabbit lung than liver, as is glucuronyl transferase. The non-ciliated bronchiolar epithelial cell, or **Clara cell** has the highest level of cytochromes P-450, and for some isozymes the activity is greater than that in the liver. It is this cell which is responsible for the metabolic activation of ipomeanol (see Chapter 7) and other lung toxins such as **naphthalene** and **3-methylfuran**.

There are over 40 different cell types in the respiratory system, and those particularly susceptible are the lining cells of trachea and bronchi, endothelial cells and the interstitial cells (fibroblasts and fibrocytes).

Some toxic substances such as sulphur dioxide, nitrogen dioxide and ozone cause acute, direct damage to lung tissue, whereas others, such as **nickel carbonyl** may lead to the formation of tumours.

The types of toxic response shown by the lungs are briefly discussed below.

Irritation

Volatile irritants such as ammonia and chlorine initially cause constriction of the bronchioles. These two gases are water soluble, are absorbed in the aqueous secretions of the upper airways of the respiratory system and may not cause permanent damage. Irritant damage may however lead to changes in permeability and oedema, the accumulation of fluid. Some irritants such as arsenic compounds cause bronchitis.

Lining-cell damage

Direct damage to cells of the respiratory tract, either by airborne compounds or those in the blood, may cause necrosis and changes in permeability leading to oedema. This may be due to the action of the parent compound such as **nitrogen dioxide** or **ozone** or to a metabolite as is the case with ipomeanol (see Chapter 7). The former two compounds cause peroxidation of cellular membranes (see below), and **phosgene** destroys the permeability of the alveolar cell membrane following hydrolysis to carbon dioxide and hydrogen chloride in the aqueous environment of the alveolus. Phosgene will also react directly with cellular constituents. The result is a change in permeability and extensive oedema. This also occurred with the highly reactive industrial chemical **methyl isocyanate** which caused death and injury to thousands in Bhopal in India when it leaked from a factory. In the case of ipomeanol, metabolic activation in the Clara cell results in necrosis of these cells and oedema of the lung tissue. Similarly, the solvent **tetrachloroethylene** undergoes metabolic activation in the lung and causes necrosis and oedema. In contrast, **pyrrolizidine alkaloids** such as monocrotaline (figure 6.7) are metabolically activated in the liver, and then cause damage in the lungs after being transported there by the blood.

In contrast to compounds such as nitrogen dioxide and ozone, compounds which are very water soluble will tend to damage the cells higher up in the system such as in the trachea or bronchi.

Fibrosis

This type of response is caused by substances such as **silica** (silicosis) and **asbestos** (asbestosis). It is thought to involve phagocytic uptake of particles by macrophages. The particles are taken up by lysosomes which then rupture at some stage and release hydrolytic enzymes, resulting in the digestion of the macrophage and release of the particles. The cyclical process then starts again. The result is aggregation of lymphoid tissue and stimulation of collagen synthesis leading to fibrotic lesions.

Other types of lung damage may also lead to fibrosis, however, such as that caused by paraquat.

Stimulation of an allergic response

A variety of particles and chemicals may stimulate an allergic response in the lungs when there is *repeated* exposure via the lungs. Micro-organisms, spores, dusts such as cotton dust, and certain chemicals may produce an allergic-type response with pulmonary symptoms. For instance, **toluene di-isocyanate**, a chemical used in the plastics industry, and **trimellitic anhydride**, another industrial chemical, cause an allergic-type reaction when inhaled. It is thought that these compounds form protein conjugates in the respiratory system which then act as antigens (see below).

Pulmonary cancer

Many different types of materials produce lung tumours after inhalation, and also after exposure via the bloodstream. Thus **asbestos, polycyclic aromatic hydrocarbons, cigarette smoke, nickel carbonyl, nitrosamines, chromates** and **arsenic** may cause lung tumours after inhalation. Other compounds cause lung tumours after systemic administration, such as **hydrazine** and **1,2-diethylhydrazine**. Although the activity of the drug metabolizing enzymes is in general less than that in the liver, for certain isoenzymes of cytochrome P-450 this is not the case. So the lung has the ability to metabolize foreign compounds and therefore can metabolically activate compounds to which the organ is exposed. Thus, the polycyclic aromatic hydrocarbon benzo[*a*]pyrene is metabolically activated by lung tissue as described in more detail in Chapter 7, and thus *in situ* activation may be responsible for the initiation of tumours.

Detection of lung damage

Unlike liver and kidney damage there are currently no general biochemical tests for lung damage. Lung damage may be detected pathologically by light and electron microscopy of lung tissue and simple measurements such as the lung wet weight/dry weight ratio will detect oedema. Bronchoscopy can be employed for the detection of gross changes and samples may be taken at the same time for histopathology. Lung dysfunction is detected by physiological tests such as **forced expiratory volume** (FEV) over a particular time, and **forced vital capacity**. Measurement of pulmonary mechanics such as **flow resistance** may also be useful for detection of lung irritants or pharmacologically active agents. These measurements are more difficult to perform in conscious animals than in humans, but determination of lung function is possible in experimental animals.

Other target organs

Although the liver, kidney and lung are perhaps the most important and most common target organs, many other organs and tissues in the body may be specifically damaged by toxic chemicals. It is beyond the scope of this book to consider all of these, but a few other examples are briefly highlighted and the reasons given where known. Some further examples are also considered in the final chapter.

Nervous system

The nervous system, both peripheral and central, is a common target for toxic compounds, and the cells which make up the system are particularly susceptible to changes in their environment. Thus, anoxia, lack of glucose and other essential metabolites, restriction of blood flow, and inhibition of intermediary metabolism, may all underlie damage to cells of the nervous system as well as direct, cytotoxic damage. The nervous system is a highly complex network of specialized cells, and damage to parts of this system may have permanent and serious effects on the organism as there is little capacity to regenerate and little reserve functional capacity. **Peripheral neuropathy** is a toxic response to a variety of foreign compounds such as organophosphorous compounds, methyl mercury and isoniazid for example.

The 'designer drug' contaminant, **1-methyl-4-phenyl-1,2,3,6-tetrahydropyridine** (MPTP) causes specific damage to the dopamine containing cells in the substantia nigra area of the brain. MPTP is highly lipophilic and readily enters the brain. Once there it is metabolized to a toxic metabolite and is taken up by dopamine neurones and hence damages these particular cells. Thus, the chemical structure of the compound and its physicochemical characteristics determine its toxicity, and the specific uptake system determines the target organ and cell. This example is considered in more detail in the final chapter. A similar example discussed in Chapter 7 is 6-hydroxydopamine.

The solvent **hexane** causes a different type of neurotoxicity, involving swelling and degeneration of motor neurones. This leads to paraesthesia and sensory loss in the hands and feet, and weakness in toes and fingers. Hexane has been widely used in industry as a solvent, and there have been many cases of neuropathy reported from different parts of the world. The toxicity is due to the metabolite 2,5-hexanedione which arises by ω-1 oxidation at the 2- and 5-position to 2,5-hexanediol, and then further oxidation to the diketone (figure 4.9). The 2,5-hexanedione then reacts with protein to form pyrrole adducts. The γ-diketone structure is important, as 2,3- and 2,4-hexanedione are not neurotoxic. **Methyl *n*-butyl ketone** also causes similar neurotoxic effects and is also metabolized to 2,5-hexanedione. The lipophilicity of the molecule allows distribution to many tissues including the nervous system. Thus, chemical structure and metabolism are important prerequisites for this toxicity. Exposure to the solvent **carbon disulphide** in industry causes neuronal damage in the central and peripheral nervous system. The mechanism may involve chelation of metal ions essential for enzyme activity by the oxothiazolidine and dithiocarbamate metabolites of carbon disulphide, which are formed by reaction with glutathione and glycine, respectively. However, reactive metabolites may also be formed such as active sulphur (figure 4.26) which might play a role in the various toxic effects.

Gonads

A number of compounds cause specific damage to the male reproductive system. Thus **cadmium, 2-methoxyacetic acid, dibromochloropropane** and **diethylhexylphthalate** all cause different types of damage to the testis. The spermatocytes (germ cells) may be susceptible to different compounds at different stages of development, such as the pachytene stage which is specifically damaged by **2-methoxyethanol**. Also, the Sertoli and Leydig cells may be damaged such as by **1,3-dinitrobenzene** and **ethanedimethylsulphonate**, respectively. The testis is a target organ for several reasons. It has rapidly growing and dividing tissue and thus compounds such as anti-cancer drugs may damage the testis. It has a limited blood supply and hence is susceptible to a reduction in this supply. For example, cadmium is believed to cause ischaemic damage to the testis by reducing blood flow. There are particular biochemical pathways, such as the utilization of lactate as an energy source, which may suffer interference. Thus, the testicular toxicity of **2-methoxyethanol** and **2-methoxyacetic acid** are thought to involve interference with the metabolism of lactate.

The **female reproductive system** may also be a target for damage by foreign compounds. Thus, the oocytes may be specifically damaged or destroyed by compounds such as **polycyclic aromatic hydrocarbons, bleomycin** and **cyclophosphamide**. Studies in mice have suggested that the immature oocyte is particularly susceptible. With the polycyclic aromatic hydrocarbons, metabolic activation appears to be necessary. The uterus may be affected by foreign compounds, and in turn this can affect fertility. Thus, **DDT** can cause an oestrogenic response in rats leading to thickening of the endometrium and increase in uterus weight. This is due to competitive inhibition of the binding of oestradiol to receptor sites in the uterus. A number of other pesticides may cause similar effects. Exposure to the solvent **carbon disulphide** can cause menstrual disorders and reduced fertility. The compound most well known for its toxic effects to the female reproductive system is **diethylstilboestrol** which causes vaginal cancer as described later in this chapter.

Heart

The heart is occasionally a target for toxic compounds. Thus, **allyllamine** specifically damages heart tissue. This is partly due to metabolism of this compound by amine oxidases present in cardiac tissue. The product is allyl aldehyde (acrolein), which is highly reactive and toxic (figure 4.31) as already mentioned in the context of the hepatotoxicity of allyl alcohol. Some compounds such as **hydralazine** (figure 7.54) may cause myocardial necrosis as a result of their pharmacological action. Heavy metals such as **cobalt** cause cardiomyopathy when given repeatedly to animals. The lesions seen are similar to those detected in humans who were exposed to cobalt as a result of heavy consumption of beer which contained cobalt in a stabilizing additive. The toxic effect of cobalt and certain other heavy metals is due to interference with Ca^{2+} in the muscle tissue. The anti-cancer drug **adriamycin** is cardiotoxic, possibly for several reasons. It has been suggested that it has a high affinity for, and hence binds to, complex lipids such as cardiolipin, present in the membranes of cardiac cells. Also, it has been suggested that the heart has reduced protective mechanisms (see below) such as superoxide dismutase and catalase. Adriamycin is believed to be toxic as a result of cyclical oxidation–reduction processes involving the adriamycin–iron complex which results in the production of superoxide. Occupational exposure to **carbon disulphide** is known to be a factor in coronary heart disease. Other components of the vascular system may also be specifically damaged.

Pancreas

The compounds **alloxan** and **streptozotocin** both specifically damage the pancreas and are used experimentally to induce diabetes. It is the β-cells which are particularly susceptible to these two compounds.

Olfactory epithelium

This tissue is known to contain a high level of cytochromes P-450 activity. Clearly it is similar to the lung in exposure. An example is the industrial chemical **trifluoromethylpyridine** which specifically damages the olfactory epithelium in animals exposed to it in the inspired air. The compound is metabolized by *N*-oxidation and the *N*-oxide product (or a nitroxide radical) is believed to be responsible for the toxicity (figure 4.20). The activity for this metabolic pathway is particularly high in the olfactory epithelium.

Blood

As almost all foreign compounds are distributed via the bloodstream, the components of the blood are exposed at least initially to significant concentrations of toxic compounds. Damage to and destruction of the blood cells results in a variety of sequelae such as a reduced ability to carry oxygen to the tissues if red blood cells are destroyed. Aromatic amines such as aniline and the drug **dapsone** (4,4-diaminodiphenyl sulphone) are metabolized to hydroxylamines, and in the latter case the metabolite is concentrated in red blood cells. Also, nitro compounds such as **nitrobenzene**, which can be reduced to hydroxylamines, are similarly toxic to red blood cells. These hydroxylamines are often unstable and can be further oxidized to reactive products, in the presence of oxygen in the red cell, and thereby damage the haemoglobin. The haemoglobin may be oxidized to methaemoglobin, and thus be unable to carry oxygen. Irreversible damage to haemoglobin can occur as a result of oxidation of thiol groups in the protein, with the subsequent denaturation of the protein and haemolysis of the red cell. **Phenylhydrazine** is another compound which damages red cells both *in vitro* and *in vivo* causing haemolysis.

The production of blood cells which takes place in the bone marrow is also a target for foreign compounds as the cells in the bone marrow are actively dividing. Damage to the bone marrow can result in reduced numbers of red and/or white blood cells, as is the case with the solvent **benzene**. This compound destroys the bone marrow cells which form both the red and white blood cells, giving rise to the condition of aplastic anaemia.

Eye

The eye may be a target organ as a result of its external position in the organism and direct exposure, but also from systemic exposure. Thus, the various components of the eye may be specifically damaged. The presence of pigments in the eye such as melanin has been suggested as the cause of **chloroquine** toxicity to the retina.

Methanol ingestion causes blindness by damage to the optic nerve, as discussed in more detail in Chapter 7. The drug practolol caused a serious keratinization of the cornea by an unknown mechanism and had to be withdrawn (see Chapter 7). Paracetamol when given in large doses to certain strains of mice causes opacification of the cornea by a similar mechanism to that involved in the hepatic damage (see final chapter). The lens may be damaged with the formation of cataracts such as by **2,4-dinitrophenol** and **naphthalene**, and the latter will also damage the retina. Carbon disulphide causes various effects on the eye in exposed humans.

Ear

The auditory system may be damaged by systemic exposure to foreign compounds. Perhaps the most well-known example is the damage to the cochlear system by aminoglycoside antibiotics such as **gentamycin**. The hair cells in this part of the ear seem to be particularly sensitive and may be destroyed by these drugs. The cells can be readily seen by electron microscopy, and the dead or missing cells can be counted, giving a *quantifiable* indication of ototoxicity. Exposure to carbon disulphide causes hearing loss to high-frequency tones in humans.

Skin

Although the skin is usually the target for external foreign compounds, systemic exposure can also lead to damage such as the photosensitization caused by use of the drug **benoxaprofen** (Opren) and the skin reactions following immune responses (see below). Exposure to **organochlorine** compounds gives rise to the condition of chloracne which may occur after oral as well as cutaneous exposure. The sebaceous gland ducts become keratinized and are replaced with a keratinous cyst. As with some of the other toxic effects, chloracnegenicity is dependent on the lateral symmetry and position of the chlorine atoms and correlates with ability to induce the mono-oxygenase enzymes. Thus the physicochemical characteristics of the compound are important in this toxic effect.

For further examples of target organs the reader is referred to the systemic toxicology section of *Casarett and Doull's Toxicology, Target Organ Toxicity*, Vols I and II, and the *Target Organ Toxicity* series (see Bibliography).

Detection of organ damage

There are various ways of detecting damage to organs, in some cases using specific *biochemical tests*, in others *function tests* are utilized. *Histopathology* is generally used at some stage, however, for all types of damage.

Thus, for damage to the nervous system and ear, functional tests and histopathology are utilized. For damage to the eye and skin, gross observation may be useful before microscopy. For damage to the reproductive system, hormonal changes may be detected. For testicular damage a urinary marker, creatine, has been proposed, and changes in the plasma levels of **lactate**

dehydrogenase isoenzyme C4 is also utilized diagnostically. Damage to blood cells is readily detected by light microscopy and automated counting techniques.

For damage to the heart and pancreas, specific enzymes may be measured in the plasma or serum such as **creatine kinase** and **amylase**, respectively.

Mechanism and response in cellular injury

Toxic compounds can damage cells in target organs in a variety of ways and the cellular injury caused by such compounds leads to a complex sequence of events. The eventual response may be reversible injury or an irreversible change leading to the death of the cell. The processes leading to cell death and the 'point of no return' for a cell are not yet clearly understood and are the subject of considerable research, speculation and some controversy. However, some of the key elements are known, and a picture of at least part of the sequence of cellular changes is emerging.

It is convenient for this discussion to divide the stages following exposure of a cell to a toxic compound into primary, secondary and tertiary events.

The **primary events** are those which result in the initial damage, the **secondary events** are cellular changes which follow from that initial damage and the **tertiary events** are the observable and final changes.

(1) Primary events. As already mentioned many compounds are toxic following metabolism to reactive metabolites. These reactive metabolites may then initiate one or more primary events. For example, paracetamol- and bromobenzene-induced liver damage results from metabolic activation, discussed in more detail in Chapter 7. In other cases metabolic activation is not necessary, and the parent compound or a stable metabolite initiates the primary event. For example, cyanide is cytotoxic as a result of inhibition of crucial enzymes and carbon monoxide deprives the cell of oxygen (see Chapter 7 for more details).

The major primary events are:

> lipid peroxidation
> covalent binding to macromolecules
> changes in thiol status
> enzyme inhibition
> ischaemia

In some cases several of these primary events may occur and may be interrelated; in others perhaps only one occurs. Each of these are discussed in more detail below and will also be considered in the final chapter as part of detailed discussions of specific examples.

(2) Secondary events. These are the changes which may occur in cells exposed to toxic compounds following the primary events. When a cell is damaged a number of changes may be detected both biochemically and structurally and some of these are interrelated. However, some of these changes are

consequences of damage rather than causal, and result from the loss of cellular control and the inability of the cell to compensate.

The major secondary events are:

changes in membrane structure and permeability
changes in the cytoskeleton
mitochondrial damage
inhibition of mitochondrial function
depletion of ATP and other cofactors
changes in Ca^{2+} concentration
DNA damage and poly ADP-ribosylation
lysosomal destabilization
stimulation of apoptosis
damage to the endoplasmic reticulum.

In some cases these may be primary events; in other situations they may follow primary events or follow sequentially from another secondary event.

(3) Tertiary events. These are the final, observable manifestations of exposure to a toxic compound. Several may occur together or they may occur sequentially. The major tertiary events are:

steatosis/fatty change
hydropic degeneration
blebbing
apoptosis
necrosis.

All of these except the last two are potentially reversible.

Primary events

LIPID PEROXIDATION

As already discussed in Chapter 4 and mentioned above, many foreign compounds are metabolically activated to reactive intermediates which are responsible for initiating toxic effects. One particular type of reactive intermediate is the free radical, and these are implicated in the hepatotoxicity of carbon tetrachloride and **white phosphorus**, the pulmonary damage caused by paraquat, the destruction of pancreatic β-cells by **alloxan** and the destruction of nerve terminals by **6-hydroxydopamine**. The role of free radicals in carbon tetrachloride hepatotoxicity is discussed in more detail in Chapter 7. Free radicals arise by either homolytic cleavage of a covalent bond in a molecule, the addition of an electron as in the case of carbon tetrachloride, or by the abstraction of a hydrogen atom by another radical. They have an unpaired electron, but are usually uncharged and centred on a carbon, nitrogen, sulphur or oxygen atom. Consequently, free radicals are extremely reactive, electrophilic species which can

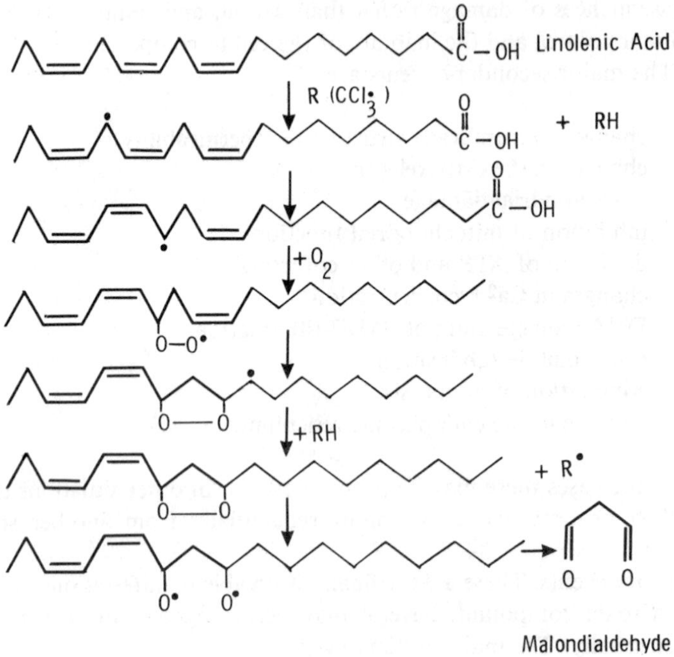

FIGURE 6.9. Peroxidative destruction of a polyunsaturated lipid initiated by a free radical attack such as by the trichloromethyl radical ($CCl_3^·$).
Adapted from Slater, T. F. (1972) *Free Radicals in Tissue Injury* (London: Pion).

react with cellular components. Alternatively, molecular oxygen may accept electrons from a variety of sources to produce the oxygen free radical superoxide. This can then produce further reactive molecules such as singlet oxygen and hydroxyl radical. The process of lipid peroxidation is initiated by the attack of a free radical on unsaturated lipids, and the resulting chain reaction is terminated by the production of lipid breakdown products: lipid alcohols, aldehydes or smaller fragments such as **malondialdehyde** (figure 6.9). Thus there is a *cascade* of peroxidative reactions which ultimately leads to the destruction of the lipid (figure 6.9) and possibly the structure in which it is located:

$$L· + O_2 \rightarrow LO_2$$
$$\textbf{Propagation} \quad LO_2 + LH \rightarrow LOOH + L·$$

This cascade however may be propagated throughout the cell unless terminated by a protective mechanism (see below) or a chemical reaction such as disproportionation which gives rise to a non-radical product. Polyunsaturated fatty acids, found particularly in membranes, are especially susceptible to free radical attack. The effects of lipid peroxidation are many and various. Clearly the structural integrity of membrane lipids will be adversely affected. In the lipid radical produced the sites of unsaturation may change, thereby altering the fluidity of the membrane (see Chapter 3). Lipid radicals may interact with other lipids and macromolecules such as proteins to cause cross linking, again potentially altering membrane structure and function. These membrane effects could lead to increased permeability of either the plasma membrane or the membranes of organelles. Thus, the lysosomal membrane may be destabilized leading to further damage to the cell from the loss of the hydrolytic contents. Some of the mitochondrial enzymes and those of the endoplasmic reticulum are membrane bound, and consequently membrane damage will compromise the function of these organelles leading to changes in cellular homeostasis (see below). Sulphydryl-containing enzymes are particularly susceptible to inhibition by lipid peroxidation. As well as the destruction of lipids and possibly of membranes, lipid peroxidation gives rise to breakdown products which themselves have biological activity (see figure 7.7). One cytotoxic breakdown product in particular which is known to be formed is **4-hydroxynonenal**. This causes a number of the effects seen with toxic compounds which are responsible for lipid peroxidation. Such reactive carbonyl compounds may have a variety of cellular effects such as changing membrane permeability, inhibiting enzymes and hence causing the disruption of the cell via such effects as changes in ion levels and energy production. Thus **4-hydroxynonenal** is known to be a potent inhibitor of Ca^{2+} sequestration by the endoplasmic reticulum. However, after low doses of carbon tetrachloride, Ca^{2+} sequestration is inhibited in the absence of lipid peroxidation, perhaps due to a direct effect of carbon tetrachloride on the endoplasmic reticulum. There is evidence that lipid peroxidation, at least in some cases, may be a consequence rather than a cause of cellular injury. Thus, the compound **diquat** causes lipid peroxidation via active oxygen species, yet cell death requires depletion of cellular glutathione. However, depletion of cellular glutathione does not occur with all cytotoxic compounds. In other cases protection against lipid peroxidation does not prevent cell death. Certainly there are many toxic compounds which do not cause lipid peroxidation, such as the hepatotoxins **hydrazine** and **thioacetamide**. Therefore, although lipid peroxidation may be an important part of cellular injury for *some* compounds, it is clearly not a general mechanism underlying *all* toxic responses.

OXIDATIVE STRESS

When active oxygen species are produced this may lead to a cycle of oxidation and reduction (redox cycle) with electrons being donated to oxygen to yield superoxide. This is the case with paraquat and also a number of cytotoxic quinones. Thus, there is a cyclic process which produces superoxide by adding electrons from paraquat (see Chapter 7) or a semiquinone to oxygen (figure 6.10). The superoxide produced may then be metabolized to hydrogen peroxide by superoxide dismutase, which is further metabolized to water (see below) or protonated to the hydroperoxy radical, $HOO\cdot$.

$$2\,O_2\cdot^- + 2H^+ \rightarrow H_2O_2 + O_2$$

However, under certain circumstances, such as in the presence of transition metal

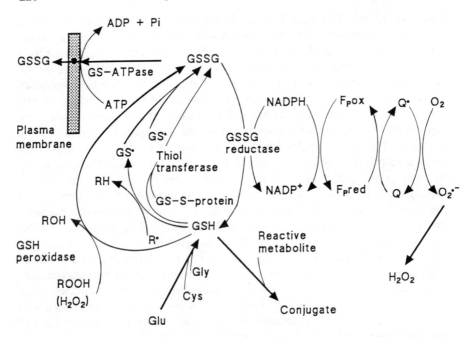

FIGURE 6.10. The various protective roles of reduced glutathione (GSH) and relationship with oxidized glutathione (GSSG). $F_p$red and $F_p$ox are the reduced and oxidized forms of a flavoprotein respectively. Q and Q· represent a quinone and semiquinone respectively. The NADPH may be regenerated by, for example, the pentose phosphate shunt, but this is not shown.

ions, hydroxyl radicals may result from either the **Haber–Weiss reaction** or the **Fenton reaction** which cause lipid peroxidation and cell injury.

Haber–Weiss reaction:

$$O_2\cdot^- + Fe^{3+} \rightarrow O_2 + Fe^{2+}$$
$$2O_2\cdot^- + 2H^+ \rightarrow O_2 + H_2O_2$$

Fenton reaction:

$$H_2O_2 + Fe^{2+} + H^+ \rightarrow HO\cdot + Fe^{3+} + H_2O$$

Hydroxyl radicals are cytotoxic and can be involved in the production of further active oxygen species such as singlet oxygen.

If a large amount of the toxic substance is present then the detoxication processes present are overwhelmed. Excess superoxide is produced, reduced glutathione and NADPH are depleted and hydroxyl radicals and singlet oxygen are formed. This is the condition known as oxidative stress. As well as causing lipid peroxidation, reactive oxygen species will also cause DNA damage and damage to proteins.

COVALENT BINDING TO MACROMOLECULES

As well as free radicals, other reactive intermediates, usually electrophilic species, may be produced by metabolism. These reactive intermediates can interact with proteins and other macromolecules and bind covalently to them. Studies have shown that there is a correlation between both the amount and site of binding and tissue damage. Therefore it was suggested that covalent binding to critical macromolecules was a possible cause of the cellular injury. While this may be true with regard to mutations and tumour induction where the target macromolecule, DNA, is known, with other types of toxic response the target molecule and the mechanism is less clear. Rather, it may simply be that covalent binding to macromolecules is an indication of the production of a reactive intermediate which is also a necessary step in the mechanism of cytotoxicity. Studies with **paracetamol** have shown that treatments which protect against cytotoxicity do not alter covalent binding in hepatocytes, and that more covalent binding may occur with non-toxic analogues of paracetamol. However, the measurements of binding are relatively non-specific as the target molecule and intracellular site are mostly unknown. Studies with paracetamol *in vitro* have shown that the binding to protein is mainly directed towards sulphydryl groups. Investigation of the covalent binding of the hepatotoxic compound **acetylhydrazine** to protein *in vivo* revealed that binding to cellular organelles show different profiles and that these changed after pretreatments which increased the toxicity. Therefore, although covalent binding to cellular protein is a useful marker of activation, unless the specific target(s) are known it cannot be assumed that this is a direct cause or indicator of cytotoxicity. Clearly, binding to critical sites on proteins could conceivably alter the function of those proteins such as inhibiting an enzyme or damaging a membrane, but equally binding may be to non-critical sites and be of no toxicological consequence.

THIOL STATUS

As already discussed in Chapter 4, reactive intermediates can react with reduced glutathione (GSH) either by a direct chemical reaction or by a glutathione transferase-mediated reaction. If excessive, these reactions can deplete the cellular glutathione. Also, reactive metabolites can oxidize glutathione and other thiol groups such as those in proteins and thereby cause a change in thiol status. When the rate of oxidation of glutathione exceeds the capacity of glutathione reductase, then oxidized glutathione (GSSG) is actively transported out of the cell and thereby lost (figure 6.10). Thus, reduced glutathione may be removed reversibly by oxidation or formation of mixed disulphides with proteins and irreversibly by conjugation or loss of the oxidized form from the cell. Thus, after exposure of cells to quinones such as **menadione**, which cause oxidative stress, glutathione conjugates, mixed disulphides and GSSG are formed, all of which will reduce the cellular GSH level.

The role of glutathione in cellular protection (figure 6.10 and see below) means that if depleted of glutathione the cell is more vulnerable to toxic compounds. The loss of reduced glutathione from the cell leaves other thiol groups such as those in critical proteins, vulnerable to attack with subsequent oxidation, cross-linking, formation of mixed disulphides or covalent adducts. The sulphydryl groups of proteins seem to be the most susceptible nucleophilic targets for attack, as shown by studies with paracetamol (see Chapter 7), and are often crucial to the function of enzymes. Consequently, modification of thiol groups of enzyme proteins such as by **mercury** and other heavy metals, often leads to inhibition of the enzyme function. Such enzymes may have critical endogenous roles such as the regulation of ion concentrations, active transport or mitochondrial metabolism. There is evidence that alteration of protein thiols may also be a crucial part of cell injury. Thus, changes to isolated hepatocytes caused by paracetamol can be prevented, and even reversed, by the use of dithiothreitol which reduces oxidized thiols. Similarly, data on the use of *N*-acetylcysteine as an antidote for paracetamol poisoning *in vivo* indicates the importance of thiol status in cell injury (see Chapter 7 for a more detailed discussion).

ENZYME INHIBITION

Although inhibition of enzymes may be a common consequence of exposure to toxic compounds, there are examples of where the inhibition of a single endogenous enzyme is the critical primary event. Thus, the inhibition of cytochrome aa_3 by cyanide leads to cell death by blocking cellular respiration, and the blockade of Krebs' cycle by the inhibition of aconitase by fluorocitrate similarly inhibits respiration (see Chapter 7 for more details). Obviously there will be various sequelae to this enzyme inhibition, such as depletion of ATP and other vital endogenous molecules. The effect of the inhibition will depend on the importance of the enzyme to the particular cell type. Thus, inhibition of cellular respiration by cyanide is especially critical in cells of the CNS. There are a number of compounds which inhibit protein synthesis, either specifically at particular points such as **tetracycline** or more indiscriminately by damage to the endoplasmic reticulum such as carbon tetrachloride. This may cause cellular injury, especially steatosis (see below), but cell death is not an inevitable consequence. Several enzymes may be inhibited by one toxic compound such as **hydrazine** for example, exposure to which inhibits phosphoenol pyruvate carboxy kinase, and many pyridoxal phosphate requiring enzymes such as the transaminases which will cause a variety of effects on intermediary metabolism. Another example is **bromobenzene** which inhibits Na^+/K^+ ATPase, GSSG reductase and glucose-6-phosphate dehydrogenase. Clearly, inhibition of each of these enzymes will have a different effect; inhibition of the Na^+/K^+ ATPase will affect the active transport of ions into and out of the cell leading to an imbalance; inhibition of GSSG reductase will exacerbate depletion of reduced glutathione, and inhibition of glucose-6-phosphate dehydrogenase – the first enzyme in the pentose phosphate pathway – will reduce the production of NADPH and hence influence the reduction of GSSG and metabolic pathways such as those catalysed by cytochromes P-450.

ISCHAEMIA

The cessation or reduction of the supply of blood containing oxygen and nutrients such as glucose to a tissue leads to damage and cell death if prolonged.

The lack of oxygen, **hypoxia** (or **anoxia** if total) may be a specific effect, as in the case of methaemoglobinaemia caused by **nitrite** for example, or carbon monoxide poisoning discussed in detail in Chapter 7, or may simply be due to a reduction of blood flow. The latter is believed to underlie the testicular necrosis which follows acute doses of **cadmium**. The blood supply to the testis arrives via the vessels of the pampiniform plexus and if these are restricted by vasoconstriction, as caused by cadmium or by surgical ligation, then the testis suffers ischaemic necrosis. Ischaemia may also be a secondary event and occur due to the swelling of cells in tissue with subsequent reduction in blood flow. For example, damage to the sinusoidal cells of the liver caused by **pyrrolizidine alkaloids** leads to swelling and reduction in liver blood flow, and hence ischaemic necrosis. The centrilobular region is especially susceptible as it receives less oxygenated blood.

Secondary events

CHANGES IN MEMBRANE STRUCTURE AND PERMEABILITY
Both the plasma membrane and internal membranes associated with organelles may be damaged by toxic compounds. As already discussed, this may be due to lipid peroxidation which alters and destroys membrane lipids. As many enzymes and transport processes are membrane bound this will affect the function of the organelle as well as the structure. Sulphydryl groups in membranes may be targets for **mercuric** ions in kidney tubular cells and for **methyl mercury** in the CNS for example. The result is changes in membrane permeability and transport and subsequent cell death. Structural damage can be detected by electron microscopy as disruption of the endoplasmic reticulum, for example, or swelling of the mitochondria. The result of mitochondrial damage could be alterations in intracellular Ca^{2+} concentrations (see below). The result of damage to the plasma membrane might be an influx or efflux of ions and other endogenous substances, leading to swelling of the cell due to an influx of water for example (see 'Hydropic degeneration', below). However, before permeability changes occur, reversible structural changes such as blebbing of the cell surface can be detected. This structural change is visible by light microscopy in isolated cells *in vitro* and is also detectable *in vivo*. It can be considered a tertiary change and is discussed more fully below. Changes in the plasma membrane can be detected *in vitro* as leakage of ions such as K^+, leakage of enzymes such as **lactate dehydrogenase** and uptake of dyes such as **Trypan Blue**. Changes in the lysosomal membrane may have important sequelae, as this may lead to leakage of degradative enzymes (see below).

CHANGES IN THE CYTOSKELETON
The cytoskeleton may be specifically damaged by toxic compounds such as **phalloidin** and the **cytochalasins**. Thus, phalloidin which is a toxic component of certain poisonous mushrooms, binds to actin filaments and stabilizes them. They are thus unable to depolymerize, and this in some way leads to release of Ca^{2+} from an intracellular compartment. As the cytoskeleton is associated with the plasma membrane, damage to it may underlie the appearance of cell surface blebs

which rapidly occur after exposure of hepatocytes to phalloidin. The intracellular level of calcium also seems to be an important factor in the functioning of the cytoskeleton (see below). An increase in the level of Ca^{2+} causes a dissociation of actin microfilaments from α-actinin, a protein involved with binding the actin cytoskeletal network to the plasma membrane. Actin-binding proteins are also involved in the binding of the cytoskeleton to the plasma membrane, and these may be cleaved by proteases activated by calcium. The consequence of these changes will be to weaken the attachment of the plasma membrane to the cytoskeleton and hence blebs may form. However, not all cytoskeletal changes are due to changes in Ca^{2+} homeostasis.

MITOCHONDRIAL DAMAGE AND INHIBITION OF MITOCHONDRIAL FUNCTION

The mitochondrion is often a target for toxic compounds and effects on the organelle can be seen with the electron microscope. Thus, **hydrazine** causes swelling of mitochondria after acute doses and prolonged treatment leads to the formation of megamitochondria. As structure and function are closely linked, compounds which affect function may cause changes in structure such as swelling or contraction. This may in some cases be due to fluid changes in the organelle. If intake of fluid is excessive and it moves into the inner compartment from the outer, then this inner compartment expands and the cristae unfold. The membranes may then rupture. Contraction of the mitochondria is associated with increases in the ADP : ATP ratio, such as occurs after anoxia or with uncouplers of oxidative phosphorylation. Prolonged contraction leads to deterioration of the inner membrane and high amplitude swelling results. Eventually contraction becomes impossible, and rupture and deterioration of the membranes results. The mitochondria are crucial to the cell, and inhibition of the electron transport chain such as caused by **cyanide** leads to rapid cell death. Damage to the mitochondria may also allow release of Ca^{2+} and so increase the levels in the cytosol, although the role of the mitochondrion in Ca^{2+} homeostasis may only be a minor one. Swelling of mitochondria and disruption of structure often occurs before cell death and necrosis. Some compounds, such as **2,4-dinitrophenol**, act by uncoupling oxidative phosphorylation, that is the production of ATP is uncoupled from electron transport. Other compounds affect mitochondrial function by inhibiting the electron transport chain at one or more specific sites, such as the toxic metabolite of MPTP which inhibits complex I (see Chapter 7). The toxic metabolite of **hexachlorobutadiene** is believed to be nephro-toxic due to inhibition of mitochondria function in the proximal tubular cells.

DAMAGE TO THE ENDOPLASMIC RETICULUM

As the smooth endoplasmic reticulum is the site for the oxidative metabolism of many foreign compounds, it is vulnerable to damage from reactive metabolites such as epoxides and free radicals. Short-lived, reactive intermediates with only a narrow radius of action will obviously damage the immediate vicinity. Thus, with **carbon tetrachloride**, damage to both smooth and rough endoplasmic reticulum occurs leading to disruption of functions, such as metabolism and protein synthesis, of the whole organelle. One particular function of the endoplasmic reticulum which is important in terms of cellular homeostasis is calcium sequestration. Compounds which damage the endoplasmic reticulum such as carbon tetrachloride are known to inhibit this function of sequestrating calcium.

DEPLETION OF ATP AND OTHER COFACTORS

Depletion of ATP is caused by many toxic compounds and this will result in a variety of biochemical changes. Although there are many ways for toxic compounds to cause a depletion of ATP in the cell, interference with mitochondrial oxidative phosphorylation is perhaps the most common. Thus, compounds such as **2,4-dinitrophenol** which uncouple the production of ATP from the electron transport chain will cause such an effect, but also inhibition of electron transport or depletion of NADH. Excessive utilization of ATP or sequestration are other mechanisms, the latter being more fully described in relation to ethionine toxicity in Chapter 7. Also, DNA damage which causes the activation of poly(ADP-ribose)polymerase may lead to ATP depletion (see below). A lack of ATP in the cell means that active transport into, out of and within the cell, is compromised or halted with the result that the concentration of ions such as Na^+, K^+ and Ca^{2+} in particular compartments, will change. Also, various synthetic biochemical processes such as protein synthesis, gluconeogenesis and lipid synthesis will tend to be decreased. At the tissue level this may mean that hepatocytes do not produce bile efficiently and proximal tubules do not actively reabsorb essential amino acids and glucose.

Depletion of other cofactors such as UTP, NADH and NADPH may also be involved in cell injury either directly or indirectly. Thus, the role of NADPH in maintaining reduced glutathione levels means that excessive glutathione oxidation such as caused by certain quinones which undergo redox cycling may in turn cause NADPH depletion (figure 6.10). Alternatively, NADPH may be oxidized if it donates electrons to the foreign compound directly. However, NADPH may be regenerated by interconversion of NAD^+ to $NADP^+$. Some quinones such as **menadione, 1,2-dibromo-3-chloropropane** (DBCP) and **hydrogen peroxide** also cause a depletion of NAD, but probably by different mechanisms. Thus, with menadione the depletion may be the result of synthesis of NADPH at the expense of NAD; with hydrogen peroxide and DBCP the depletion is thought to be a consequence of poly(ADP-ribose) polymerase activation.

CHANGES IN Ca^{2+} CONCENTRATION

Perhaps the most important cellular changes described recently in terms of mechanisms of cell injury and death are those involving changes in the concentration of calcium. It has long been known that calcium is in some way involved in cell death, and that it accumulates in damaged tissue. More recently changes in the intracellular distribution of this ion have been implicated in the mechanisms underlying the cytotoxicity of many different, although not all, toxic compounds in different tissues. Thus, the toxicity of **carbon tetrachloride** and **paracetamol** to liver cells, the toxicity of **lead** and **methyl mercury** and the toxicity of **TCDD** to cardiac and thymus cells are all believed to involve Ca^{2+}.

The level of calcium in the cell is low, about $0·1\,\mu M$ compared with approximately $1·3\,mM$ outside the cell. The intracellular concentration is kept low by the activity of transport systems which transport calcium out of the cell and sequester it in the mitochondria and endoplasmic reticulum. The plasma membrane thus houses a Ca^{2+} transporting ATPase, as does the endoplasmic reticulum and nucleus. Calcium may also be stored in the mitochondria, and the intracellular protein calmodulin will bind calcium. Interference with any of these processes may be caused by toxic compounds and can alter calcium homeostasis. This can allow an influx of Ca^{2+}, inhibition of export of Ca^{2+} out of the cell, or a release of Ca^{2+} from compartments within the cell. The result of each of these will be a rise of intracellular Ca^{2+} which can cause a variety of damaging events.

Interference with Ca^{2+} homeostasis may occur in a variety of ways:

(i) inhibition of Ca^{2+} ATPases. Compounds which react with sulphydryl groups will inhibit the ATPase which are involved in calcium transport. Thus, intracellular movement and efflux of calcium will be altered. Consequently if the plasma membrane Ca^{2+} transporting ATPase is inhibited Ca^{2+} will tend to enter the cell due to the high concentration outside, but will not be effectively pumped out. Similarly if the ATPase in the endoplasmic reticulum is inhibited this organelle will fail to sequester Ca^{2+} and so the free cytosolic concentration will again rise. The reactive aldehyde product of lipid peroxidation, **4-hydroxy-nonenal**, will inhibit sequestration of Ca^{2+} by the endoplasmic reticulum giving another mechanism whereby lipid peroxidation may lead to cell injury and death. **Carbon tetrachloride** and **bromobenzene** cause release of Ca^{2+} from the endoplasmic reticulum. Quinones and peroxides inhibit the efflux of Ca^{2+} from the cell.

(ii) damage to the plasma membrane allows increased influx of Ca^{2+}. The result of this will be the same as inhibition of the efflux, a rise in cytosolic free calcium. Paracetamol and carbon tetrachloride cause this increased influx.

(iii) depletion of ATP will also lead to a rise in intracellular Ca^{2+}, presumably as a result of the reduction in activity of the ATPases and a reduction in other metabolic activity.

The cytosolic free calcium may also rise due to release from mitochondrial stores as caused by cadmium, MPTP and uncouplers.

It seems that the crucial event is a sustained rise in cytosolic free calcium and there are various consequences which arise from this:

(a) alterations in the cytoskeleton. The cytoskeleton depends on the intracellular Ca^{2+} concentration which affects actin bundles, the interactions between actin and myosin and α-tubulin polymerization. The effect of increases in Ca^{2+} on the cytoskeletal attachments to the plasma membrane and the role of the cytoskeleton in cellular integrity have

already been mentioned (see above). If the cytoskeleton is d̶
disrupted or its function altered by an increase in Ca^{2+}, then
protrusions appear on the plasma membrane (see below). As well a̶
increase in Ca^{2+}, oxidation of, or reaction with sulphydryl groups, such as
alkylation or arylation for example, may disrupt the cytoskeleton as thiols
are important for its integrity. Similarly direct interaction with the
cytoskeleton as occurs with **phalloidin** will also lead to cellular damage.

(b) Ca^{2+}-activated phospholipases. These enzymes are found in biological
membranes where they catalyse the hydrolysis of membrane phospholipids and Ca^{2+} is required as a cofactor. Phospholipase A_2 enzymes are
involved in detoxifying phospholipid hydroperoxides, which may occur
during lipid peroxidation. However, prolonged action of these enzymes
leads to breakdown of the membrane and release of compounds such as
lysophospholipids, which are cytotoxic, and arachidonic acid which may
be further metabolized to substances which are inflammatory mediators.
Prolonged increases in cytosolic Ca^{2+}, therefore, will activate these
enzymes and lead to membrane breakdown and cytotoxicity.

(c) Ca^{2+}-activated proteases. These enzymes, also known as calpains, are
extra-lysosomal. They are involved in normal cell functions such as
enzyme activation and membrane remodelling. When activated by
increases in cytosolic Ca^{2+} the proteases modify the cytoskeleton and also
membrane protein with cytotoxic results as already indicated above.

(d) Ca^{2+}-activated endonucleases. These enzymes are involved with the
normal process of apoptosis or 'programmed cell death' (see below) and
cleave DNA into fragments. This destruction of DNA results in cell death.
Other enzymes involved in DNA fragmentation may also be activated by
Ca^{2+}.

Increases in cytosolic Ca^{2+} may have other cellular effects such as inhibiting
mitochondrial function.

DNA DAMAGE AND POLY ADP-RIBOSYLATION

Compounds such as hydrogen peroxide and alkylating agents such as **dimethyl
sulphate** which cause single strand breaks in DNA cause the activation of the
enzyme poly(ADP-ribose)polymerase. This enzyme catalyses ADP-ribosylation
which is post-translational protein modification, involving the addition of ADP-
ribose to amino acids. It is also involved in polymerization reactions and, through
modulation of DNA ligase activity, with DNA repair. The result of activation of
this enzyme is the cleavage of the glycosidic link in NAD to yield ADP-ribose and
to release nicotinamide. Therefore, there is a loss of NAD, and with severe DNA
damage this depletion of NAD is sufficient to lead to cell death. This may indeed
be a protective mechanism, a form of cellular euthanasia which ensures that cells
with extensively damaged DNA do not replicate and so cannot pass on any errors
which might be encoded in the altered DNA.

ƆN

that release of lysosomal enzymes was the mechanism
d death. However, although this occurs during cell
te event occurring after the point of no return has been
r than a cause of damage. There are situations, however,
e may be an initiating event. For example, lysosomal
the toxicity of silica particles which are engulfed by
ungs and are then taken up into lysosomes, and with
gentai... taken up into lysosomes in the proximal tubular cells. In
both cases the pr.. ce of the foreign substance leads to rupture of the lysosomes
and the hydrolytic enzymes released damage and destroy the cell.

STIMULATION OF APOPTOSIS

Apoptosis is the process of 'programmed cell death' which occurs as part of the normal maintenance and renewal of tissues. The process is distinct from necrosis (see below) which may be caused by a toxic compound. However, it has been suggested that some foreign compounds such as TCDD may stimulate apoptosis. In this case it is suggested that TCDD acts through a receptor to stimulate the synthesis of a protein which mediates the influx of Ca^{2+} into the cell and thereby activates endonucleases which initiates the apoptotic process. This occurs in thymocytes which suffer atrophy in animals exposed to TCDD. It is possible, but not yet proven, that this occurs in other cell types exposed to cytotoxic compounds.

Tertiary events

STEATOSIS/FATTY CHANGE

The accumulation of fat is a common cellular response to toxic compounds which is normally reversible. Usually it is triglycerides which accumulate, although sometimes phospholipids accumulate as occurs after exposure to the drug chlorphentermine (see Chapter 2). Steatosis is particularly common in the liver as this organ has a major role in lipid metabolism (figure 6.11). The lipid may appear in the cell as many small droplets or as one large droplet. Interference with lipid metabolism can occur at several points:

(i) inhibition of excretion of lipid from the cell. This is the most common cause of steatosis, although there are several possible reasons for the inhibition. Thus, inhibition of protein synthesis will block the synthesis of the lipid acceptor protein (apoprotein) required for the transport of lipid out of the cell as Very Low Density Lipoprotein (VLDL) (figure 6.11). There are numerous points at which protein synthesis can be inhibited: at the point of DNA transcription; during RNA translation and during assembly of the apoprotein. Thus, **puromycin** and **tetracycline** both cause steatosis by inhibiting protein synthesis. **Carbon tetrachloride** causes steatosis by damaging the endoplasmic reticulum and Golgi apparatus where protein synthesis and assembly take place.

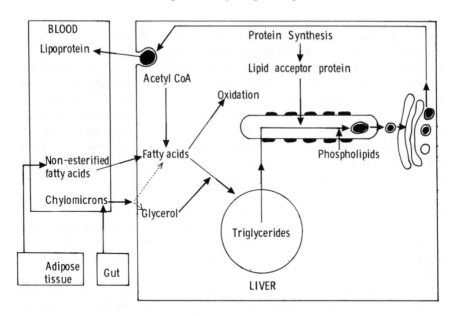

FIGURE 6.11. Pathways involved in lipid metabolism and distribution.
Adapted from Trump, B. F. and Arstila, A. U. (1980) In *Principles of Pathobiology*, edited by M. F. La Via and R. B. Hill (New York: Oxford University Press).

Ethionine (see Chapter 7), which depletes ATP, may cause steatosis by inhibiting protein synthesis as ATP is required for this. Also, ATP will be required for the transport of the lipoprotein complex out of the hepatocyte and the lack of ATP will compromise this process.

(ii) increased synthesis of lipid or uptake. Increased synthesis of lipid may be the cause of fatty liver after **hydrazine** administration as this compound increases the activity of the enzyme involved in the synthesis of diglycerides. **Hydrazine** also depletes ATP and inhibits protein synthesis however. Large doses of **ethanol** will cause fatty liver in humans and it is believed that this is partly due to an increase in fatty acid synthesis. This is a result of an increase in the NADH : NAD ratio and therefore of the synthesis of triglycerides. Changes in the mobilization of lipids in tissues followed by uptake into the liver can also be another cause of steatosis.

(iii) decreased metabolism of lipids. Decreased mitochondrial oxidation of fatty acids is another possible cause of ethanol induced steatosis. Other possible causes are vitamin deficiencies and the inhibition of the mitochondrial electron transport chain.

HYDROPIC DEGENERATION
This term describes the swelling of a cell due to the intake of water. It is a reversible change but may often precede irreversible changes and cell death. The osmotic balance of the cell is maintained by active processes which control the

influx and efflux of ions. The intracellular Na^+ and Ca^{2+} concentrations are maintained at a lower level than that in the extracellular fluid by an active process requiring ATP, whereas K^+ is maintained at a higher concentration inside the cell. However, the system depends on the integrity of the membrane and transport systems and on the supply of ATP. Inhibition of the ATPase enzyme involved in the active transport or reduction of the supply of ATP by metabolic inhibitors will upset the balance allowing the concentration of ions in the cell to increase. The osmotic pressure thus created will cause water to enter the cell with swelling. This will lead to damage to the cell and organelles if it is excessive. Thus, the mitochondria and endoplasmic reticulum will swell and may become damaged.

BLEBBING OF THE PLASMA MEMBRANE

This process is an *early morphological change* in cells often seen in isolated cells *in vitro* but also known to occur *in vivo*. The blebs, which appear before membrane permeability alters, are initially reversible. However, if the toxic insult is sufficiently severe and the cellular changes become irreversible, the blebs may rupture. If this occurs vital cellular components may be lost and cell death follows. The occurrence of blebs may be due to damage to the cytoskeleton which is attached to the plasma membrane as described above. The cause may be an increase in cytosolic Ca^{2+}, interaction with cytoskeletal proteins or modification of thiol groups.

NECROSIS

When *irreversible damage* occurs a cell may reach the point of no return and cell death ensues. This is followed by a series of degenerative changes including hydrolysis of cellular components and denaturation of proteins. This cellular necrosis, which is the death of cells forming part of living tissue, is characterized by changes such as swelling of the mitochondria and the endoplasmic reticulum, the appearance of vacuoles and the accumulation of fluid. These changes may occur prior to necrosis and be reversible, or they may presage cell death and necrosis. Abrupt and extensive expansion of the mitochondria with disruption of the structure often occurs immediately prior to necrotic change. The endoplasmic reticulum may also undergo dilation and disruption. Which organelle shows damage first may depend on the mechanism. Thus, in hypoxia the mitochondria show damage before the endoplasmic reticulum, whereas with carbon tetrachloride-induced cellular injury the endoplasmic reticulum is damaged first. Cells undergoing necrosis also show changes in the nucleus in which the chromatin becomes condensed (pyknosis) and may then become fragmented (karyorrhexis). Alternatively in some cases the nucleus simply fades and dissolves (karyolysis).

In contrast to necrosis, in cells undergoing apoptosis, the mitochondria remain normal in appearance. The cell contracts away from its neighbours and blebs may appear on the plasma and nuclear membranes. The condensation of chromatin that occurs may reflect the action of endonucleases.

Protective mechanisms

Biological systems possess a number of mechanisms for protection against toxic foreign compounds, some of which have already been mentioned. Thus, metabolic transformation to more polar metabolites which are readily excreted is one method of detoxication. For example, conjugation of paracetamol with glucuronic acid and sulphate facilitates elimination of the drug from the body and diverts the compound away from potentially toxic pathways (see Chapter 7). Alternatively, a reactive metabolite may be converted into a stable metabolite. For example, reactive epoxides can be metabolized by epoxide hydrolase to stable dihydrodiols.

GLUTATHIONE (GSH)

Probably the most important protective mechanisms involve the tripeptide glutathione (figure 4.56). This compound is found in most cells and in liver cells it occurs at a relatively high concentration, about 5 mM or more. It has a nucleophilic thiol group and it can detoxify substances in one of three ways:

   (i) conjugation catalysed by a glutathione transferase
  (ii) chemical reaction with a reactive metabolite to form a conjugate
 (iii) donation of a proton or hydrogen atom to reactive metabolites or free radicals, respectively.

The first *two* of these are discussed in Chapter 4 and there are specific examples in Chapter 7. The products are either excreted directly into the bile or further metabolized and excreted into the urine as cysteine or *N*-acetylcysteine conjugates. There are, however, examples of glutathione conjugates being involved in toxicity as indicated in Chapters 4 and 7.

The *third* protective role of glutathione involves the reduction of reactive metabolites, the glutathione becoming oxidized in the process (figure 6.10). If the reactive metabolite is a peroxide such as hydrogen peroxide or an organic peroxide, the glutathione becomes oxidized and the peroxide is reduced to an alcohol. This reaction is catalysed by glutathione peroxidase which is a selenium-containing enzyme located in the mitochondria and cytosol. Other reactive metabolites may chemically oxidize GSH to GSSG and in turn are reduced. Alternatively, glutathione may donate a hydrogen atom to a free radical intermediate and be converted into a glutathione radical (thiyl radical), which may then react with *another* glutathione radical to form GSSG or abstract a hydrogen atom from other substances to form new radicals and GSH (figure 6.10). Thus, the result of these types of reactions may be the oxidation of glutathione which may then suffer one of two fates. Under normal conditions oxidized glutathione (GSSG) is reduced back to GSH via GSSG reductase, an enzyme which requires NADPH (figure 6.10). However, in conditions of excess oxidation of GSH, such as oxidative stress induced by a toxic chemical, there may be insufficient NADPH and the enzyme-mediated reduction will be unable to cope with the production of GSSG. Then GSSG is exported out of the cell in an active process which utilizes ATP and a translocase enzyme located in the plasma membrane (figure 6.10). Glutathione conjugates can also be transported out of the cell by this mechanism. The export of GSSG under these conditions of excessive oxidation can lead to depletion of glutathione in the cell requiring resynthesis.

SUPEROXIDE DISMUTASE

Free radical intermediates or other reactive intermediates may donate electrons to oxygen-forming active oxygen species such as superoxide anion radical, O_2^- which can cause cellular damage (see above). Mammalian systems have an enzyme, superoxide dismutase, which removes superoxide and produces hydrogen peroxide. This reactive product can then removed by the action of catalase or glutathione peroxidase (see figure 6.10):

$$2\,O_2^- + 2\,H^+ \rightarrow H_2O_2 + O_2$$

$$2\,H_2O_2 + \text{Catalase} \rightarrow 2H_2O + O_2$$

As indicated above however, hydrogen peroxide may yield hydroxyl radicals in the presence of metal ions such as Fe^{2+}.

VITAMIN E

Lipid radicals and other radicals may be removed by a number of endogenous compounds as well as glutathione. One is vitamin E (α-tocopherol), which reacts with lipid hydroperoxide radicals to yield the hydroperoxide and α-tocopherol quinone:

$$LO_2\cdot + \alpha - TH \rightarrow LOOH + \alpha - T.$$

$$LO_2\cdot + \alpha T\cdot \rightarrow LOOH + \alpha - TQ$$

Other compounds which may have a protective role are vitamin K, cysteine and ascorbate. These compounds act as alternative hydrogen donors in preference to the allylic hydrogen atoms of unsaturated lipids.

Lipid hydroperoxides can be removed by reaction with glutathione catalysed by glutathione peroxidase. The enzyme phospholipase A_2 has been proposed to have a role in the detoxication of phospholipid hydroperoxides by releasing fatty acids from peroxidized membranes.

Pharmacological, physiological and biochemical effects

These three types of toxic effect will be considered together as they are often all closely interrelated. Thus, toxic foreign compounds can affect the homeostasis of an organism by altering basic biochemical processes. These effects may often be reversible if inhibition of an enzyme or binding to a receptor is involved. A well defined dose–response relation is often observed with such effects, and the end-point may be death of the animal if a process of central importance is affected. These types of response include the **exaggerated** or **unwanted pharmacological responses** to clinically used drugs which are the most commonly observed types of drug toxicity in humans.

Two distinct bases for these types of effects may be distinguished: pharmacokinetic and pharmacodynamic. **Pharmacokinetic**-based toxic effects are due to an increase in the concentration of the compound or active metabolite at the target site. This may be due to an increase in the dose, altered metabolism or saturation of elimination processes for example. An example is the increased hypotensive effect of **debrisoquine** in poor metabolizers where there is a genetic basis for a reduction in metabolic clearance of the drug (see Chapter 5).

Pharmacodynamic-based toxic effects are those where there is altered responsiveness of the target site perhaps due to variations in the receptor. For example, individual variation in the response to **digitoxin** means that some patients suffer toxic effects after a therapeutic dose (see below Chapter 7). The inhibition of enzymes, blockade of receptors or changes in membrane permeability which underlie these types of effects often rely on reversible interactions. These are dependent on the concentration of the toxic compound at the site of action, and possibly the concentration of an endogenous substrate if competitive inhibition is involved. Therefore, with the loss of the toxic compound from the body, by the processes of metabolism and excretion, the concentration at the site of action falls and the normal function of the receptor or enzyme returns. This is in direct contrast to the type of toxic effect in which a cellular structure or macromolecule is permanently damaged, altered or destroyed by a toxic compound. In some cases, however, irreversible inhibition of an enzyme may occur, which if not fatal for the organism will require the synthesis of new enzyme, as is the case with organophosphorus compounds which inhibit cholinesterases.

It also should be noted that a profound physiological disturbance such as prolonged anoxia may result in irreversible pathological damage to sensitive tissue. There are many different types of response within this category with a variety of mechanisms, and it is beyond the scope of this book to examine all of these in detail. However, a brief illustration of some of the various types of response will be given and some examples will be considered in more detail in Chapter 7.

Anoxia

Lack of oxygen in the tissues may be due to respiratory or circulatory failure or absence of oxygen. Thus, the first situation may arise if a toxic compound affects breathing rate via central control, such as the drug **dextropropoxyphene** when taken in overdoses, or by effecting respiratory muscles such as **Botulinum toxin**. The second situation arises when a toxic compound inhibits oxygen transport. The classic example of this is carbon monoxide which binds to haemoglobin in place of oxygen (see Chapter 7 for more details). Another example is the oxidation of haemoglobin by **nitrite**; the methaemoglobin produced does not carry oxygen.

Inhibition of cellular respiration

This type of effect can occur in all tissues, and is caused by a metabolic inhibitor such as **azide** or cyanide which inhibits the electron transport chain. Inhibition of one or more of the enzymes of the tricarboxylic acid cycle such as that caused by fluoroacetate (figure 7.39) also results in inhibition of cellular respiration (for more details of cyanide and fluoroacetate see Chapter 7).

Respiratory failure

Excessive muscular blockade may be caused by compounds such as the cholinesterase inhibitors. Such inhibitors, exemplified by the organophosphate insecticides such as **malathion** (figure 5.10 and see also Chapter 7), and **nerve gases** (e.g. isopropylmethylphosphonofluoridate) cause death by blockade of respiratory muscles as a result of excess acetylcholine accumulation. This is due to inhibition of the enzymes normally responsible for the inactivation of this neurotransmitter (see page 327). Respiratory failure may also result from the inhibition of cellular respiration by cyanide for example or central effects caused by drugs such as **dextropropoxyphene**.

Disturbances of the central nervous system

Toxic substances can interfere with normal neurotransmission in a variety of ways, either directly or indirectly, and cause various central effects. For example, cholinesterase inhibitors such as the **organophosphate insecticides** cause accumulation of excess acetylcholine. The accumulation of this neurotransmitter in the CNS in humans after exposure to toxic insecticides leads to anxiety, restlessness, insomnia, convulsions, slurred speech and central depression of the respiratory and circulatory centres.

Hyper/hypotension

Drastic changes in blood pressure may occur as a toxic response to a foreign compound, such as the hypotension caused by **hydrazoic acid** and **sodium azide**. There may be various mechanisms involved such as vasodilation, β-adrenoceptor blockade or altered water balance. Anti-hypertensive drugs will clearly cause dangerous hypotension if given in large doses, but such an exaggerated response may also be due to reduced metabolism in some individuals, as in the case of debrisoquine (see Chapter 5).

Hyper/hypoglycaemia

Changes in blood sugar concentration can be caused by foreign compounds and this may involve a variety of mechanisms. Drugs such as **tolbutamide,** a sulphonylurea, are used therapeutically to lower blood sugar levels. **Streptozotocin,** which destroys the pancreatic β-cells which produce insulin, causes hyperglycaemia indirectly by reducing insulin levels. **Hydrazine** the industrial chemical causes first hyperglycaemia as a result of glycogen mobilization due to the hepatic effects, and then hypoglycaemia as glycogen stores are depleted and gluconeogenesis is inhibited. Similarly, salicylate poisoning causes first hyper- then hypoglycaemia as a result of mobilization of glycogen due to an increase in adrenaline followed by a depletion of available stores (see Chapter 7 for more details).

Anaesthesia

The induction of unconsciousness may be the result of exposure to excessive concentrations of toxic solvents such as carbon tetrachloride or vinyl chloride, as occasionally occurs in industrial situations (solvent narcosis). Also, volatile and non-volatile anaesthetic drugs such as halothane and thiopental, respectively, cause the same physiological effect. The mechanism(s) underlying anaesthesia are not fully understood, although various theories have been proposed. Many of these have centred on the correlation between certain physicochemical properties and anaesthetic potency. Thus the oil : water partition coefficient, the ability to reduce surface tension and the ability to induce the formation of clathrate compounds with water are all correlated with anaesthetic potency. It seems that each of these characteristics are all connected to hydrophobicity and so the site of action may be a hydrophobic region in a membrane or protein. Thus, again, physicochemical properties determine biological activity.

Changes in water and electrolyte balance

Certain foreign compounds may cause the retention or excretion of water. Some, such as the drug **furosemide,** are used therapeutically as diuretics. Other compounds causing diuresis are ethanol, caffeine and certain mercury compounds such as mersalyl. Diuresis can be the result of a direct effect on the kidney, as with **mercury** compounds which inhibit the reabsorption of chloride, whereas other diuretics such as ethanol influence the production of antidiuretic hormone by the pituitary. Changes in electrolyte balance may occur as a result of excessive excretion of an anion or cation. For example, salicylate-induced alkalosis leads to excretion of Na^+ and HCO_3^-, and ethylene glycol causes the depletion of calcium, excreted as calcium oxalate.

Ion transport

Alteration of the movement of ions, such as potassium, in heart tissue may be a toxic response to some foreign compounds. For example, the **digitalis glycosides** cause changes in tissue potassium and this may lead to serious cardiac effects. Diuretics may also cause low plasma potassium with potentially dangerous effects on heart function.

Failure of energy supply

Toxic compounds which interfere with major pathways in intermediary metabolism can lead to depletion of energy-rich intermediates. For example, fluoroacetate blocks the tricarboxylic acid cycle, giving rise to cardiac and central nervous system effects which may be fatal (see Chapter 7). Another example is cyanide (see Chapter 7).

Changes in muscle-contraction/relaxation

This type of response may be caused by several mechanisms. For instance, the muscle relaxation induced by succinylcholine, discussed in more detail in Chapter 7, is due to blockade of neuromuscular transmission. Alternatively, acetylcholine antagonists such as **tubocurarine** may compete for the receptor site at the skeletal muscle end plate, leading to paralysis of the skeletal muscle. **Botulinum toxin** binds to nerve terminals and prevents the release of acetylcholine; the muscle behaves as if denervated and there is paralysis.

Hypo/hyperthermia

Certain foreign compounds can cause changes in body temperature which may become a toxic response if they are extreme. Substances such as **2,4-dinitrophenol** and salicylic acid will raise body temperature as they uncouple mitochondrial oxidative phosphorylation. Thus, the energy normally directed into ATP during oxidative phosphorylation is released as heat.

Substances which cause vasodilation may cause a decrease in body temperature.

Heightened sensitivity

Certain foreign compounds may cause an increase in sensitivity of a particular tissue to endogenous substances such as catecholamines. For instance exposure to some **halogenated hydrocarbons,** such as the fluorinated and chlorinated compounds used in aerosol spray cans and fire extinguishers, may cause heightened sensitivity to catecholamines. The heart tissue is sensitized by exposure to the halogenated solvents, and then if exposure is followed by a fright or other situation – when catecholamines such as adrenaline are released into the body – the heart may fail. A number of cases of sudden death from heart failure in teenagers and young, and otherwise healthy adults, have occurred due to this type of toxicity. This type of toxic effect became known particularly as it occurred during glue sniffing or **solvent abuse.** However, only certain halogenated solvents cause this type of toxicity.

Teratogenesis

Introduction

Teratogenesis involves interference with the normal development of either the embryo or foetus *in utero*, giving rise to abnormalities in the neonate. This interference may take many forms and there is therefore no general mechanism underlying this type of response. Many of the toxic effects described elsewhere in this book may be teratogenic in the appropriate circumstances.

Teratogenic agents may be drugs taken during pregnancy, radiation, both ionizing and non-ionizing, environmental pollutants, chemical hazards in the workplace, dietary deficiencies and natural contaminants.

Although mutations occurring in germ cells may give rise to abnormalities in the neonate, such as Down's syndrome, teratogenicity is normally confined to the effect of foreign agents on somatic cells within either the developing embryo or foetus and the consequent effects on that individual, rather than inherited defects. However, the effects of foreign compounds on germs cells will also be considered within the context of teratogenesis in this section.

Embryogenesis is a very complex process involving cell proliferation, differentiation, migration and organogenesis. This sequence of events (figure 6.12) is controlled by information transcribed and translated from DNA and RNA respectively (figure 6.22) in a time-dependent manner.

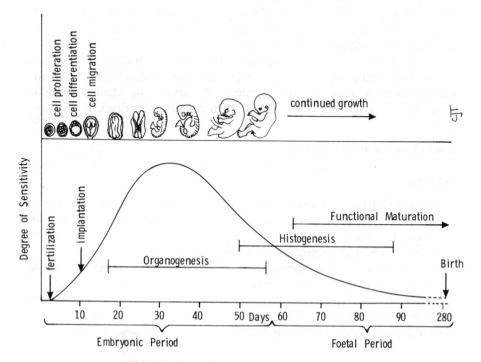

FIGURE 6.12. The stages of embryogenesis.

Although embryogenesis is not fully understood and is a highly complex process, with the knowledge available certain predictions may be made regarding teratogenesis:

(a) The sequence of events in embryogenesis could be easily disturbed, its interrelationships readily disrupted and such interference could be very specific.
(b) The timing of the interference with the process of embryogenesis would be very important to the final expression of the teratogenic effect.

Both of these predictions are found to be correct, and the teratogenic effect may be out of all proportion to the initiating event, if this event occurs at a critical time. This teratogenic effect might simply be a slowing down in the development of one group of cells, which are then out of synchrony with the overall process of embryonic development. For example, a deficiency of folic acid in pregnant rats may cause reno-uteral defects in the neonate because migration of certain primordial cell groups is disturbed by this deficiency. The cell groups are then not able to attain their correct position at a later stage.

A further prediction might be that any effect or foreign compound which interfered with biochemical pathways, particularly transcription and translation, damaged macromolecules, caused a deficiency of essential cofactors, caused a reduction in energy supply or produced direct tissue damage, would probably be teratogenic. Thus ionizing radiation, which causes mutations and other effects, will readily cause birth defects if the pregnant animal is irradiated at the appropriate time. Pre-existing mutations derived from the maternal or paternal germ cells may also result in mutations, of course.

The mechanisms underlying teratogenesis are therefore many and varied, and true understanding will only come with greater understanding of the remarkable but complex process of embryogenesis. However, the characteristics of teratogenesis can be examined and general mechanisms described.

Characteristics of teratogenesis

Selectivity

Teratogens interfere with either embryonic or foetal development, but often do not affect the placenta or maternal organism. They are therefore selectively either embryotoxic or foetotoxic, giving rise to manifestations of such toxicity up to and including death, with subsequent abortion. The most potent teratogens may be regarded as those with no observable toxicity to the maternal system, but causing malformations in the foetus rather than death, such as the now infamous drug thalidomide. Clearly, most teratogens will cause foetal death at high enough doses and probably maternal toxicity also. Certain compounds, such as **colchicine**, which cause foetal death do not cause malformations, however.

Genetic influences

Observations that species and strain differences exist in the susceptibility to certain teratogens suggest that genetic factors may be involved in teratogenesis. Similarly, it seems clear that in some cases at least a teratogen may increase the frequency of a naturally occurring abnormality. These genetic factors may simply be differences in the maternal metabolism or distribution of the compound which lead to variation in the exposure to the ultimate teratogenic agent.

It is also clear from the previous discussion, however, that the instructions for the complex series of events which constitute organogenesis are probably coded in DNA, and consequently mutations or damage to DNA will be expected to cause certain abnormalities if that information is transcribed and translated. Also, genetic susceptibility to mutation or chromosome breakage may be factors. For example, the vitamin antagonist **6-aminonicotinamide** causes cleft palate and chromosomal abnormalities in mice. The chromosomal abnormalities are found in the somatic cells of a number of tissues, leading to faulty mitosis. This is supported by the fact that in some cases only one or two embryos are malformed out of several in multiparous animals, and this indicates that the embryonic genotype may sometimes be an important factor. However, although many mutagens and carcinogens are teratogenic, not all are, suggesting that genetic factors are not always involved in the underlying mechanisms of teratogenesis.

Susceptibility and development stage

The susceptibility of both an embryo and a foetus to a teratogen is variable, depending on the stage of development when exposure occurs. For gross anatomical abnormalities the critical periods of organogenesis are the most susceptible to exposure, whereas other types of abnormality may have other critical periods for exposure.

After fertilization, the cells divide, giving rise to a blastocyst. During this time there is little morphological differentiation of cells, except that some are located on the surfaces and others internally. The development of the blastocyst gives an internal cavity (the blastocoele) and hence further surfaces and positional differences. However, there are few specific teratogenic effects which occur at this time, the major effects being death or overall developmental retardation.

The appearance of the embryonic germ layers, the ectoderm, endoderm and mesoderm is the next stage, with the gross segregation of cells into groups. Damage at this stage may be associated with specific effects.

The chemical determination which precedes structural differentiation is still not understood but it is probable that this period is sensitive to interference.

Organogenesis, which is the segregation of cells, cell groups and tissues into primordia destined to be organs is particularly sensitive to teratogens although not exclusively so. Histological differentiation occurs concurrently with, and continues after, organogenesis, as does acquisition of function. Both of these stages may be susceptible to teratogens, and exposure to them may lead to defects, although generally these defects are not gross structural ones.

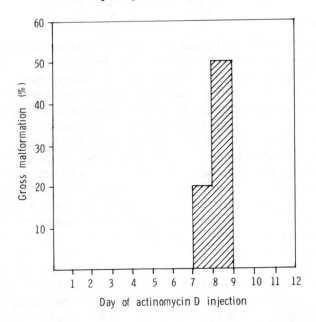

FIGURE 6.13. Critical timing of the teratogenic effects of actinomycin D in the rat. The histogram shows the percentage of gross malformations among surviving foetuses after a dose of 75 μg/kg, i.p.
Adapted from Tuchmann-Duplessis and Mercier-Parot (1960) *Congenital Malformations*, Ciba Foundation Symposium, edited by G. E. W. Wolstenholme and C. M. O'Connor (Boston: Little Brown).

The sensitive period for induction of malformations is the 5–14 day period in the rat and mouse and the 3rd week to the 3rd month in humans. This is illustrated in figure 6.13 for the teratogen **actinomycin D**. The later period of foetal development, like the initial proliferative stage, is less susceptible to specific effects and an 'all or none' type of response is usually seen, such as either death or no gross effect.

Specificity

Different types of teratogens may give similar abnormalities if given during the same critical periods and conversely the *same teratogen* given at *different* times may produce *different effects*. Therefore, the particular abnormality may represent interference with a specific developmental process (figure 6.14). Although there is a wide variety of possible biochemical effects or structural changes at the molecular level, these may be manifested as relatively few types of abnormal embryogenesis and hence similar defects may result from different teratogens. Thus, excessive cell death, incorrect cellular interaction, reduced biosynthesis of RNA, DNA or protein might be the primary manifestations of the teratogenic effect, but all give rise to the impaired migration of cells or cell groups. The eventual effect might conceivably be too few cells or cell products at a particular site for normal morphogenesis or maturation to proceed.

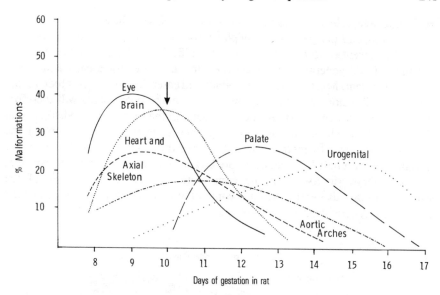

FIGURE 6.14. Periods of peak sensitivity to teratogens in the rat. A teratogen administered at the time shown by the arrow would cause a mixture of malformations. It would particularly damage the eyes and brain but would have little or no effect on the palate.
Adapted from Wilson, J. G. (1973) *Environment and Birth Defects* (New York: Academic Press).

Manifestations of abnormal development

The final consequences of abnormal development are death, malformation, growth retardation and functional disorder. Each of these consequences follows exposure at different stages. Thus foetal death may result from exposure at an early or late stage of development to high concentrations of the toxin, without evidence of malformation. This may be due to substantial damage to the undifferentiated cells in the early stages or interference with some physiological mechanism in the late stage of development.

Malformations tend to occur following exposure during organogenesis, whereas functional disorders would be expected at later stages. The period of foetal growth is susceptible and agents acting at this time may cause growth retardation. Growth retardation may reflect a variety of functional deficiencies and both of these may occur without any sign of malformations.

Access to the embryo and foetus

For foreign compounds, in contrast to ionizing and other radiation or mechanical force, the route of access is via the maternal body. This is either through the fluids surrounding the embryo or via the blood, following the formation of the placenta. The blood from the maternal circulation enters a sinus into which the vessels of the embryo protrude like fingers (figure 6.15). Consequently, there is rapid and efficient exchange between maternal and embryonic blood for most foreign substances, provided they are not bound to large proteins and have a molecular weight less than about 1000. Some proteins do, however, seem to cross the placenta, and phagocytosis may play a role here. Lipophilic compounds will cross the placenta most readily by passive diffusion, whereas those ionized at plasma pH will generally not unless they are substrates for a carrier-mediated transport system. Most foreign compounds will enter the embryonic bloodstream by passive diffusion, and the exposure of the embryo or foetus will therefore depend on the concentration of the compound in the maternal bloodstream and the blood flow which increases as pregnancy progresses. The ability of the maternal organism to metabolize and excrete the compound, reflected in the plasma level and half-life, will therefore be a major determinant of the exposure of the embryo or foetus to the compound. Similarly, metabolites of the compound may also enter the embryonic bloodstream provided they have the necessary physicochemical characteristics.

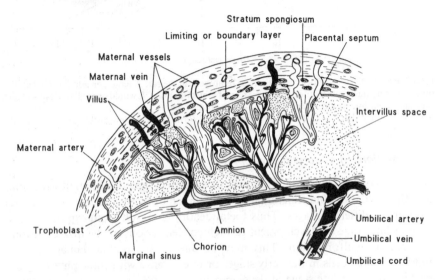

FIGURE 6.15. The structure and blood supply of the mammalian placenta.
Adapted from Gray, H., *Anatomy of the Human Body*, edited by C. D. Clemente (Philadelphia: Lea & Febiger), 30th Edition. 1985.

However, the embryo or foetus may have the ability to metabolize the compound itself or further metabolize metabolites. This is especially the case in humans where the fetal liver at mid-gestation has 20–40% of adult activity for phase 1 reactions. If the metabolites produced by the embryo or foetus are polar or large, they may become trapped inside the embryonic system as they will be unable to diffuse across the placenta into the maternal bloodstream. Similarly, if the compound or metabolites entering the embryo bind to or react with proteins or other macromolecules, this process may effectively entrap the compound inside the embryonic system. Either of these processes may cause a concentration gradient to exist across the placenta and will prolong embryonic or foetal exposure to the compound.

Dose–response

It is generally accepted that there is a true no-effect level for teratogenesis, certainly as far as death and malformations are concerned, because of the presence and influence of the maternal organism. Thus, it is clear that a critical number of cells may have to be damaged before the lethality or malformation becomes apparent. For potent teratogens the dose–response curve will be steep and displaced from the dose–response curve for the maternal organism. Three different types of dose–response relationship can be identified depending on the end-point or manifestation which is measured. Thus, in the first type, malformation, growth retardation and lethality may each show dose–response curves which are displaced (figure 6.16a), each being a separate manifestation, dependent on a different mechanism. For example, thalidomide is only embryolethal at doses several times higher than those required to produce malformations. Alternatively, as in the second type, one or more of these manifestations may show a similar dose–response curve (figure 6.16b), as is the case with **actinomycin D** where lethality and malformations occur at the same doses (figure 7.48). Thus, the responses may be different degrees of manifestation of the initial insult. In the third type, malformations may not occur at all, simply growth retardation and lethality, the latter showing a steeper dose–response curve than the former (figure 6.16c). As with the first type, these two responses are separate manifestations of the insult. Thus, the drugs **chloramphenicol** and **thiamphenicol**, which inhibit mitochondrial function, either cause growth retardation or embryolethality, which shows a very steep dose–response curve. There is no basis for selective effect, and therefore malformations in particular organs and tissues do not occur. However, it is clear that the relationships between lethality, malformation and growth retardation are complex and will vary with the type of teratogen, dose and time of dosing. Most foreign compounds will be embryotoxic if given in sufficient doses at an appropriate time.

Mechanisms of teratogenesis

Because the process of development of the embryo and then the foetus is a complex series of events it is clear that many different mechanisms may cause the disruption that underlies the teratogenic effects of foreign compounds. In many cases the initiating events probably occur at the subcellular or molecular level in the embryo, but may be only detectable as malformations or cell death.

The various known and postulated underlying mechanisms are discussed below.

Cytotoxicity

Direct damage to cells leading to cell death and necrosis is one suggested cause of teratogenesis. Thus, cytotoxic compounds such as the alkylating agent N-methyl-N'-nitro-N-nitroso-guanidine (MNNG) are teratogenic. This compound causes a spectrum of malformations, embryolethality and growth retardation

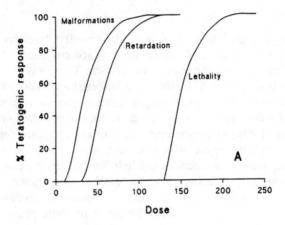

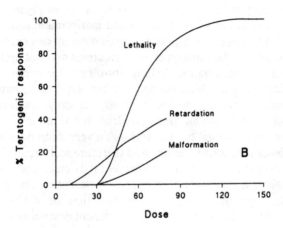

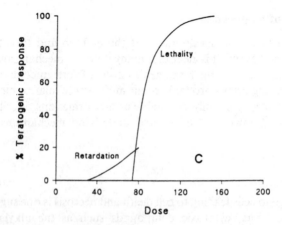

FIGURE 6.16. Dose-response relationship patterns for different types of teratogen.
Adapted from Manson, J.M., Teratogens, in *Casarett and Doull's Toxicology*, edited by C.D.
Klaassen, M.O. Amdur and J. Doull (New York: Macmillan), 3rd Edition, 1986.

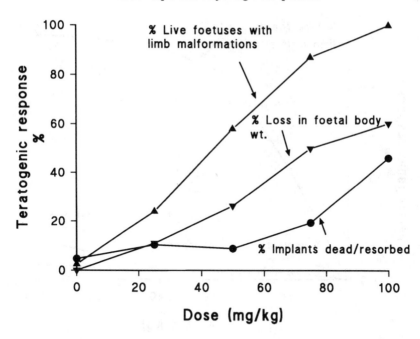

FIGURE 6.17. The dose-response patterns for teratogenicity caused by the cytotoxic agent *N*-methyl-*N'*-nitro-*N*-nitroso-guanidine (MNNG). Pregnant mice were treated on day 11 with MNNG.
Data from Manson, J.M., Teratogens, in *Casarett and Doull's Toxicology*, edited by C.D. Klaassen, M.O. Amdur and J. Doull (New York: Macmillan), 3rd Edition, 1986.

(figure 6.17). Thus, cell death and reduced proliferation specifically in the target tissue, which may be a limb bud or tissue destined to be an organ, will give rise to a malformation in that tissue if there is insufficient time for replacement. There are many underlying causes of cytotoxicity as already discussed. Generalized cell death is more likely to lead to embryolethality if extensive, or growth retardation if the embryo is at an early stage where cells can be replaced by compensatory hyperplasia. After a single exposure to MNNG, foetuses within a single litter may show all three outcomes, and this therefore is an example of the second type of dose–response relationship (see above). In the case of **MNNG** the necrosis of limb bud cells, or interference with or damage to DNA may be involved in the teratogenicity.

Receptor-mediated teratogenicity

These are highly specific mechanisms which will not involve extensive embryolethality or growth retardation, but a well-defined structural malformation. Thalidomide might be an example of such a teratogen, but the mechanism is unknown. An example where the receptor and mechanism are partly understood are the **glucocorticoids**, which are involved in normal growth and differentiation of embryonic tissue. After doses of glucocorticoids leading to

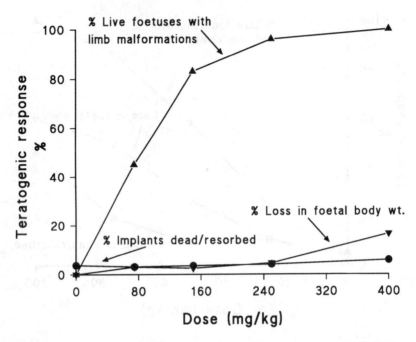

FIGURE 6.18. The dose-response patterns for teratogenicity caused by the herbicide nitrofen (2,4-dichloro-4'-nitro diphenyl ether). Pregnant rats were treated on day 11 with nitrofen.
Data from Manson, J.M., Teratogens, in *Casarett and Doull's Toxicology*, edited by C.D. Klaassen, M.O. Amdur and J. Doull (New York: Macmillan), 3rd Edition, 1986.

exposure of the embryo to high levels relative to the normal physiological concentration, malformations such as cleft palate occur. There is a cytosolic receptor for glucocorticoids, and the level of these receptors correlates with the susceptibility of the animal to teratogenicity.

A foreign compound which may act in this way is the herbicide **nitrofen** (2,4-dichloro-4'-nitro diphenyl ether) which causes a variety of malformations lethal to the neonate. There is no growth retardation or embryolethality at doses which are teratogenic, however (figure 6.18), and therefore this exhibits the first type of dose–response relationship (see above). The mechanism may involve the production of a metabolite which interferes with the thyroid hormone status. The metabolite, which results from reduction and hydroxylation of the parent compound, was detected in embryonic tissue and cross-reacted with antibodies to tri-iodothyronine (T_3). Thus the metabolite may have T_3 activity and exert this in the embryo.

Mutation

It is clear that mutations are the basis of many errors of development; maybe 20–30% of those occurring in humans are due to heritable mutations in the germ line. Mutations may occur in the genetic material of either the germ cells or the somatic cells. Germ cell mutations are heritable, whereas effects in somatic cells are only apparent in the foetus exposed. Somatic cell mutations will probably not lead to gross manifestations of damage such as malformations, unless a sufficiently large number of progency cells are affected. Consequently, somatic cell mutations are probably an infrequent cause of foetal abnormality. Some somatic mutations, however, may be incompatible with the viability of the affected cell or group of cells. This might be less damaging than a viable mutant cell which was the ancestor of an important organ or structure, however.

Not all mutagens are teratogens, but the cytotoxic alkylating agent MNNG mentioned above is a potent teratogen and causes replication-dependent mutations. As cell proliferation is important in embryogenesis, a mutation which is induced in a sufficient number of critical cells and results in cell death or altered growth characteristics for example, is potentially able to cause a malformation. Clearly the nature of the mutation and the number and position of the cells affected will all be important factors.

Chromosomal aberrations

Aberrations in chromosomes or chromatids which are sometimes microscopically visible may arise during mitotic division when newly divided chromosomes fail to separate or do so incorrectly. The absence of a chromosome is usually lethal, and an excess is often poorly tolerated, giving rise to serious defects. Aberrations of the sex chromosomes are more readily tolerated, however. Chromosome aberrations may be caused by foreign compounds as indicated in the section on mutagenesis (see page 259). However, those cells with aberrations seem to be rapidly eliminated and so may contribute to cell death rather than a heritable mutation.

Mitotic interference

As embryogenesis involves extensive cellular proliferation, any interference with this process is potentially teratogenic. Interference with spindle formation, inhibition or arrest of DNA synthesis or the incorrect separation of chromatids all fall into this category. Inhibition of DNA synthesis, such as that caused by cytosine arabinoside, slows or arrests mitosis, which cannot progress beyond the S-phase. Thus those areas of the embryo which show extensive cellular proliferation are the most susceptible to both necrosis and subsequent malformation from cytotoxic compounds such as **cytosine arabinoside**.

Inhibition of spindle formation such as that caused by **vincristine or colchicine** stops separation of chromosomes at anaphase (see page 268). Proper separation of chromatids may not occur because of 'stickiness' or bridging between the chromatids. Clearly interference with mitosis, and hence cell proliferation, is an important cause of teratogenic effects.

Interference with nucleic acids

A number of antineoplastic and antibiotic drugs which interfere with nucleic acid function are teratogenic. Such effects as changes in replication, transcription, the incorporation of bases and translation occurring in somatic cells are non-heritable and may be embryotoxic. However inhibition of DNA synthesis or DNA damage does not seem to be as important as cell death and significant inhibition of DNA synthesis alone usually seems to be insufficient to cause malformations. Interference with protein synthesis is generally lethal to the embryo rather than malformation-causing.

Examples of teratogenic compounds which interfere with nucleic acid metabolism are **cytosine arabinoside**, which inhibits DNA polymerase, **mitomycin C**, which causes cross-linking, and **6-mercaptopurine**, which blocks the incorporation of the precursors, adenylate and guanylate. **Actinomycin D**, a well known teratogen, intercalates with DNA and binds deoxyguanosine, interfering with RNA transcription and causing erroneous base incorporation. **8-Azaguanine** is a teratogenic analogue of guanine.

Compounds such as **puromycin** and **cycloheximide**, which block rRNA transfer, and **streptomycin** and **lincomycin**, which cause misreading of mRNA, block protein synthesis and are therefore often embryolethal.

Substrate deficiency

If the requirements for the growth and development of the foetus are withheld, a disruption to these processes may occur and damage may ensue. Deficiencies in essential substrates, such as **folic acid**, may be caused by dietary lack or by substrate analogues.

Failure in the placental transport of essential substrates may be teratogenic and can be caused by certain compounds such as azo dyes. This has, however, only been demonstrated in rodents because of the inverted yolk sac type of placenta such animals have.

Deficiency of vitamins, such as folic acid, is highly teratogenic, as essential synthetic metabolic pathways are blocked or reduced. This may be caused by the administration of specific vitamin analogues or antagonists as well as by a failure in supply.

Deficiency of energy supply

The rapidly proliferating and differentiating tissue of the embryo would be expected to require high levels of energy, and therefore interference with its supply, not surprisingly, may be a teratogenic action. A deficiency of glucose due to dietary factors or due to hypoglycaemia, which may be induced by foreign compounds, is teratogenic. Interference in glycolysis, such as that caused by **6-aminonicotinamide**, inhibition of the tricarboxylic acid cycle as caused by fluoroacetate (see page 338), and impairment or blockade of the terminal electron transport system as caused by hypoxia or cyanide, all cause abnormal foetal development. For example, **chloramphenicol** and **thiamphenicol** both interfere with the mitochondria and cause ATP depletion, inhibition of mitochondrial respiration and cytochrome oxidase activity. They cause embryolethality and growth retardation rather than malformations. This may be because they affect all cells non-specifically and hence either slow growth overall, or at higher doses are lethal to a sufficient number of cells to lead to death of the embryo. The dose–response curve for embryolethality is very steep (0–100% mortality between 100 and 125 mg/kg for thiamphenicol) depending on a threshold of critical energy depletion. This is the third type of dose-response relationship, as discussed where malformations do not normally occur.

Inhibition of enzymes

Teratogens in this category are also included in some of the previous categories. Enzymes of central importance in intermediary metabolism are particularly vulnerable, such as dihydrofolate reductase, which may be inhibited by folate antagonists giving rise to a deficiency in folic acid. **6-Aminonicotinamide**, which inhibits glucose-6-phosphate dehydrogenase, an important enzyme in the pentose-phosphate pathway, is a potent teratogen.

5-Fluorouracil (figure 3.6) an inhibitor of thymidylate synthetase, and **cytosine arabinoside**, which inhibits DNA polymerase, are both teratogens which have already been referred to.

Changes in osmolarity

Various conditions and agents may change the osmolarity within the developing embryo and thereby disrupt embryogenesis. Thus, induction of hypoxia may cause hypo-osmolarity, which leads to changes in fluid concentrations. This causes changes in pressure and the consequent disruption of tissues may result in abnormal embryogenesis. Other agents causing osmolar changes are hypertonic solutions, certain hormones and compounds such as the azo dye **Trypan blue**, and **benzhydryl piperazine**. Trypan blue has been widely studied and causes fluid changes in rodent embryos leading to malformations of the brain, eyes, vertebral column and cardiovascular system. It may also have other effects, however, such as inhibition of lysosomal enzymes, which may interfere with release of nutrients from the yolk sac, impairing foetal nutrition.

Benzhydryl piperazine causes orofacial malformations in rat embryos, possibly the result of the oedema induced in the embryo. It has also been suggested that changes in *embryonic pH* may underlie the teratogenicity of certain compounds.

Membrane permeability changes

Changes in membrane permeability might be expected to lead to osmolar imbalance and foetal abnormality. This is hypothetical, as there are no real examples of such a mechanism, although high doses of **vitamin A** are teratogenic and may cause ultrastructural membrane damage.

Changes in maternal and placental homeostasis

As well as direct effects on the embryo or foetus, foreign compounds can also be teratogenic by influencing the maternal organism or the placenta. Thus, maternal malnutrition or protein deficiency, or a deficiency in one or more vitamins or minerals may lead to effects on the embryo and foetus. As might be expected maternal malnutrition will tend to cause growth retardation. However, **vitamin A** and folic acid deficiencies cause malformations and embryolethality as well as growth retardation. **Trypan Blue** is believed to be teratogenic to animals with a yolk sac placenta by interfering with the nutrition of the embryo (see above).

A permanent, but partial, decrease in placental blood flow will lead to growth retardation in the embryo. A total cessation of blood flow for a short time (3 h) during organogenesis will cause malformations. Thus, foreign substances which cause vasoconstriction in the placenta may give rise to effects on the embryo indirectly. Exposure of pregnant rabbits to **hydroxyurea** rapidly results in a large reduction in placental blood flow (77%) which may be the cause of craniofacial and cardiac haemorrhages.

Role of metabolic activation

There is clear evidence from many different sources that the metabolism of compounds may be involved in their teratogenic effects, as will be seen in the final chapter in the discussion of thalidomide and diphenylhydantoin teratogenicity. The embryo and foetus of some species clearly have metabolic activity towards foreign compounds which may be inducible by other foreign compounds. Thus, foetal liver from primates has a more well-developed metabolic system for xenobiotics than does that from rodents and rabbits for example. This may be due to the late development of the smooth endoplasmic reticulum and therefore of cytochrome(s) P-450 in the latter species. The use of metabolic inducers and inhibitors *in vivo* and the use of metabolizing systems with embryo or limb bud culture *in vitro*, have all indicated that for some teratogens, metabolism is involved. Metabolic activation in the maternal organism or in the embryo itself may both occur and have different roles in the eventual toxicity.

A good example of a compound which is a teratogen and which requires metabolic activation is the anti-cancer drug **cyclophosphamide**, which has been studied extensively both *in vivo* and *in vitro*.

However, just as with other toxic effects, either the parent drug or a metabolite may be responsible for embryotoxicity, but it is often difficult to predict which, without substantial metabolic and biochemical data being available.

Transplacental carcinogenesis

Brief mention should be made of the phenomenon of transplacental carcinogenesis. This is the induction of cancer in the offspring by exposure of the pregnant female. It may not occur in the foetus or even in the neonate, but may be evident only in adulthood. The best known example of this is the appearance of

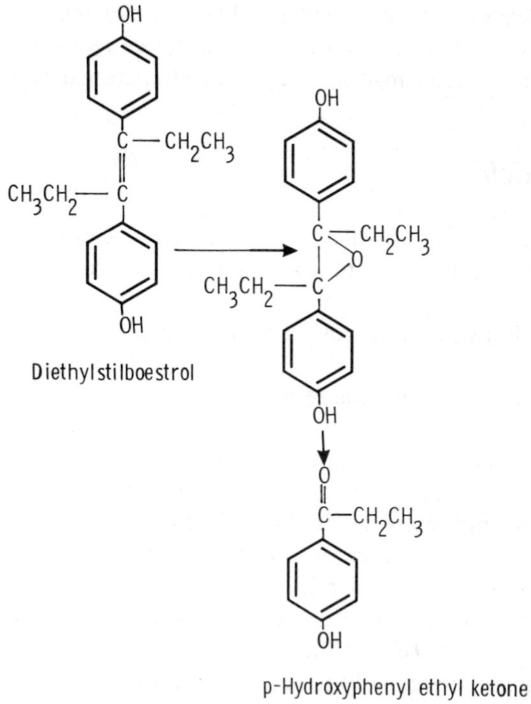

FIGURE 6.19. Metabolism of diethylstilboestrol via an epoxide intermediate. This potentially reactive intermediate may show an affinity for the oestradiol receptor and thereby accumulate in oestrogen target organs. This may facilitate reaction with DNA in these organs.
From Metzler and McLachlan (1979) *Archs. Toxicol*, Suppl. 2, 275.

vaginal cancer in human females born of mothers given the drug **diethylstilboestrol** (figure 6.19) during pregnancy. The vaginal cancer did not appear in the female offspring until puberty.

Male-mediated teratogenesis

This is a controversial area with regard to humans where there is currently little hard data. Theoretically it is possible for a foreign compound to cause mutations in male germ cells which result in malformations or the development of abnormal offspring. This is similar to the situation in which inherited mutations or chromosomal aberrations lead to the birth of abnormal offspring, such as occur in Down's syndrome, for example, where an extra chromosome occurs (Trisomy 21).

Although this has been shown to occur in experimental animals after exposure of males to foreign compounds such as **cyclophosphamide**, there is only inconclusive evidence that this occurs in man. Thus, studies of exposure of human males to **vinyl chloride, dibromochloropropane** and anaesthetic gases, for example, have revealed only equivocal evidence of developmental toxicity in the offspring.

There now seems to be some evidence that the leukaemia occurring in children, which appears to be clustered around nuclear fuel reprocessing plants such as Sellafield in the United Kingdom, may be due to paternal exposure to radiation.

Immunotoxicity

There are several ways in which toxic compounds can react with the immune system, but the most important for the purposes of this chapter are:

(a) interaction with a component of the system to inhibit or depress immune function

(b) elicitation of an immune response.

It is beyond the scope of this book to consider the subject of immunology in any detail, but a brief introduction will be useful. Further details can be found in the books, chapters and references in the Bibliography.

The immune system functions to protect the organism against infection, foreign proteins and neoplastic cells. It is organized into primary lymphoid tissue such as bone marrow and thymus, and secondary lymphoid tissues such as spleen and lymph nodes. The cells of the immune system are leucocytes and a specific type produces immunoglobulins. Leucocytes are derived from stem cells in the bone marrow. These cells become functionally mature cells: granulocytes, lymphocytes and macrophages. Lymphocytes divided into T-cells (thymus derived) and B-cells (bursa equivalent) depending on the site of maturation. The T-cells may be subdivided into T-helper and T-cytotoxic cells. The B-lymphocytes give rise to plasma cells which are responsible for the production of antibodies (immunoglobulins). The immune system utilizes two mechanisms, a non-specific resistance and specific, acquired immunity. The non-specific mechanism does not require prior contact and involves macrophages and granulocytes which are phagocytic cells. The specific mechanism involves the lymphocytes and also macrophages and requires prior contact with the foreign substance. Following this initial contact and the establishment of memory of the foreign substance or organism, the second contact evokes an immune response. However, there are links between the specific and non-specific mechanisms such as the presentation of antigens by lymphocytes to macrophages and the activation of macrophages by lymphokines produced by sensitized T-lymphocytes.

There are two main types of immune response: cell-mediated and humoral. Cell-mediated immunity involves specifically sensitized thymus-dependent lymphocytes. Humoral immunity involves the production of antibodies (immunoglobulins) from bursa equivalent lymphocytes or plasma cells. The mechanisms will be discussed in more detail below.

Immunosuppression

Drugs and other foreign compounds can inhibit the functioning of the immune system, an activity which has been used in clinical medicine for the suppression of graft rejection. However, it has become clear that this activity may be considered as a toxic response to certain chemicals as it causes secondary effects such as increased susceptibility to infection. This toxic effect has been observed in both experimental animals and humans after exposure to environmental pollutants and industrial chemicals for example. There may be many mechanisms underlying immunosuppression, but damage to primary and secondary lymphoid organs is certainly involved.

It is well known that immunosuppression is caused by a variety of hydrocarbons such as **benzene, polycyclic aromatic hydrocarbons** and halogenated aromatic hydrocarbons. Thus, human exposure to **polychlorinated biphenyls** (PCBs) in Japan (Yusho accident) and China has been associated with increased respiratory infections and decreased levels of immunoglobulins in serum. In animals exposed to these compounds there is atrophy of both primary and secondary lymphoid organs, lower circulating immunoglobulins and decreased antibody responses after exposure to antigens. Similarly, the exposure of both humans and farm animals to **polybrominated biphenyls**, which occurred in Michigan in 1973, resulted in depressed immune responses.

With **benzene**, the target organ seems to be the bone marrow and aplastic anaemia results from benzene exposure. This has several effects, one of which is a reduction in the lymphocyte population, an increased susceptibility to infection and inhibited immunoglobulin production. These effects have been detected in both experimental animals and humans occupationally exposed to benzene. Benzene seems to be metabolized in the bone marrow, but the toxic metabolite is not yet unequivocally established, although a reactive quinone is one candidate.

Dibenzodioxins, including **TCDD** (dioxin), have been demonstrated to cause severe thymic atrophy with resulting effects on the immune system. The target seems to be the epithelial cells in the thymus, and it is interesting that thymus cells contain the TCDD receptor. Thymic atrophy correlates with the presence of this receptor, and other compounds which compete with TCDD for this receptor also cause thymic atrophy. The basis of the toxicity of TCDD to the thymus may be an effect on T-cell maturation and differentiation. The result of exposure of animals to TCDD is depressed antibody responses, increased susceptibility to infectious agents, and depressed T-cell function. In humans exposed to TCDD occupationally, decreased serum levels of some immunoglobulins and depressed lymphocyte responses to mitogens were reported.

There are many other compounds which have been shown to cause immunosuppression such as **organophosphorus compounds, ozone**, metals, **organotin** compounds and various drugs such as alkylating agents, for example **cyclophosphamide**, used in anti-cancer chemotherapy. In the case of such alkylating agents the target organs are those with rapidly proliferating cells such as the bone marrow and lymphoid tissue. **Dialkyltin** compounds seem selectively toxic to the thymus and hence affect T-cell function without affecting B-cell function or causing myelotoxicity or toxicity to non-lymphoid tissue.

Stimulation of immune responses

This type of response to a foreign compound is often termed an allergic or hypersensitivity reaction. There are some such responses which seem to be non-immunologically mediated, but these are poorly understood. The response can be manifested in a variety of ways and various organ systems may be involved. The response might be a minor one such as the development of a temporary skin rash, or it may be fatal as in the case of anaphylactic shock. However, the underlying mechanisms and characteristics of immune-mediated responses do have common features which can be considered together.

Most immunotoxic responses can be considered as a response of the organism to a foreign substance. The mechanisms may be complex and give rise to a variety of symptoms, but the basis is a specific altered reactivity of the organism involving either humoral or cellular systems. The basis therefore is the recognition of a substance as an antigen. Antigens are:

(a) normally large molecules with molecular weights greater than 3000–5000 Da

(b) recognized as foreign or 'not-self' by immunologically competent lymphocytes.

Thus, foreign proteins such as those associated with bee stings, pollen grains or present in rat urine are potential antigens and are examples of substances known to cause allergies by immune mechanisms.

Clearly, however, many foreign compounds which are known to induce immune responses, are too small to be antigens. Such compounds, known as haptens, must be covalently linked to a larger carrier molecule in order to stimulate an immune response. Reversible binding to proteins such as plasma proteins is not sufficient to stimulate an immune response. Therefore small, foreign molecules which act as haptens are often chemically reactive and so are able to react with proteins or other macromolecules. The hapten may be the foreign compound to which the animal is exposed, or it may be a reactive metabolite. Although several different antigens might be produced, possibly only one with a particular tertiary structure or number of haptens (epitopes) attached to the carrier molecule, will stimulate an immune response. The epitope density is believed to be an important factor in immunogenicity. Alternatively, a foreign compound may cause the destruction or alteration of an endogenous macromolecule such as to make the fragments antigenic. This may precipitate an autoimmune response where normal body constituents are altered, and then become targets for attack by the immune system.

Once an antigen has been produced it can then interact with the immune system. This occurs in two stages:

(a) sensitization stage

(b) elicitation stage.

The structural requirements for antigens in the two stages may be different. For example, after the immune system of an animal has been sensitized, other antigens may be sufficiently similar to elicit a response. A divalent antigen is normally required and for some types of reaction a trivalent antigen is required.

The sensitization stage involves exposure to a foreign compound, which if not large enough to be recognized as foreign is either metabolized to a reactive metabolite or reacts directly with a protein or other macromolecule. The resulting antigen is recognized as foreign by the immune system (figure 6.20). The immune system responds by producing antibodies (immunoglobulins; humoral system) or activating T-cells (cell-mediated system). As mentioned above there may be collaboration between the specific and non-specific parts of the system. Thus macrophages and T- and B-cells may collaborate to give rise to specific memory cells which synthesize antibodies or to sensitize lymphocytes.

Antibodies are immunoglobulins, proteins with a molecular weight of 150 000 (IgD, IgG) to 890 000 (IgM) which are synthesized in response to an antigen. There are five classes of immunoglobulins some of which are detectable in plasma: IgA, IgD, IgE, IgG, IgM. All of the immunoglobulins have certain common structural features, the light and heavy polypeptide chains which have regions of variable and constant structure. The variable region retains the same number of amino acids, but the particular amino acids vary. The antibody

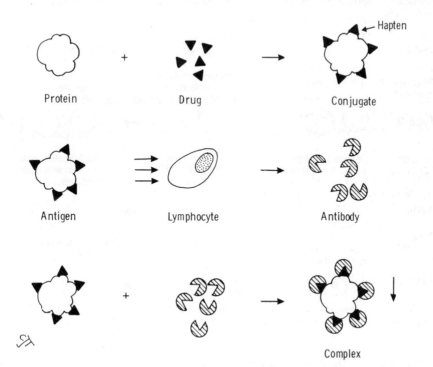

FIGURE 6.20. Formation of an antigenic drug-protein conjugate and subsequent stimulation of antibody production. The antibody so produced combines with the conjugate and inactivates it, in some cases by causing precipitation of the complex.

molecule is structurally symmetrical and has binding sites for antigens on the variable sections of both light and heavy chains. More than one antibody may be produced in response to a single antigen. The plasma immunoglobulin fraction is often raised after exposure to an antigen, and is one indicator that an immune response has occurred.

The elicitation stage follows a further exposure when the presence of the antigen starts the immune response which may be cell-mediated or humoral or both. The response of the animal which occurs in this stage varies, but there are four basic types shown in figure 6.21. Types I–III are humoral responses and type IV is a cell-mediated response. The antibodies involved in the reactions may be different, and more than one type of reaction may occur in response to a particular antigen. One type of reaction may prevent another by removing the antigen. For example, with some immune responses, IgG and IgM antibodies may remove the antigen before eliciting a response-mediated by IgE which would be more serious. The IgG and IgM antibodies in this case are known as blocking antibodies. Following the interaction of the antigen with antibody or T-cell, a series of events takes place which depends on the type of reaction (figure 6.21).

Thus, with **type I reactions** (anaphylaxis, immediate hypersensitivity), where IgE is often involved, binding of the antigen to the antibody on the surface of the

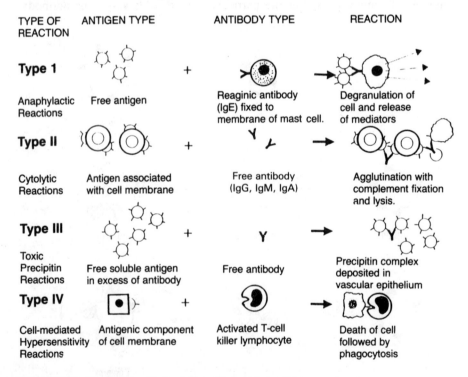

FIGURE 6.21. Mechanisms of stimulation of immune responses.
From Timbrell, J. A., *Introduction to Toxicology*. Taylor and Francis, London, 1989. (Adapted from Bowman, W. C. and Rand, M. J. *Textbook of Pharmacology*, 2nd Edition, Blackwell Scientific Publishers, Oxford).

mast cell releases mediators (figure 6.21). The mediators released such as histamine, slow reacting substance of anaphylaxis (SRSA), serotonin and heparin cause a variety of physiological effects such as inflammation, oedema and vasodilation. The antibody, a homocytotropic antibody, circulates in the bloodstream, but has a high affinity for the surface of mast cells and binds to receptors on the surface. This type of reaction which occurs quickly after re-exposure, underlies reactions in the respiratory system (asthma; rhinitis), skin (urticaria), gastrointestinal tract (food allergies) and vascular system (anaphylactic shock). Type I reactions can be severe, causing difficulty in breathing, loss of blood pressure, anoxia, oedema in the respiratory tract and bronchospasm which may prove fatal.

Type II reactions (cytolytic) involve free antibodies (IgG, IgA, IgM) and antigens bound to cells such as leucocytes, erythrocytes and platelets which are agglutinated by the antibodies (figure 6.21). The target is often the cells in the blood and the result can be haemolytic anaemia, leucocytopoenia and thrombocytopoenia. The agglutination may also stimulate the complement cascade. This is a cascade of physiologically active proteins which results in the release of kinins, histamine and lysosomal enzymes which may cause lysis of cells. Consequently, the result can be severe if significant loss of blood cells occurs either through lysis or removal of agglutinated cells by the spleen. For example, administration of the drug **aminopyrine** can cause severe granulocytopoenia as a result of agglutination and removal of leucocytes.

Type III (Arthus) reactions are those involving both free antigen and free antibody (mainly IgG) which interact to form an immune complex which can precipitate in blood vessels (figure 6.21). Again the complement system may be activated. The deposition of complexes and complement activation can give rise to thrombosis and inflammation and damage to the vasculature. The main targets are skin, joints (rheumatoid arthritis), kidneys (glomerulonephritis), lungs and circulatory system. The antigens involved may be 'self' as opposed to foreign protein or structures (see hydralazine, Chapter 7).

Type IV reactions (delayed hypersensitivity) are cell-mediated (figure 6.21) in which an altered cell membrane is the target for an activated T-cell or killer lymphocyte which attacks and destroys the antigenic cell. Again, various mediators are released such as lymphokines, which may cause the aggregation of lymphocytes, macrophages and basophils. Any organ or tissue may be the target including the liver as occurs in halothane hepatitis (see Chapter 7), and if the target organ is a major one, the consequences can be severe.

Characteristics of immunological reactions

Dose–response relationship

There is usually *no* dose–response relationship for immune responses, as the magnitude of the response is dependent on the type of reaction of the endogenous immune system, not on the concentration of the foreign compound. However, there may be a relationship between the frequency of occurrence of an immunotoxic response and the exposure as occurs with hydralazine (see Chapter 7).

Exposure

Repeated or chronic exposure is required for immunotoxic reactions as the immune system must first be sensitized as described above. The exposure to the compound need not be continuous and in fact repeated, discontinuous exposure is often more potent. There is no generalized, characteristic exposure time or sequence for immunotoxic reactions. In **halothane** hepatitis (see Chapter 7) the number of exposures seems to be important, with about four being optimal. With the **hydralazine**-induced Lupus syndrome exposure for a considerable length of time (average of 18 months), during which there is repeated exposure, is generally required for the development of the immune response (see Chapter 7). With some types of immune response exposure to minute amounts of the antigen, such as might occur with an environmental pollutant, may be sufficient to elicit a severe response in a sensitized individual.

Specificity

The nature and extent of the immunological reaction does not generally relate directly to the chemical structure of the foreign compound which is acting as a hapten. Thus, many different compounds may elicit the same type of reaction. For example, **nickel, *p*-phenylenediamine, *p*-aminobenzoic acid** and **neomycin** all cause dermatitis. The antigen produced may have some bearing on the type of response, but there are many other factors. Thus, it can be stated that chemically similar foreign compounds may produce very different immunotoxic reactions and conversely different types of substance may elicit the same type of reaction. However, once sensitized to a particular foreign substance, other chemically similar substances may show cross-reactivity and precipitate the same immune response.

    However, the exact chemical structure is often a crucial factor in the antigenicity, and small changes in structure may therefore remove antigenicity.

Site of action

Various target organs may be involved in the immune response, but this usually depends on the type of reaction rather than on the distribution of the foreign substance. The many substances which cause immune responses may cause anaphylactic reactions giving rise to asthma and various other symptoms as described above. The site of exposure to the foreign compound may not necessarily be the lungs, however. Similarly, a common immune response is urticaria, or the formation of wheals on the skin which can occur when exposure has been via the oral route. Thus, the target organ is generally due to the particular response rather than the circumstances of exposure or distribution of the compound. However, there are exceptions to this such as the type IV cell-mediated immune reactions where the cell is altered by the foreign compound and is then a target. In the case of halothane hepatitis the liver is the target for metabolic reasons (see Chapter 7 for more detail). Another exception is the situation where the foreign compound is chemically reactive and reacts with proteins in the skin or respiratory system. These altered proteins may then become targets for an immune response at the site of exposure, although there may well be other sites involved as well. Thus the final response is determined by the type of immune mechanism involved.

Thus, immune-mediated responses can be immediate or delayed, localized or widespread. The response can be restricted to the area of exposure or can be systemic. Similar compounds may cross-react or produce very different responses. Many different foreign compounds can cause an immunotoxic response: drugs such as penicillin, halothane and hydralazine, industrial chemicals such as **trimellitic anhydride** and **toluene di-isocyanate**, natural chemicals such as **pentadecylcatechol** found in **poison ivy**, food additives such as **tartazine** and food constituents such as egg white (albumen). However, these may not be immunotoxic in all exposed individuals, and sometimes chemically similar compounds are not immunotoxic. Also, some highly reactive compounds and reactive metabolites of compounds such as paracetamol (see Chapter 7) do not seem to be immunotoxic despite reacting with protein. These are currently areas of obscurity in immunotoxicology, but the importance of this aspect of toxicology is increasing in view of the growing realization that possibly many adverse effects in humans are mediated by the immune system.

Genetic toxicity

Introduction

A mutation is a heritable change produced in the cell genotype. Such a change may be induced by a variety of agents, including foreign compounds, and its occurrence implies that the DNA double helical molecule, the source of genetic information, has been changed. The interaction of toxic chemicals with the genetic material may be divided into three types of effect: **aneuploidization**, **clastogenesis** and **mutagenesis**. Aneuploidy is the acquisition or loss of a complete chromosome. Clastogenesis is the loss, addition or rearrangement of parts of chromosomes. Mutagenesis is the loss, addition or alteration of a small number of base pairs.

The genetic code incorporated in the DNA molecule consists of pairs of bases arranged in triplets. The bases utilized are the purine bases adenine (A) and guanine (G), and the pyrimidine bases, thymine (T) and cytosine (C). The basic mechanism of the transfer of the information encoded in DNA is shown in figure 6.22. The information in the DNA molecule may be *replicated* in another identical molecule, *transcribed* into RNA and *translated* into protein.

The three-base code may specify an amino acid in the protein for which the particular gene is responsible or it may specify an instruction such as start or stop

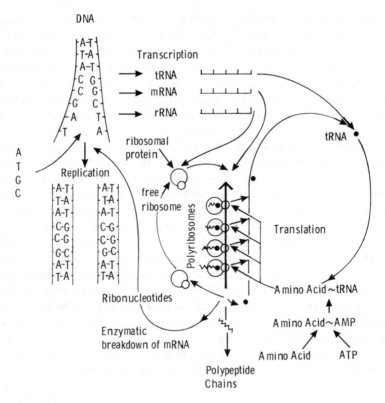

FIGURE 6.22. Replication of DNA and the transcription and translation of its encoded information into RNA and thence protein. Modified from Watson, J. D. (1965) *The Molecular Biology of Gene* (New York: W. A. Benjamin).

translating the code. The DNA molecule, the double helix, is replicated at cell division, during mitosis, for passage to the daughter cells and during meiosis in germ cells for passage to the next generation. This replication is made possible and takes place readily as a result of the base-pairing arrangement within the DNA molecule. The consequence of this specific base pairing is a complementary code on each strand of the molecule, with a purine always paired with a pyrimidine.

Unwinding of the double helical DNA molecule allows each strand to act as a template for new synthesis and guarantees the accurate transmission of information. Alteration of the DNA molecule, in particular a change in the base code, may therefore result in a mutation, although this is not always the result. Such alterations in the base code are of various kinds, each having particular consequences, as described below.

BASE-PAIR TRANSFORMATIONS
In this type of change, a base, either a purine or a pyrimidine, may be replaced by another. If the replacement base is of the same type, i.e. if a purine is replaced by

another purine, this is known as a base-pair transition. If a purine replaces a pyrimidine this is known as a base-pair transversion.

BASE-PAIR ADDITION OR DELETION

The complete removal or incorporation of a base-pair is more serious than the foregoing and may result in a frameshift mutation. Here the order of the genetic code is altered and hence the reading from that base code is shifted.

LARGE DELETIONS AND REARRANGEMENTS

Breakage of the complete DNA molecule may occur at one or more sites. This may then be followed by reconstitution of the molecule. One result of this may be erroneous re-constitution, leading to a large change in the genetic code. This type of mutation may also occur at the level of the chromosome when segments may become inverted (clastogenesis).

UNEQUAL PARTITION OR NON-DISJUNCTION

The unequal division of chromosomes during mitosis may sometimes occur (aneuploidization). This may be the result of exposure of the cell to agents which damage or disturb the spindle fibres or interfere in some way with the process of cell division. The result is a cell with more or less chromosomes than normal, which may or may not be viable.

Thus there are two types of mutagen: those acting directly on DNA and those acting on the replication or the partition of chromosomes. Mutagens of the latter type may therefore only be effective at certain times in the cell cycle. Also, some mutagens may not be able to cross the nuclear membrane and are therefore only active at mitosis, when the nuclear material is in the cytoplasm.

Replication of DNA shows great fidelity, due to the repair of errors which occur during the formation of complementary DNA strands and new DNA molecules. Therefore, many mutations which have occurred in the original molecule and those that occur during the replication process are recognized and corrected. Consequently, fewer mutations are actually present in the replicated DNA molecules than originally occurred. The erroneous bases which are the result of transformations or chemical modification are excised by repair enzymes before the mistake is encoded in the daughter molecule. If the mutation is not recognized and repaired, however, the false information is transcribed into RNA, and hence the error or mutation is expressed as an altered protein resulting from perhaps a single amino acid change. Such a slight change may drastically alter a protein if it occurs at a crucial site in the molecule. For example, in some of the inherited metabolic anaemias this is the case, with a single amino acid change in the haemoglobin molecule being sufficient to impair its function drastically. However, it is clear that only mutations recognized as errors will be repaired and therefore not transcribed. A mutation in the DNA molecule, if not recognized, may therefore be transcribed into mRNA, and the wrong amino acid will be specified and hence translated into a mutant protein. This will occur provided the original mutation is compatible with the transfer RNA codes. Alternatively, the protein may be terminated prematurely and be too short and therefore unable to function. The change may be of no consequence or may be crucial, depending on the position of the amino acid(s) in the protein specified by the gene in question. Mutations may of course lead to an improvement in the gene product.

If a mutation or major chromosomal change occurs and is not repaired it may not be compatible with the continued existence of the cell and as already discussed earlier in this chapter, cell death may occur, possibly as part of a programmed process to remove such damaged genetic material. However if the mutation is not lethal and is not repaired, when the cell divides and the DNA replicates, the mutation or genetic error will be 'fixed' in the genome of the daughter cell. Thus the mutation is passed on to the daughter cells if not recognized at replication. However, if the mutation is simply an altered base at replication, it will be corrected in one half of the molecule but compounded in the other. For a simple mutation, therefore, involving small changes in the code, the consequence will be of variable severity because:

(a) The code for amino acids is degenerate. This means that there may be several codes for any one amino acid and consequently a mutation may be hidden and no change in the amino acid will be seen.

(b) A change in the code may change an amino acid in an inconsequential position such as in a structural or terminal portion of the protein.

(c) There are nonsense codes which do not specify amino acids but specify instructions such as 'start' or 'stop' transcription or translation. A mutation in such a code might therefore have a profound effect. Alternatively, mutations in the base sequence may give rise to such nonsense codes, causing erroneous instructions to be carried out such as the premature termination of a protein. Obviously, mutations of this kind may have a much more profound effect than a single change in an amino acid.

Base-pair transformations

The smallest unit of mutation is the transformation of a single base pair. There are two mechanisms by which this may occur:

(a) the chemical modification of a base

(b) the incorporation of a mutagen into the DNA molecule or a mutagen causing erroneous base pairing.

If the transformation involves the same type of base; that is a purine or pyrimidine is transformed into another purine or pyrimidine, this is termed a transition whereas if a purine is transformed into a pyrimidine or vice versa, this is termed a transversion.

The first type of base-pair transformation may occur at any stage of the cell cycle. An example of such a transformation is the deamination of the bases adenine and cytosine, caused by nitrous acid. Adenine is deaminated to hypoxanthine, cytosine to uracil (figure 6.23).

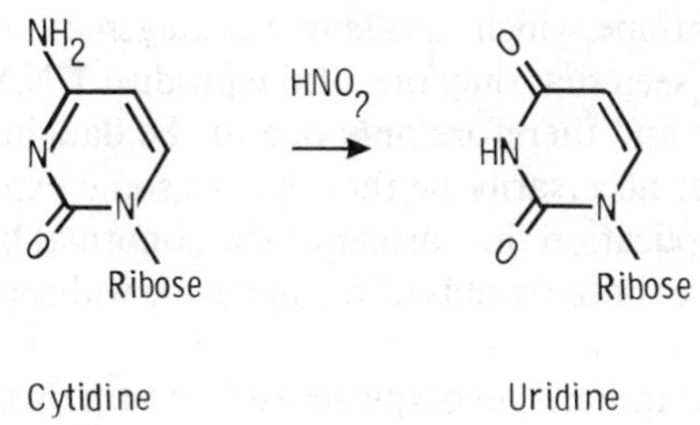

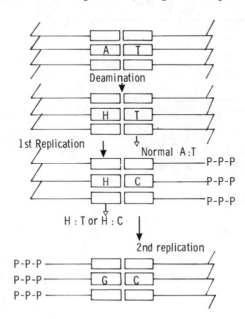

FIGURE 6.23. Deamination of cytidine to uridine by the action of nitrous acid.

For instance, if such a transformation is induced in the tobacco mosaic virus genome, changes in the viral protein can be detected which correlate with the mutation. This has been made possible because the viral protein has been characterized. Thus it is found that the amino acid threonine is replaced by isoleucine. It is also clear that several triplet codes in the nucleic acid specify the amino acid threonine, namely ACA, AGG, ACC and ACU, emphasizing the degeneracy of the code.

Changing the cytosine to uracil, such as by deamination, in these codes gives AUC, AUA and AUU, the codes which specify isoleucine. This mutation is therefore a C→U transition. The effect of a base-pair transformation on replicating DNA can be seen in figure 6.24, using the example of the deamination

FIGURE 6.24. Mechanism of a mutagenic transformation by deamination of adenine to hypoxanthine. Hypoxanthine (figure 4.29) pairs like guanine and this results in a transition A:T→G:C.

Adapted from Goldstein *et al.* (1974) *Principles of Drugs Action: The Basis of Pharmacology* (New York: John Wiley).

of adenine to hypoxanthine which results in a mutagenic transition of the type A:T→G:C. It can be seen that only one of the product DNA molecules has the altered base pair, GC, and therefore only one of the daughter cells is affected. However, this may not necessarily be the case, as some experimental evidence suggests that before replication the partner of the abnormal base may be excised and replaced. Therefore, both strands of the molecule will contain the mutagenic transition.

The expression of the mutant genotype depends in turn on certain factors such as the turnover rate of the product proteins, the rapidity of cell division and, in diploid organisms, the dominance or recessivity of the mutation.

Similar types of mutation are caused by mutagens which alkylate bases, such as the **alkyl nitrosamines**, which alkylate the N^7 and O^6 of guanine (figure 6.31). This may lead to a G:C→A:T transition. Another example is afforded by the mutagen and carcinogen **vinyl chloride** which is discussed in more detail in Chapter 7. This compound causes base-pair transformations, as a result of alkylation of N^1 of adenine and the N^3 of cytidine. The alkylation, addition of the etheno moiety, gives rise to an imidazole ring. This causes adenine to pair like

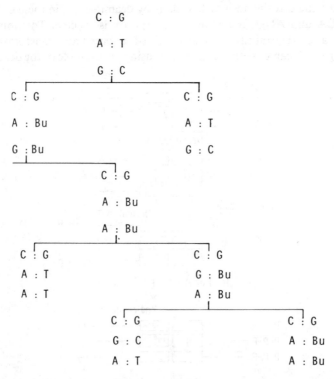

FIGURE 6.25. Replication of a nucleic acid in the presence of the base analogue bromouracil (Bu). After the first replication, Bu enters opposite guanine (G), giving a G:C→A:T transition. If G enters opposite Bu already incorporated, an A:T→G:C transition results.
Adapted from Goldstein *et al.* (1974) *Principles of Drug Action: The Basis of Pharmacology* (New York: John Wiley).

cytidine and cytidine to become similar to adenine. The result of these alkylations are A–T→C–G and C–G→A–T transversions.

The second type of base-pair transformation, incorporation of an abnormal base analogue, only occurs at replication of the DNA molecule. An example of such a mutagen is the thymine analogue, **5-bromouracil** (Bu). This compound is incorporated extensively into the DNA molecule during replication, being inserted opposite adenine in place of thymine. 5-Bromouracil first forms the deoxyribose triphosphate derivative, a pre-requisite for incorporation. If 5-bromouracil is incorporated opposite A, it seems that the A:Bu pair will function as A:T, and the organism continues to grow and divide. Therefore, this change does not seriously affect the function of the DNA. Bu may pair with G, however, as if it were C (figure 6.25), with the consequence that G:C→A:T and A:T→G:C transitions result. The first type of transition occurs because of erroneous incorporation, the second results from erroneous replication against the incorporated Bu. Other base analogues, some of which are used as anticancer drugs, may produce lethal mutations in mammalian cells, particularly in cancerous cells which are involved in rapid DNA synthesis.

Frameshift mutations

This type of mutation arises from the addition or deletion of a base to the nucleic acid molecule. This puts the triplet code of the genome out of sequence. Consequently, transcription of the information is erroneous when it is carried out distal to the mutant addition or deletion. In illustration of this, the following code makes sense when read in groups of three:

HOW CAN THE RAT EAT

After the deletion or insertion of a base, however, the code makes no sense:

HOW CAN THR ATE AT . . .

The transcription of information is therefore generally seriously affected by a frameshift mutation, as large parts of the code may be out of register:

ACA	AAG	AGU	CCA	UCA
threonine	lysine	serine	provaline	serine

becomes

ACA	AGA	GUC	CAU
threonine	lysine	valine	histidine

It is clear that in a long-chain protein produced from such a template, substantial errors may occur.

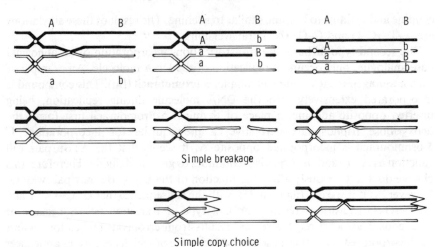

FIGURE 6.26. Crossover at meiosis. Two homologous chromosomes are shown aligned in the top line. Each consists of two chromatids, joined together at the centromere. A and B represent genes on one chromosome, and a and b the corresponding alleles on the other chromosome. After recombination two new gene combinations are apparent. The middle and bottom lines represent other possible methods of recombination.
Adapted from Srb *et al*. (1965) *General Genetics* (San Francisco: W. H. Freeman), after Goldstein *et al*. (1974) *Principles of Drug Action: The Basis of Pharmacology* (New York: John Wiley).

However, if the addition of a base is followed by a deletion in close proximity or vice versa, production of proteins with partial function may occur.

Frameshift mutations may be produced by various mechanisms. For instance, intercalation of planar molecules, such as acridine within the DNA molecule, allows base insertions at replication. **Acridine**-type compounds may also induce mutations of the frameshift type by interfering with the excision and repair processes which correct errors occurring during unequal crossover between homologous chromosomes at meiosis. Such unequal crossover gives rise to errors of the frameshift type.

Crossover, which involves the exchange of homologous segments between chromatids at meiosis and between homologous chromatids at mitosis, occurs normally in diploid organisms, by the breaking and reunion of chromatids or by switching of the replication process, a mechanism known as copy choice (figure 6.26). The broken ends of the chromatids which are formed during this process tend to reunite with other broken ends. Any disruption to this system may therefore potentially cause a frameshift mutation. Also, gross alterations of chromosome structure and the complete loss of sections of DNA may occur if the broken section is not attached to the centromere and the chromatids do not move properly at meiosis.

Chromosome breakage and deletions

This type of effect, clastogenesis, is similar to the foregoing but on a larger scale. Certain compounds cause the breakage of chromosomes, including many of those able to induce base-pair transformations. Alkylating agents, both monofunctional and difunctional, cause chromosome breakage, the latter type probably by causing cross-linking across the DNA molecule. Inhibition of DNA repair may be another cause of this type of mutation.

Chromosome breakage may therefore by linked to other mechanisms of mutagenesis in some fundamental manner. The correction of errors involves breakage of the DNA molecule for both the 'cut and patch' type of repair and for sister strand exchange (figure 6.27). Interference with these processes could clearly result in breakage in one or both chains of the DNA molecule, which it is now accepted forms the backbone of the chromatid. Large deletions may also occur after breakage of the chromosome if the repair processes are compromised.

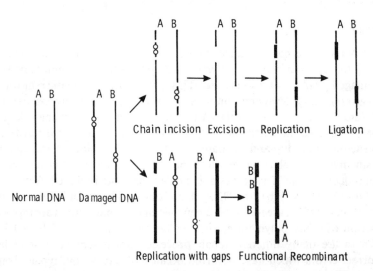

FIGURE 6.27. Representation of two mechanisms of repair of DNA damage such as ultraviolet light-induced dimerization. The upper line represents cut and patch repair, the lower sister strand exchange. Thick lines represent newly synthesized DNA.
Adapted from Goldstein *et al*. (1974) *Principles of Drug Action: The Basis of Pharmacology* (New York: John Wiley).

Metaphase poisoning

This type of mutation involves interference with the processes of mitosis and meiosis. Interference with the processes of mitosis and meiosis can lead to major changes in chromosome number (aneuploidization). During mitosis, sister chromosomes partition prior to separation of the daughter cells and in meiosis this occurs with homologous chromosomes. The partitioning of chromosomes during these processes requires positioning of the centrioles, then arrangement of the chromosomes in the correct positions at metaphase, formation of spindle fibres and involvement of these in the partition of the chromosomes. Interference with one or more aspects of this overall process constitutes metaphase poisoning.

The spindle fibres are attached to the chromosomes at the centromere and align them on the equatorial plate. The spindle arrangement contains microtubules composed of the protein tubulin. **Colchicine** is a specific spindle poison, which binds to tubulin, and inhibits its polymerization. Consequently, colchicine blocks mitosis, causing polyploidy, the unequal partition of chromosomes and metaphase arrest.

Other compounds causing such effects are **podophyllotoxin** and the vinca alkaloids, **vincristine** and **vinblastine**.

The unequal partition of chromosomes, or non-disjunction, is a serious effect if the affected daughter cells survive. Down's syndrome or Trisomy 21 is the result of chromosome non-disjunction in man, there being 47 chromosomes instead of 46. This unequal partitioning of chromosomes can occur at mitosis in germ cells or during meiosis in the production of sperm or ova.

DNA repair

If the DNA molecule has sustained damage from a toxic chemical, it can be repaired by the action of various enzymes. The cell is able to recognize and then repair damaged DNA by **excision repair**. This may involve enzymatic removal of the damaged region of DNA and synthesis of new DNA using the opposite strand as a template. The new section of DNA is then inserted into the damaged DNA molecule. With guanine methylated on the O^6 position a methyltransferase removes the methyl group and the damage is repaired in a single step. Another mechanism involves a glycosylase which breaks the bond between the purine moiety and deoxyribose. This leaves a gap or apurinic site in the DNA which is restored in three stages. First, the DNA chain is broken by an endonuclease and three to four nucleotides are removed. Then the removed nucleotides are replaced by the action of DNA polymerase, and finally the strand is closed by a DNA ligase. There are many different repair processes which remove mismatched bases, correct errors which have occurred because of replication, repair strand breaks, cross-linking and other damage.

Provided repair occurs before cell division and occurs without error, the original damage will not have caused a heritable mutation. If, however, the repair has not been carried out before the cell divides, the damaged area may cause a mutation by one of the mechanisms already described. Foreign compounds which affect the repair process itself may also cause mutations. Thus directly affecting the repair process may introduce errors, and increasing the rate of cell division may also lead to more errors in the DNA. The effect of a lack of ability to repair errors in DNA is seen in those patients with xeroderma pigmentosum. These patients are unable to excise thymidine dimers from DNA caused by exposure to ultraviolet light and have an extremely high incidence of skin cancer.

Mutagenesis in mammals

The effects of mutagens on mammalian cells *in vivo* are less clearly defined than their effects on bacterial cells. When mutagens act on mammalian germ cells, inherited defects may arise, and when somatic cells are exposed to mutagens, cellular organization may be disrupted, giving rise to tumours. It now seems clear that many mutagens are carcinogens and most, but not all, carcinogens are mutagenic (see below).

Major mutagenic changes in somatic cells will clearly lead to total disruption of DNA-directed cellular organization, possibly cell death but probably an inability to grow and divide.

Mutagenic attack in rapidly dividing cells, such as in embryonic tissue, will obviously be particularly serious, giving rise to defects in growth and differentiation, leading to birth defects. Consequently, mutagens are often also teratogens, although some are directly cytotoxic, and it may be this rather than their mutagenicity which causes the teratogenic effect.

However, both carcinogenesis and teratogenesis are discussed separately, and therefore attention will be confined to the susceptibility of the germ cell line.

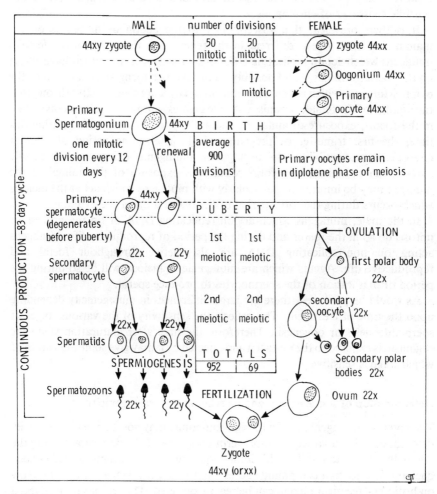

FIGURE 6.28. Gametogenesis in humans showing the number of nuclear divisions giving rise to the gametes. Each *average* fertilizing spermatozoon is the result of 900 divisions of the primary spermatogonium.

, the primordial germ cells are differentiated from about the sixth
n and are consequently susceptible from then onwards. In the
ion of the primary oocytes, which involves the first meiotic
...urs in foetal life. These primary oocytes do not mature into ova until
puberty, with the second meiotic division yielding one ovum from each primary
oocyte (figure 6.28).

In the male, the stem cells or primary spermatogonia undergo mitotic division
regularly, one of the products becoming the primary spermatocyte which, after
two meiotic divisions, gives rise to four spermatozoa. This continues throughout
childhood, but before puberty the sperm degenerate. In the female organism,
chromosomes replicate about 70 times in the production of an ovum while in the
male 950 replications take place in the production of a sperm. In both cases,
however, only two continuous meiotic divisions have occurred, involving one
possible crossover of chromatids.

It is therefore clear that susceptibility to the action of mutagens on the
mammalian germ line is different in the female and the male. In the female,
mutagens which act on replicating DNA or on the process of mitosis have effects
on the foetal germ tissue and possibly on meiosis occurring at maturation of the
ovum. Mutagens which act on non-replicating DNA are clearly active throughout
the lifetime of the female organism. As the greatest number of mitoses take place
in the foetus, exposure to mutagens should be particularly avoided during this
time, the first trimester of pregnancy. This is also the period of greatest
teratogenic susceptibility. In testing for mutagenic effects in the female
mammalian organism, therefore, continuous exposure of the animal to the
mutagen may be important, particularly with mutagens which act at the meiosis
which occurs during the short period of the maturation of the ovum.

In the male, mutagens which act on replicating DNA or on the process of
mitosis do so in the foetus and during the period of reproductive life. Mutagens
acting on non-replicating DNA are also active throughout foetal and
reproductive life. Agents which are mutagenic at meiosis will be so during the
period of maturation of the spermocytes to produce sperm.

As would be predicted, there are large differences in mutagenicity depending
upon the time of treatment. This reflects the sensitivity of the various stages of
spermatogenesis or oogenesis. Therefore, the time after maturation when the
organism is exposed to the mutagen and affected by it may give some indication of
which stage is sensitive.

Determination of mutagenicity and its relation to carcinogenicity

The potential mutagenicity of a foreign compound may now be rapidly and easily
determined using a variety of short-term tests, one of the best known being the
Ames test. This test utilizes histidine-requiring mutant strains of *Salmonella*
bacteria. If a foreign compound causes a mutation, the bacteria may then grow
on histidine-free medium and can be readily detected. The test has been adapted
to utilize liver homogenate, usually from rats or other experimental animals, so as
to identify mutagens which require metabolic activation. There are now many

different *in vitro* mutagenicity tests employing different strains of bacteria and also *in vivo* tests designed to detect mutations in higher organisms and mammalian system.

Major changes in chromosomes as in aneuploidization and clastogenesis can be detected by microscopy (see Bibliography).

As discussed in the preceding section, there is a relationship between mutagenesis and carcinogenesis and the Ames technique is used to predict potential carcinogens. However, although it is clear that many mutagens, as detected in the Ames test, are also carcinogens, there are discrepancies. There are some mutagens which are not carcinogens and some substances shown to be carcinogenic in animals which are not mutagenic. This area of research is controversial and very active.

The test is set up to be extremely sensitive to mutagenic activity and the incorporation of the liver homogenate allows for metabolically activated mutagens. However, the major route of metabolism of a mutagen *in vivo* may be detoxication rather than activation, which may account for some of the false positives, and tumour induction almost certainly requires the occurrence of more events than a mutation, as discussed in the section on carcinogenicity. Consequently, the results of short-term mutagenicity tests must be interpreted with caution as regards potential carcinogenicity.

Chemical carcinogenesis

Introduction

Study of the induction of cancer by chemicals, carcinogenesis, is a large and complex area in toxicology which is rapidly advancing. Because cancer is usually a progressive disease, and although some types are treatable many are fatal, the study of cancer is of major importance. Thus, the biochemical changes associated with exposure to carcinogens have been studied in detail, and consequently the target site and the interactions of carcinogens with it is more clearly understood than in other areas of toxicology. It is, however, beyond the scope of this book to attempt more than a brief overview of the field and to consider some specific carcinogens in Chapter 7.

Partly as a result of epidemiological evidence it is now generally accepted that the majority of human cancers result from exposure to environmental carcinogens; these include both *natural* and *man-made chemicals*, *radiation* and *viruses*. This concept follows on from many observations and work in the early part of the century which showed that coal tar, and then the **aromatic hydrocarbons** derived from it, could induce skin cancer in animals. Indeed, the relationship between such substances and cancer was suggested as early as the 18th century, when in 1775 Sir Percival Pott related scrotal cancer in chimney sweeps to exposure to soot and coal tar.

The number of chemical substances known to be carcinogenic in animals including man is now large. It is striking how diverse the substances are in terms of structure, ranging from metals to complex organic chemicals (figure 6.29), and

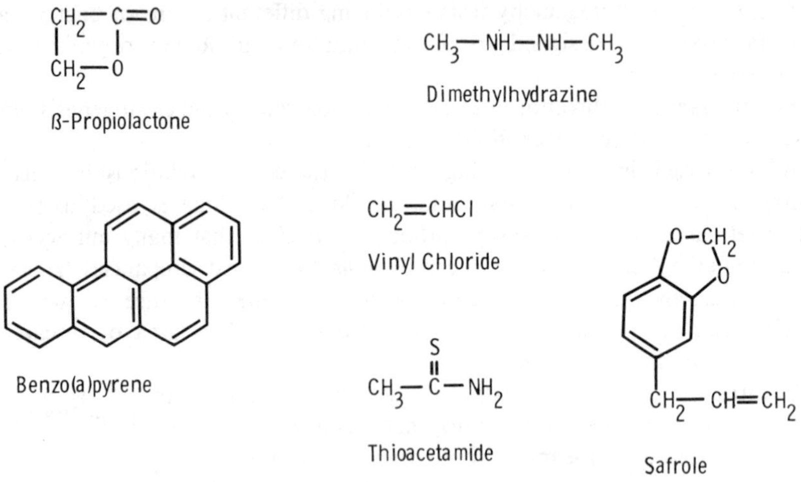

$$CH_2-C=O$$
$$CH_2-O$$

ß-Propiolactone

$$CH_3-NH-NH-CH_3$$

Dimethylhydrazine

Benzo(a)pyrene

$$CH_2=CHCl$$

Vinyl Chloride

$$CH_3-\overset{\overset{\displaystyle S}{\|}}{C}-NH_2$$

Thioacetamide

Safrole

$$CH_2-CH=CH_2$$

FIGURE 6.29. Structures of various carcinogens.

there is large variation in potency. For example, the artificial sweetener **saccharin** may cause tumours in experimental animals after large doses ($TD_{50} = 10\,g$), whereas **aflatoxin B$_1$** is an exquisitely potent carcinogen ($TD_{50} = 1\,\mu g$). The variation in structure and potency suggests that more than one mechanism is involved in carcinogenesis.

It is also clear that apart from exposure to carcinogens, other factors such as the *genetic predisposition* of the organism exposed may also be important. Thus, patients with the genetic disease xeroderma pigmentosum are more susceptible to skin cancer. It has already been mentioned that the incidence of bladder cancer is significantly higher in those individuals who have the slow acetylator phenotype, especially if they are exposed to **aromatic amines**.

Cancer is the unrestrained, malignant proliferation of a somatic cell. The result is a tumour which is a progressively growing mass of abnormal tissue. The term cancer usually applies to malignant tumours rather than benign ones as the latter do not normally invade healthy tissue. A characteristic of cancer cells is the loss of growth control which allows uncontrolled proliferation. It is clear that a permanent heritable transformation has occurred in the cell because, on division, the daughter cells possess the same characteristics. Thus, cancer cells have undergone a malignant transformation in which they lose their specific function and acquire the capacity for unrestrained growth. The process of development of a tumour after exposure is typically quite long, and there are modifying factors such as the frequency of exposure, sex of the animal, genetic constitution, age and the host immune system. Chemical carcinogens may cause:

(a) the appearance of unusual tumours
(b) an increase in the incidence of normal tumours
(c) the appearance of tumours earlier
(d) increased multiplicity.

Table 6.1. Types of carcinogens.

Type	Example
1. Genotoxic carcinogens	
Primary, direct acting	
Alkylating agents	Dimethylsulphate
	Ethylene imine
	β-Propiolactone
Procarcinogens	
Polycyclic aromatic	
hydrocarbons	Benzo[a]pyrene
Nitrosamines	Dimethylnitrosamine
Hydrazines	1,2-Dimethylhydrazine
Inorganic	Cadmium,
	plutonium
2. Epigenetic carcinogens	
Promoters	Phorbol esters
	Saccharin, bile acids.
Solid-state	Asbestos, plastic
Hormones	Oestrogens
Immunosuppressants	Purine analogues
Cocarcinogens	Catechol
3. Unclassified	
Peroxisome proliferators	Clofibrate,
	phthalate esters

Carcinogens may be divided into three classes as shown in table 6.1 The genotoxic carcinogens are those which can react with nucleic acids. They may be direct acting, or primary carcinogens if they are sufficiently chemically reactive to react directly with cellular constituents. Alternatively, they may be procarcinogens which require metabolic activation to proximate and then ultimate carcinogens. The epigenetic carcinogens are those which are not genotoxic and there are others which are not classified such as **peroxisome proliferators**.

The carcinogenic process

It is now generally recognized that cancer is a *multistage process*, with at least three easily recognizable stages:

(a) initiation
(b) promotion
(c) progression.

a matter of debate exactly how many stages are involved, but the
[fa]ctor is that it is a multistage process.

[experimen]tal studies have revealed certain features of the carcinogenic

1. The initiator must be given before the promoter.
2. Repeated doses of an initiator may cause tumours in the absence of a promoter.
3. The initiator produces an irreversible change.
4. The initiator is often chemically reactive and interacts with DNA whereas the promoter does not.
5. The interval between initiation and exposure to the promoter can be varied, yet the number of tumours produced will be similar.
6. Exposure to the promoter is usually repeated and the action seems to be reversible.

The first stage, *initiation*, is a normally rapid, irreversible change which is believed to involve a change in the genetic material of the cell. Thus, the cell undergoes a transformation in which it gains the potential to develop into a clone of preneoplastic cells. Most initiators are thus genotoxic; however, there are other classes of carcinogen (see below). The **somatic cell-mutation theory** depends on there being an interaction between the genetic material of the cell and the carcinogen. This theory fits much of the data available on carcinogenesis. Thus, such a mutation would account for the heritable nature of tumours, their monoclonal character, the mutagenic properties of carcinogens and the evolution of malignant cells *in vivo*. However, although many carcinogens are also mutagens, by no means all of them are. Thus, **ethionine** and **thioacetamide** are carcinogens, but do not seem to be mutagens. Conversely however, some mutagens such as **sodium azide** and **styrene oxide** do not appear to be carcinogenic, indicating that a mutation alone is not sufficient to cause a tumour to develop. Thus, it is highly likely that after the initial mutational event or other initial transformation, other coincident changes are necessary in order for the cell to proceed to a cancerous one. Indeed, some multistage models of carcinogenesis require two critical changes in a cell in order that it becomes neoplastic, each of which is due to a genetic alteration. (It should also be noted that there are many ways of detecting mutagens of various types, and it is possible that a particular compound may be negative or positive in a particular system, whereas *in vivo* other conditions may prevail.)

Consequently, epigenetic mechanisms have also been proposed. Thus the original change in the cell may not necessarily be mutational or it may be indirectly so. For example, some carcinogens may increase cell proliferation and hence decrease the time available for DNA repair. Alternatively, some compounds might interfere with the repair process itself so as to increase error-prone repair, or interact with RNA or protein rather than DNA. Reversion of tumour cells from a neoplastic to a quasi-normal state may occur and has been interpreted as indicative of an epigenetic mechanism. This would involve repression and de-repression of genetic material, or expression of the genotype in an analogous fashion to differentiation being regarded as directed repression and de-repression of particular sections of the genome concerned with the process. Therefore, the interaction of carcinogens with cellular mediators of genetic expression would be a non-mutational mode of carcinogenesis. A carcinogen might also alter the response of the cell to growth controls by non-genetic effects or decrease the efficiency of the immunological surveillance system.

Promotion is the second stage of the carcinogenic process. After the initial transformation a preneoplastic cell can remain dormant, and the necessary proliferation may be facilitated by a promoter. This second stage is characterized by an alteration in genetic expression, and the growth of the clone from the initiated cell. Promoters can act in several ways and cause several different effects. For example, they may cause de-repression and repression, cell membrane effects or inhibition of intercellular communication such as through gap junctions. The latter may be important in cellular homeostasis and suppression of latent tumour cell growth. Promoters may thus release preneoplastic cells from regulatory control. The result may be an increase in the incidence of cancer when exposure occurs after a low dose of initiator. They may also increase the number of tumours produced, and they may decrease the latency period between exposure to the initiator and the development of tumours.

Progression is the stage in which the neoplastic cells become a malignant tumour and may involve changes in the phenotype. This stage is characterized by changes in the number and/or arrangement of chromosomes. The result is an increased growth rate, invasion of healthy tissue and formation of metastases.

Mechanisms of chemical carcinogenesis

Interaction with DNA

There are numerous cellular targets with which carcinogens may interact, but attention has focused on nucleic acids, particularly DNA. It is indisputable that many carcinogens are electrophiles (direct acting or procarcinogens) which react covalently with DNA as well as other cellular macromolecules. The fact that the neoplastic transformation is heritable implies that an alteration in DNA may have occurred, and the observation that many cancers have altered gene expression and chromosomal abnormalities also supports the central role of DNA as the target.

Thus, some of the mutagenic carcinogens studied are known to result in arylation or alkylation of bases in DNA, and such covalent interactions correlate well with carcinogenicity (figure 6.30). The electrophilic ultimate carcinogens, which may be carbonium ions or nitrenium ions for example, react with nucleophilic oxygen and nitrogen atoms in the DNA molecule. The most reactive sites are the N^7 and O^6 of guanine (figure 6.31) and the N^3 and N^7 of adenine. However the important site for attack may vary with the particular carcinogen. Also the specific nature of the covalent interaction is important, possibly because of the ability of the cell to repair the altered nucleic acid, and hence avoid mispairing and the incorporation of a mutation.

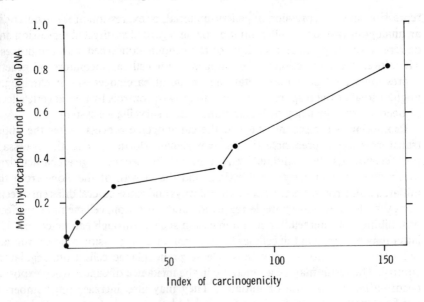

FIGURE 6.30. Relationship between the binding of carcinogens to DNA and carcinogenic potency.
Adapted from Brookes (1966) *Cancer Res.*, **26**, 1994.

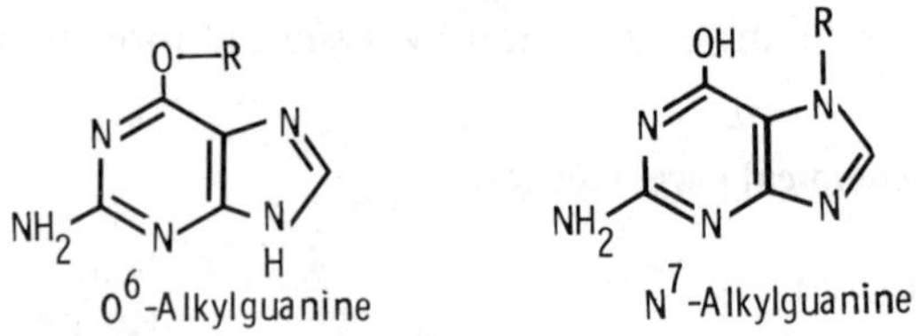

O^6-Alkylguanine N^7-Alkylguanine

FIGURE 6.31. O^6 and N^7 alkylation of guanine.

As well as alkylation or arylation of bases in DNA, intercalation of planar molecules within the double helix may also be important if it allows a subsequent reaction to take place. Also, the effect of carcinogens, such as some **metals**, on the fidelity of DNA polymerase may be another mechanism whereby altered DNA is produced. Some carcinogens, such as **hydrazine**, for example, can indirectly cause DNA methylation which may be involved in the carcinogenic process. After the initial reaction with DNA has taken place the damage is either repaired or fixed if replication takes place before repair. The result of this is a point mutation or frame shift mutation (see above) and possibly the activation of proto-oncogenes (see below).

Interaction of carcinogens with other macromolecules such as RNA and the proteins which regulate gene function may also be involved in the carcinogenic process. Interactions with RNA have been suggested as possibly leading to a permanently altered state of differentiation through changes in gene expression.

Work with the carcinogen **acetylaminofluorene** found that residues of the compound in ribosomal RNA may correlate more closely with liver tumour formation than residues in DNA. Direct interactions with the mechanisms of protein synthesis, or with DNA and RNA polymerase enzymes, can also be seen as possible mechanisms. For instance, a modification of the polymerase enzymes by a carcinogen, either directly or indirectly, could lead to the erroneous replication of DNA or RNA and hence the permanent incorporation of a mutation.

Activation of cellular oncogenes

Cellular oncogenes, termed **proto-oncogenes** in their normal state, are found in the DNA of cells and are important in the regulation of cell growth. It is now clear that the activation of cellular oncogenes, and also the inactivation of tumour **suppressor genes**, are critical events in the neoplastic transformation of a cell.

Thus, proto-oncogenes may be activated by at least five different mechanisms including point mutations, and they then produce proteins which result in the neoplastic transformation of the cell. Thus, the gene products may be growth factors, growth factor receptors, binding proteins (GTP and nuclear) and kinases. Similarly, tumour suppressor genes may be inactivated by mutations, but additionally by deletions or loss of large sections of the chromosome. Hereditary inactivation of tumour suppressor genes may be an important genetic predisposition in the development of cancer in humans.

DNA repair

The cell has the ability to recognize and repair damage to DNA as already discussed in the section on mutagenesis (see above). This must occur *before* cell division in order to be effective, otherwise there may be mispairing of bases, rearrangements and translocations of sections of DNA. The size of the DNA adducts will influence the amount of DNA repaired. Chemicals which affect the repair process may therefore increase the possibility of a neoplastic transformation. There are however differences in the rate of repair depending on the type of repair and also species and tissue differences in this rate. For example, the damage to liver DNA caused by **dimethylnitrosamine** is more slowly repaired in the hamster than in the rat, and the hamster is more susceptible to tumours from this compound than the rat. There are also tissue differences in repair within the same species which account for differences in susceptibility between tissues and organs. For example, the methylation of guanine at the O^6 position is more crucial than that at N^7, as excision repair of O^6 methylguanine is poor in certain tissues, such as the brain, in comparison with the liver and kidney. The brain is consequently more susceptible to tumour development.

Epigenetic mechanisms

Epigenetic carcinogens cause the appearance of tumours without a genotoxic effect. This includes *promotion, immunosuppression, cocarcinogenicity* and *cytotoxicity*. This group probably involves diverse mechanisms. They may act by:

(a) exposure of a pre-existing genetic abnormality
(b) impairment of DNA polymerase
(c) gene amplification
(d) alteration of chromosome composition
(e) induction of a stable, altered state of gene expression.

Other types of carcinogen include the peroxisome proliferators. These are a diverse group of chemical substances which may be potent carcinogens, but are *not* mutagens. These substances, such as the hypolipidaemic drug **clofibrate** and various **phthalate esters** used as plasticizers, when given to rodents cause an increase in the number of peroxisomes. These organelles contain various enzymes, but especially those concerned with the oxidation of fatty acids. However, unlike the β-oxidation pathway which is carried out in mitochondria, the pathway in the peroxisome produces hydrogen peroxide rather than water. Peroxisome proliferators not only cause an increase in the number of peroxisomes, especially in the liver, but also an increase in some of the enzymes in the organelle and an increase in liver size due to hyperplasia. As well as these changes there are also liver tumours, and with some of the hypolipidaemic drugs the incidence of liver tumours may be as high as 100% in experimental animals. A current theory to explain the carcinogenicity of these compounds is that the increase in some of the enzymes involved in β-oxidation, which may be 20-fold, results in a large increase in the production of hydrogen peroxide. However, removal of this hydrogen peroxide depends on catalase, but this enzyme, although present in the peroxisome, is overwhelmed by the amount of hydrogen peroxide because catalase is only increased about two-fold by peroxisome proliferators. Thus, there is an excess of hydrogen peroxide which can diffuse out of the peroxisome into the cytoplasm and may reach the nucleus. The hydrogen peroxide and further products may therefore damage the DNA and other macromolecules.

Cell proliferation

Chemically-induced cell proliferation can also be an important part of carcinogenesis and may be a mechanism of epigenetic carcinogenesis. The increase in cellular replication will lead to an increase in the frequency of spontaneous mutations and a decrease in the time available for DNA repair. During cell replication the DNA molecule is unwound allowing the transcription of certain genes, and hence the molecule is more vulnerable to attack by carcinogens. Tumour promotion and progression are associated with cell proliferation. The stimulus for cell proliferation might be direct mitogenic action due to compounds such as the peroxisome proliferators. Alternatively, cytotoxic compounds especially when administered chronically, may cause regenerative cell growth. For example, the induction of $\alpha_{2\mu}$-globulin nephropathy by a variety of chemicals is associated with the appearance of renal tumours. The protein $\alpha_2\mu$-globulin is filtered by the kidney glomerulus and reabsorbed into the proximal tubules where it is degraded enzymatically by lysosomal enzymes. Such compounds as 1,2-**dichlorobenzene, unleaded petrol** and **dimethyl phosphonate**, which all cause both renal tumours and nephropathy, make $\alpha_2\mu$-globulin more resistant to enzymatic breakdown. This results in protein droplets, lysosomal overload and the end result is necrosis of the proximal tubular cells. Thus, chronic exposure to such compounds results in increased cell proliferation. This is believed to cause promotion of neoplastic lesions which may arise by spontaneous mutation.

In conclusion, it is clear that the mechanisms underlying chemical carcinogenesis are complex, involving numerous steps and more than a single gene mutation. It is also clear that the cell may protect itself from chemical carcinogens by repair of altered macromolecules, thereby resisting potent mutagens. Perhaps failure of these repair mechanisms leading to incorporation of mutations, coupled with epigenetic effects or the derepression of growth control, are some of the major factors involved in the causation of cancer by chemicals.

Bibliography

General

KLAASSEN, C. D., AMDUR, M. O. and DOULL, J. (editors) (1986) *Casarett and Doull's Toxicology, The Basic Science of Poisons* (New York: Macmillan).

HODGSON, E. and LEVI, P. E. (1987) *A Textbook of Modern Toxicology* (New York: Elsevier).

LU, F. C. (1991) *Basic Toxicology*, 2nd edition (Washington, DC: Hemisphere).

PRATT, W. B. and TAYLOR, P. (editors) (1990) *Principles of Drug Action* (New York: Churchill Livingstone).

Basic mechanisms

ANDERS, M. W. (editor) (1985) *Bioactivation of Foreign Compounds* (New York: Academic Press).

BAKKE, J. and GUSTAFSSON, J. A. (1984) Mercapturic acid pathway metabolites and xenobiotics: generation of potentially toxic metabolites during enterohepatic circulation. *TIPS*, **5**, 517.

BRIDGES, J. W., BENFORD, D. J. and HUBBARD, S. A. (1983) Mechanisms of toxic injury. *Ann. N.Y. Acad. Sci.*, **407**, 42.

BOOBIS, A. R., FAWTHROP, D. J. and DAVIES, D. S. (1989) Mechanisms of cell death. *TIPS*, **10**, 275.

BOYD, M. R. and STATHAM, C. N. (1983) The effect of hepatic metabolism on the production and toxicity of reactive metabolites in extrahepatic organs. *Drug Metab. Rev.*, **14**, 35.

CHEESEMAN, K. H. (1984) General mechanisms of toxic liver injury with special reference to free radical reactions. In *Advances in Inflammation Research*, edited by K. D. Rainsford and G. P. Velo, Vol. 6. (New York: Raven Press).

COTGREAVE, I. A., MOLDEUS, P. and ORRENIUS, S. (1988) Host biochemical defense mechanisms against prooxidants. *Annu. Rev. Pharmac. Toxicol.*, **28**, 189.

LOCK, E. A., MITCHELL, A. M. and ELCOMBE, C. R. (1990) Biochemical mechanisms of induction of hepatic peroxisome proliferation, *Annu. Rev. Pharmacol. Toxicol.*, **29**, 145.

MITCHELL, J. R. and HORNING, M. G. (editors) (1984) *Drug Metabolism and Drug Toxicity* (New York: Raven Press).

MONKS, T. J. and LAU, S. S. (1988) Reactive intermediates and their toxicological significance. *Toxicology*, **52**, 1.

KEHRER, J. P., JONES, D. P., LEMASTERS, J. J., FARBER, J. L. and JAESCHKE, H. (1990) Mechanisms of hypoxic cell injury. *Toxicol. Appl. Pharmacol.*, **106**, 165.

KOCSIS, J. J., JOLLOW, D. J., WITMER, C. M., NELSON, J. O. and SNYDER, R. (editors) (1986) *Biological Reactive Intermediates III. Mechanisms of Action in Animal Models and Human Disease* (New York: Plenum Press).

ORRENIUS, S., McCONKEY, D. J., BELLOMO, G. and NICOTERA, P. (1989) Role of Ca^{2+} in toxic cell killing. *TIPS*, **10**, 281.

TRUMP, B. F., BEREZESKY, I. K., SMITH, M. W., PHELPS, P. C. and ELLIGET, K. A. (1989) Relation between ion deregulation and toxicity. *Toxicol. Appl. Pharmacol.*, **97**, 6.

YOST, G. S., BUCKPITT, A. R., ROTH, R. R. and McLEMORE, T. L. (1989) Mechanisms of lung injury by systemically administered chemicals. *Toxicol. Appl. Pharmacol.*, **101**, 179.

Direct toxic action: tissue lesions

BALAZS, T. (editor) (1981) *Cardiac Toxicology*, Vols I–III (Boca Raton, Florida: CRC Press).

BALAZS, T., HANIG, J. P. and HERMAN, E. H. (1986) Toxic responses of the cardiovascular system. In *Casarett and Doull's Toxicology, The Basic Science of Poisons*, edited by C. D. Klaassen, M. O. Amdur and J. Doull (New York: Macmillan).

BUS, J. S. and GIBSON, J. E. (1979) Lipid peroxidation and its role in toxicology. In *Reviews in Biochemical Toxicology*, Vol. 1, edited by E. Hodgson, J. R. Bend and R. M. Philpot (New York: Elsevier-North Holland).

CALDWELL, J. and JAKOBY, W. B. (editors) (1983) *Biological Basis of Detoxication* (New York: Academic Press).

COHEN, G. M. (editor) (1986) *Target Organ Toxicity*, Vols. I and II (Boca Raton, Florida: CRC Press).

DIXON, R. J. (editor) (1985) *Reproductive Toxicology, Target Organ Toxicology Series*, edited by R. L. Dixon (New York: Raven Press).

DIXON, R. L. (1986) Toxic responses of the reproductive system. In *Casarett and Doull's Toxicology, The Basic Science of Poisons*, edited by C. D. Klaassen, M. O. Amdur and J. Doull (New York: Macmillan).

DRILL, V. A. and LAZAR, P. (editors) (1984) *Cutaneous Toxicity*, Target Organ Toxicology Series, edited by R. L. Dixon (New York: Raven Press).

EMMETT, E. A. (1986) Toxic responses of the skin. In *Casarett and Doull's Toxicology, The Basic Science of Poisons*, edited by C. D. Klaassen, M. O. Amdur and J. Doull (New York: Macmillan).

FARBER, E. and FISHER, M. M. (editors) (1980) *Toxic Injury to the Liver, Parts A and B* (New York: Marcel Dekker).

GLAISTER, J. R. (1986) *Principles of Toxicological Pathology* (London: Taylor & Francis).

HAYES, A. W. (editor) (1985) *Toxicology of the Eye, Ear, and Other Special Senses*, Target Organ Toxicology Series, edited by R. L. Dixon (New York: Raven Press).

HOOK, G. R. (editor) (1988) *Toxicology of the Lung*, edited by D. E. Gardner, J. D. Crapo and E. J. Massaro (New York: Raven Press).

HOOK, J. B. (editor) (1981) *Toxicology of the Kidney, Target Organ Toxicology Series*, edited by R. L. Dixon (New York: Raven Press).

HOOK, J. B. and HEWITT, W. R. (1986) Toxic responses of the kidney. In *Casarett and Doull's Toxicology, The Basic Science of Poisons*, edited by C. D. Klaassen, M. O. Amdur and J. Doull (New York: Macmillan).

HOOK, J. B., McCORMACK, K. M. and KLUWE, W. M. (1979) Biochemical mechanisms of nephrotoxicity. In *Reviews in Biochemical Toxicology*, Vol. 1, edited by E. Hodgson, J. R. Bend and R. M. Philpot (New York: Elsevier-North Holland).

IOANNIDES, C. and PARKE, D. V. (1990) The cytochrome P450 I gene family of microsomal haemoproteins and their role in the metabolic activation of chemicals. *Drug Metab. Rev.*, **22**, 1.

IRONS, R. D. (editor) (1985) *Toxicology of Blood and Bone Marrow, Target Organ Toxicology Series*, edited by R. L. Dixon, (New York: Raven Press).

KULKARNI, A. P. and HODGSON, E. (1980) Hepatotoxicity. In *Introduction to Biochemical Toxicology*, edited by E. Hodgson and F. E. Guthrie (New York: Elsevier-North Holland).

LAMB, J. C. and FOSTER, P. M. B. (editors) (1988) *Physiology and Toxicology of the Male Reproductive System* (London: Academic Press).

MENZEL, D. B. and AMDUR, M. O. (1986) Toxic responses of the respiratory system. In *Casarett and Doull's Toxicology, The Basic Science of Poisons*, edited by C. D. Klaassen, M. O. Amdur and J. Doull (New York: Macmillan).

MITCHELL, C. L. (editor) (1982) *Nervous System Toxicology, Target Organ Toxicology Series*, edited by R. L. Dixon (New York: Raven Press).

NICHOLSON, J. K. and WILSON, I. D. (1989) High resolution proton magnetic resonance spectra of biological fluids. *Prog. in NMR Spec.*, **21**, 449.

NORTON, S. (1986) Toxic responses of the central nervous system. In *Casarett and Doull's Toxicology, The Basic Science of Poisons*, edited by C. D. Klaassen, M. O. Amdur and J. Doull (New York: Macmillan).

PLAA, G. L. and HEWITT, W. R. (editors) (1982) *Toxicology of the Liver, Target Organ Toxicology Series*, edited by R. L. Dixon (New York: Raven Press).

PLAA, G. L. (1986) Toxic responses of the liver. In *Casarett and Doull's Toxicology, The Basic Science of Poisons*, edited by C. D. Klaassen, M. O. Amdur and J. Doull (New York: Macmillan).

POTTS, A. M. (1986) Toxic responses of the eye. In *Casarett and Doull's Toxicology, The Basic Science of Poisons*, edited by C. D. Klaassen, M. O. Amdur and J. Doull (New York: Macmillan).

RAPPAPORT, A. M. (1976) The microcirculatory acinar concept of normal and pathological hepatic structure. *Beitr. Path.*, **157**, 215.

SCHILLER, C. M. (editor) (1982) *Intestinal Toxicology, Target Organ Toxicology Series*, edited by R. L. Dixon (New York: Raven Press).

SIES, H. and KETTERER, B. (editors) (1989) *Glutathione Conjugation. Mechanisms and Biological Significance* (London: Academic Press).

SLATER, T. F. (editor) (1978) *Biochemical Mechanisms of Liver Injury* (London: Academic Press).

SMITH, R. P. (1986) Toxic responses of the blood. In *Casarett and Doull's Toxicology, The Basic Science of Poisons*, edited by C. D. Klaassen, M. O. Amdur and J. Doull (New York: Macmillan).

SMUCKLER, E. A. (1976) Structural and functional changes in acute liver injury. *Environ. Health Perspect.*, **15**, 13.

THOMAS, J. A., KORACH, K. S. and McLACHLAN, J. A. (editors) (1985) *Endocrine Toxicology, System, Target Organ Toxicology Series*, edited by R. L. Dixon (New York: Raven Press).

TIETZE, N. W. (editor) (1987) *Fundamentals of Clinical Chemistry*, 3rd edition (Philadelphia: Saunders).

TRUMP, B. F. and ARSTILA, A. V. (1980) Cellular reaction to injury. In *Principles of Pathobiology*, edited by M. F. La Via and R. B. Hill (New York: Oxford University Press).

282   *Principles of Biochemical Toxicology*

VAN STEE, E. W. (editor) (1982) *Cardiovascular Toxicology, Target Organ Toxicology Series*, edited by R. L. Dixon (New York: Raven Press).
WORKING, P. K. (editor) (1989) *Toxicology of the Male and Female Reproductive Systems* (London: Hemisphere/Taylor and Francis)
ZIMMERMAN, H. J. (1978) *Hepatotoxicity* (New York: Appleton Century Crofts).

Pharmacological, physiological and biochemical effects

ALBERT, A. (1979) *Selective Toxicity* (London: Chapman & Hall).
ALBERT, A. (1987) *Xenobiosis* (London: Chapman & Hall).
GILMAN, A. G. (editor) (1990) *Goodman and Gilman's The Pharmacological Basis of Therapeutics*, 8th edition (New York: Pergamon).
GORROD, J. W. (editor) (1979) *Drug Toxicity* (London: Taylor & Francis).
HATHWAY, D. E. (1984) *Molecular Aspects of Toxicology* (Royal Society of Chemistry: London).
KLAASSEN, C. D., AMDUR, M. O. and DOULL, J. (editors) (1986) *Casarett and Doull's Toxicology, The Basic Science of Poisons*, (New York: Macmillan).
LEVI, P. E. (1987) Toxic action. In *A Textbook of Modern Toxicology*, edited by E. Hodgson and P. E. Levi (New York: Elsevier)
PRATT, W. B. and TAYLOR, P. (editors) (1990) *Principles of Drug Action* (New York: Churchill Livingstone).

Teratogenesis

BROWN, L. P., FLINT, O. P., ORTON, T. C. and GIBSON, G. G. (1986) Chemical teratogenesis: Testing methods and the role of metabolism. *Drug Metab. Rev.*, **17**, 221.
JUCHAU, M. R. (1990) Bioactivation in chemical teratogenesis. *Annu. Rev. Pharmacol. Toxicol.*, **29**, 165.
KIMMEL, C. A. and BUELKE-SAM, J. (editors) (1981) *Developmental Toxicology, Target Organ Toxicology Series*, edited by R. L. Dixon (New York: Raven Press).
MANSON, J. M. (1986) Teratogens. In *Casarett and Doull's Toxicology, The Basic Science of Poisons*, edited by C. D. Klaassen, M. O. Amdur and J. Doull (New York: Macmillan).
RUDDON, R. W. (1990) Chemical teratogenesis. In *Principles of Drug Action, The Basis of Pharmacology*, edited by W. B. Pratt and P. Taylor (New York: Churchill Livingstone).
SULLIVAN, F. M. and BARLOW, S. M. (1979) Congenital malformations and other reproductive hazards from environmental chemicals. *Proc. Roy. Soc. Lond.*, B, **205**, 91.
WILSON, J. G. (1973) *Environment and Birth Defects* (New York: Academic Press).
WILSON, J. G. and FRASER, F. C. (editors) (1977) *Handbook of Teratology*, Vols 1–4 (New York: Plenum Press).

Immunotoxicity

BERLIN, A., DEAN, J. H., DRAPER, M. H., SMITH, E. M. B. and SPREAFICO, F. (editors) (1987) *Immunotoxicology* (Lancaster: Martinus Nijhoff).
DEAN, J. H., LUSTER, M. I., MUNSON, A. E. and AMOS, H. (editors) (1985) *Immunotoxicology and Immunopharmacology, Target Organ Toxicology Series*, edited by R. L. Dixon, (New York: Raven Press).
DEAN, J. H., MURRAY, M. J. and WARD, E. C. (1986) Toxic responses of the immune system. In *Casarett and Doull's Toxicology, The Basic Science of Poisons*, edited by C. D. Klaassen, M. O. Amdur and J. Doull (New York: Macmillan).

GREALLY, J. F. and SILVANO, V. (editors) (1983) *Allergy and Hypersensitivity to Chemicals* (Copenhagen: WHO; Luxembourg: CEC).

LUSTER, M. I., BLNK, J. A. and DEAN, J. H. (1987) Molecular and cellular basis of chemically induced immunotoxicity. *Annu. Rev. Pharmacol. Toxicol.*, **27**, 23.

PARK, B. K. and KITTERINgHAM, N. R. (1990) Drug–protein conjugation and its immunological consequences. *Drug Metab. Rev.*, **22**, 87.

POHL, L. R., SATOH, H., CHRIST, D. D. and KENNA, J. G. (1988) The immunologic and metabolic basis of drug hypersensitivities. *Annu. Rev. Pharmacol.*, **28**, 367.

PRATT, W. B., (1990) Drug allergy. In *Principles of Drug Action, The Basis of Pharmacology*, edited by W. B. Pratt, and P. Taylor (New York: Churchill Livingstone).

DE WECK, A. L. and BUNDGAARD, H. (editors) (1983) *Allergic Reactions to Drugs* (Berlin: Springer).

ROITT, I. M. (1984) *Essential Immunology*, 5th edition (Oxford: Blackwell).

Mutagenesis

AMES, B. N. (1979) Identifying environmental chemicals causing mutations and cancer. *Science*, **204**, 587.

BRUSICK, D. (1987) *Principles of Genetic Toxicology*, 2nd edition (New York: Plenum Press).

DRAKE, J. W. (1970) *The Molecular Basis of Mutation* (San Francisco: Holden Day).

GROSCH, D. S. (1980) Genetic poisons. In *Introduction to Biochemical Toxicology*, edited by E. Hodgson and F. E. Guthrie (New York: Elsevier-North Holland).

HOLLEANDER, A. (editor) (1971–1981) *Chemical Mutagens: Principles and Methods for Their Detection*, several volumes (New York: Plenum Press).

RUDDON, R. W. (1990) Chemical mutagenesis. In *Principles of Drug Action, The Basis of Pharmacology*, edited by W. B. Pratt and P. Taylor (New York: Churchill Livingstone).

THILLY, W. G. and CALL, K. M. (1986) Genetic toxicology. In *Casarett and Doull's Toxicology, The Basic Science of Poisons*, edited by C. D. Klaassen, M. O. Amdur and J. Doull (New York: Macmillan).

VOGEL, F. and ROHRBORN, G. (editors) (1970) *Chemical Mutagenesis in Mammals and Man* (New York: Springer-Verlag).

Carcinogenesis

BRADSHAW, R. A. and PRENTIS, S. (editors) (1987) *Oncogenes and Growth Factors* (Elsevier Science Publications)

BOYLAND, E. (1980) History and future of chemical carcinogenesis. In *Chemical Carcinogenesis*, edited by P. Brookes. *British Medical Bulletin*, **36**, 5.

COOPER, C. S. and GROVER, P. L. (editors) (1989) *Chemical Carcinogenesis and Mutagenesis I. Handbook of Experimental Pharmacology*, Vol. 94 (Berlin: Springer).

COOPER, C. S. and GROVER, P. L. (editors) (1990) *Chemical Carcinogenesis and Mutagenesis II. Handbook of Experimental Pharmacology*, Vol. 94. (Berlin: Springer).

GLOVER, D. M. and HAMES, B. D. (1989) *Oncogenes* (Oxford: IRL Press/OUP).

LAFOND, R. E. (editor) (1988) *Cancer: The Outlaw Cell*, 2nd edition (Cambridge: Royal Society of Chemistry).

LAWLEY, P. D. (1980) DNA as a target of alkylating carcinogens. In *Chemical Carcinogenesis*, edited by P. Brookes, *British Medical Bulletin*, **36**, 19.

MILLER, E. C. and MILLER, J. A. (1976) The metabolism of chemical carcinogens to reactive electrophiles and their possible mechanism of action in carcinogenesis. In *Chemical Carcinogens*, edited by C. E. Searle (Washington D.C.: American Chemical Society).

RUDDON, R. W. (1990) Chemical carcinogenesis. In *Principles of Drug Action, The Basis of Pharmacology*, edited by W. B. Pratt and P. Taylor (New York: Churchill Livingstone).

SEARLE, C. E. (editor) (1984) *Chemical Carcinogens*, 2nd edition (Washington D.C.: American Chemical Society).

SIMS, P. (1980) Metabolic activation of chemical carcinogens. In *Chemical carcinogenesis*, edited by P. Brookes, British Medical Bulletin, **36**, 11,

WILLIAMS, G. M. (1980) The pathogenesis of rat liver cancer caused by chemical carcinogenesis. *Biochim. Biophys. Acta*, **605**, 167.

WILLIAMS, G. M. and WEISBURGER, J. H. (1986) Chemical carcinogens. In *Casarett and Doull's Toxicology, The Basic Science of Poisons*, edited by C. D. Klaassen, M. O. Amdur and J. Doull (New York: Macmillan).

Chapter 7

Biochemical mechanisms of toxicity: specific examples

Chemical carcinogenesis

Acetylaminofluorene

This compound is a well known carcinogen and one of the most widely studied. Research into the mechanism underlying its carcinogenicity is at present an area of high activity and therefore this discussion must confine itself to the principles rather than details.

The study of acetylaminofluorene carcinogenicity has provided insight into the carcinogenicity of other aromatic amines and also illustrates a number of other important points. Acetylaminofluorene is a very potent mutagen and a carcinogen in a number of animal species, causing tumours primarily of the liver, bladder and the kidney. It became clear from the research that metabolism of the compound was involved in the carcinogenicity. The important metabolic reaction was found to be N-hydroxylation, catalysed by the microsomal mixed function oxidases, and this was demonstrated both *in vitro* and *in vivo*. Thus N-hydroxyacetylaminofluorene (figure 7.1), the product, is a more potent carcinogen than the parent compound. The production of this metabolite *in vivo* was found to be increased nine-fold by the repeated administration of the parent compound, a finding of particular importance when considering the general use of single-dose rather than multiple low-dose studies for evaluating the toxicity of compounds. This effect is presumably the result of induction of the microsomal enzymes involved in the production of N-hydroxyacetylaminofluorene. N-hydroxylation has since proved to be an important metabolic reaction in the toxicity of a number of other compounds. The N-hydroxy intermediate in particular is of importance for the carcinogenicity of a number of aromatic amino, nitro and nitroso compounds. This intermediate may arise by reduction as well as oxidation (see page 101). However, N-hydroxyacetylaminofluorene is not the ultimate carcinogen, as it requires further metabolism in order to initiate tumour production. The N-hydroxylated intermediate is stable enough to be conjugated and the N-O glucuronide of acetylaminofluorene is an important metabolite (figure 4.51).

Thus, it has been found that conjugates of N-hydroxyacetylaminofluorene are more potent carcinogens than the parent compound, and there is a wealth of

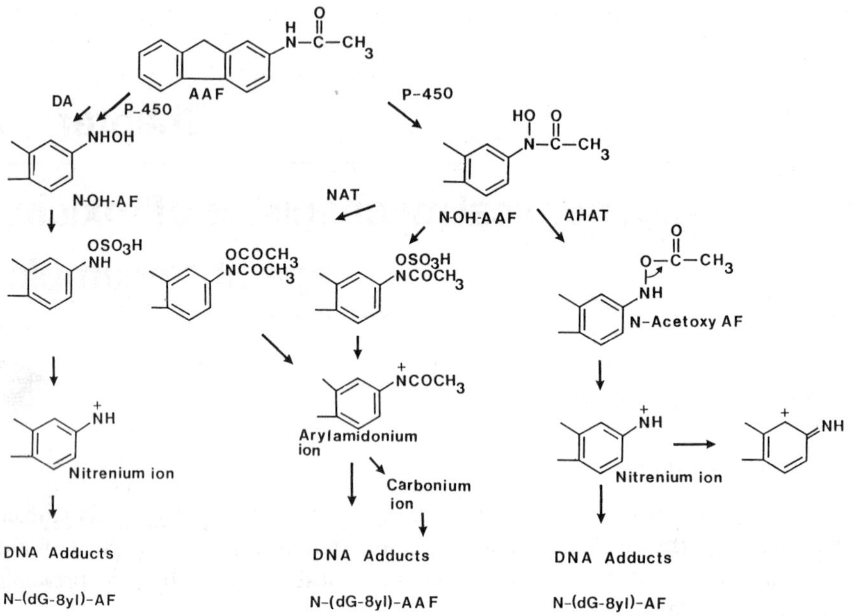

FIGURE 7.1. Some of the possible routes of metabolic activation of acetylaminofluorene (AAF). N-OH-AF: N-hydroxyaminofluorene; N-OH-AAF: N-hydroxyacetylaminofluorene; N-acetoxy AF: N-acetoxyaminofluorene; N-(dG-8yl)-AF: N-deoxyguanosinyl-aminofluorene; N-(dG-8yl)-AAF: N-deoxyguanosinyl-acetylaminofluorene. P-450: cytochrome(s) P-450; DA: deacetylase; NAT: N-acetyltransferase; AHAT: N,O-arylhydroxamic acid acyltransferase.

evidence, at times confusing and conflicting, which indicates that both the **sulphate** and **acetyl esters** of *N*-hydroxyacetylaminofluorene are involved and that both may give rise to reactive **nitrenium** ions and **carbonium** ions. Thus, the metabolism of acetylaminofluorene may involve several different pathways as shown in figure 7.1 and some of the metabolites are mutagenic and some are also cytotoxic and carcinogenic. Thus, sulphate conjugation is clearly a requirement for both the cytotoxicity and tumorigenicity of *N*-hydroxyacetylaminofluorene and results in DNA adducts. The cytotoxicity is believed to be due to binding to DNA rather than protein. However, some data suggests that deacetylation also occurs to yield *N*-hydroxyaminofluorene which is subsequently sulphated to yield the ultimate carcinogen. Deacetylation of acetylaminofluorene itself may also occur and could be an alternative route to *N*-hydroxyaminofluorene. Whether sulphation of *N*-hydroxyacetylaminofluorene is also involved is as yet unclear: the sulphate conjugate is not mutagenic when formed *in situ* in a *Salmonella in vitro* test system and has low carcinogenicity when applied to the skin of experimental animals. Similarly, the acetate ester of *N*-hydroxyacety-laminofluorene which is carcinogenic, has been suggested as giving rise to a reactive nitrenium ion which could be responsible for the DNA adduct isolated after *in vivo* exposure and the nucleoside adduct detected *in vitro*. Alternatively the *N*-acetyltransferase enzyme (*N,O*-arylhydroxamic acid acyltransferase) is known to catalyse a transfer of the acetyl group from the nitrogen atom to the oxygen to yield the *O*-acetyl ester of *N*-hydroxyaminofluorene. This compound is mutagenic, is highly reactive and spontaneously reacts with macromolecules. It can give rise to a reactive nitrenium ion and will form the same DNA adduct as the sulphate conjugate: *N*-(deoxyguanosin-8-yl)-2-aminofluorene. It is noteworthy that this adduct has lost the acetyl group and therefore deacetylation must occur at some point and may be a prerequisite. In fact guinea-pig microsomal deacetylase will metabolize both acetylaminofluorene and *N*-hydroxyacetylaminofluorene to products which bind to DNA *in vitro*. Inhibition of the deacetylase decreases the mutagenicity of *N*-hydroxyacetylaminofluorene.

The reactive nitrenium or carbonium ions postulated to be produced will react with nucleophilic groups in nucleic acids, proteins and sulphydryl compounds such as glutathione and methionine. The arylation of DNA by acetyl-aminofluorene has been demonstrated *in vivo* and *in vitro*. The involvement of sulphate conjugation brings other factors into play. Depletion of body sulphate reduces and supplementation with organic sulphate increases the carcinogenicity of acetylaminofluorene. The production of covalent adducts between acetylaminofluorene and cellular macromolecules *in vivo* can be shown to be correspondingly decreased and increased by manipulation of body sulphate levels.

Although the N-O glucuronide has a lower chemical reactivity and carcinogenicity than the sulphate conjugate, the glucuronide may be responsible for the production of bladder cancer by acetylaminofluorene. Furthermore, the N-O sulphate conjugate is not the only ultimate carcinogen, as acety-laminofluorene induces tumours in tissues without sulphotransferase activity. For example, the mammary gland has no deacetylase and does not carry out **sulphate conjugation**. However, a single application of *N*-hydroxyacetyl-aminofluorene yields tumours at the site of application. This may be due therefore to the presence of the *N,O*-acetyltransferase enzyme or another means of activation.

N-hydroxylation is not the only or major route of metabolism *in vivo*, nor is it the only reaction catalysed by the microsomal enzymes. Ring hydroxylation is the major route of metabolism, the products of which are not carcinogenic. These alternative routes of metabolism are inducible *in vivo* by pretreatment with agents such as phenobarbital and 3-methylcholanthrene (see Chapter 5). Glucuroni-dation of the resulting hydroxyl derivatives is also induced by phenobarbital pretreatment. Consequently, pretreatment of animals with both these agents reduces the carcinogenicity of acetylaminofluorene. This illustrates the difficulty of predicting the effect of environmental influences on toxicity when multiple metabolic pathways are involved.

A well documented species difference in susceptibility to acetylaminofluorene carcinogenicity is the resistance of the guinea-pig. This affords an interesting illustration of the role of species differences in metabolism as a basis for species differences in susceptibility to toxicity. The guinea-pig is not resistant to the carcinogenicity of the metabolite *N*-hydroxyacetylaminofluorene, however, indicating that this species has low activity for *N*-hydroxylation. As well as this, the guinea-pig has low sulphotransferase activity. A combination of these factors therefore confers resistance on the guinea-pig. The low production of *N*-hydroxyacetylaminofluorene followed by low sulphate conjugation result in little of the ultimate carcinogen being produced.

The study of acetylaminofluorene carcinogenesis therefore provides many insights into the factors affecting chemical carcinogenesis. A wealth of other data not discussed here is available which confirms the occurrence of covalent interactions between the ultimate carcinogenic metabolite of acetylaminofluorene and nucleic acids to yield covalent conjugates (figure 7.1). It remains to be seen how these conjugates initiate the process of carcinogenesis.

Clearly the mechanism(s) underlying the carcinogenesis of acetyl-laminofluorene is very complex and the sometimes apparently conflicting and confusing data may reflect the fact that several different metabolic pathways are involved which are more or less predominant in different tissues and in different animal species. It has been suggested for instance that the sulphate ester of *N*-hydroxyacetylaminofluorene is responsible for cytotoxicity, cell death and cell proliferation, and therefore promotion, whereas an *N*-hydroxyaminofluorene metabolite is responsible for initiation. Thus, both pathways would be necessary (figure 7.1).

Benzo[*a*]pyrene

The polycyclic aromatic hydrocarbons constitute a large group of compounds, which includes a number of carcinogens found originally in coal tar but which have since been detected in cigarette smoke, the exhaust fumes from internal combustion engines, and smoke from other processes involving the burning of organic material. Benzo[*a*]pyrene is one of the most intensely studied, as it is an extremely potent carcinogen. Although chemically stable, *in vivo* polycyclic aromatic hydrocarbons undergo a wide variety of metabolic transformations catalysed by the microsomal mixed-function oxidases as illustrated for benzo[*a*]pyrene (figure 7.2). These are mainly hydroxylations occurring at the various available sites on the aromatic rings, and conjugations of the hydroxyl groups with glucuronic acid or sulphate. The majority of these hydroxylation reactions probably proceed through an epoxide intermediate, as discussed in Chapter 4. The prostaglandin synthetase system may also be involved in the production of quinols and phenols via free radicals.

Initially, particular attention was focused on the epoxides of the so-called K region. As in the case of benzo[*a*]pyrene and certain other polycyclic aromatic hydrocarbons, these were more carcinogenic than the parent compound. The **K region** had attracted particular interest, as it is electronically the most reactive portion of the polycyclic aromatic hydrocarbon molecule. However, with other carcinogenic polycyclic aromatic hydrocarbons this was not found to be the case. It now seems that the ultimate carcinogen is an epoxide of a dihydrodiol metabolite where the epoxide is adjacent to the so-called **bay-region** (figure 7.2).

Although a number of the epoxides and **diol epoxides** are mutagenic, the 7,8-dihydrodiol, 9,10-oxide, shown in figure 7.2 is believed to be the ultimate

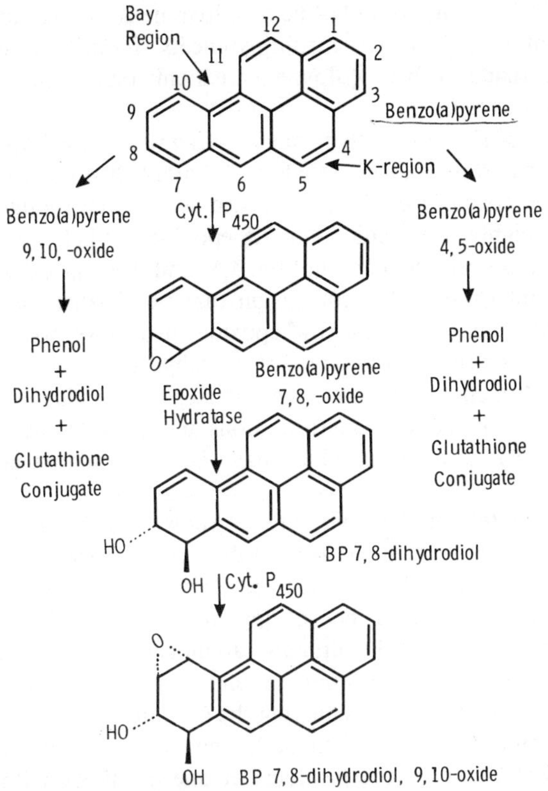

FIGURE 7.2. Part of the metabolism of benzo[*a*]pyrene. The major sites for hydroxylation are the 4, 5, 7, 8 and 9, 10 positions, giving rise to numerous phenols and their conjugates and also to dihydrodiols and glutathione conjugates. Formation of the epoxide of the 7,8-dihydrodiol is thought to be the crucial step in the carcinogenesis.

carcinogen. It should be noted that there are several possible diastereoisomers of this metabolite. However, as the action of epoxide hydrolase yields a *trans* dihydrodiol, and the epoxide ring produced by the action of cytochromes P-450 may be *cis* or *trans*, two **diastereoisomers** are produced metabolically. The diastereoisomer believed to be the ultimate carcinogen is shown in figure 7.2 and the benzylic carbon of the epoxide ring is the electrophilic site. It may be that the cytochromes P-450 system involved in this metabolic activation is present in the nucleus as well as the endoplasmic reticulum, and the epoxide hydrolase necessary may be active in the cytosol as well as in the endoplasmic reticulum.

It is postulated that this ultimate carcinogen reacts covalently with nucleic acids, producing nucleic acid adducts. It has been demonstrated that benzo[*a*]pyrene reacts covalently with nucleic acids *in vitro*, provided that the microsomal enzyme systems necessary for activation are present, and also in whole cell systems. The 7,8-dihydrodiol metabolite of benzo[*a*]pyrene binds more extensively to DNA after microsomal enzyme activation than does benzo[*a*]pyrene or other benzo[*a*]pyrene metabolites, and the nucleoside adducts formed from the 7,8-dihydrodiol of benzo[*a*]pyrene are similar to those obtained from cells in culture exposed to benzo[*a*]pyrene itself. Furthermore, the synthetic 7,8-diol-9,10-epoxides of benzo[*a*]pyrene are highly mutagenic in mammalian as well as in bacterial cells.

Other studies of DNA adducts formed *in vivo* with benzo[*a*]pyrene and using cells in culture have also indicated that the 7,8-diol-9,10-epoxides are responsible for most of the covalent binding to nucleic acids, even though the K region, 4,5-epoxide is highly mutagenic. The 7,8-epoxide for benzo[*a*]pyrene and the 7,8-dihydrodiol are carcinogenic, but the 4,5- and 9,10-epoxides are not. These findings all point towards the conclusion that the further metabolism of the 7,8-epoxide of benzo[*a*]pyrene by epoxide hydrolase to the dihydrodiol metabolite and then further oxidation of the dihydrodiol by the mixed function oxidases to the diol-epoxide are the necessary steps for production of the ultimate carcinogen. These metabolic transformations are illustrated in figure 7.2. Other possible reactive intermediates which have been postulated are radicals and radical cations.

Other metabolites of benzo[*a*]pyrene known to be produced are the **3,6-quinone** and **semiquinone**. These metabolites are cytotoxic and cause DNA strand breaks, and are also mutagenic. They also inhibit the cytochromes P-450-mediated oxidation of benzo[*a*]pyrene and of the dihydrodiol, thus inhibiting the production of the ultimate carcinogen. Reduction and subsequent glucuronidation removes the quinone and semiquinone and consequently decreases their cytotoxicity. By removing the quinone however, glucuronidation increases the mutagenicity of benzo[*a*]pyrene and the covalent binding of metabolites to DNA. Thus the products of one metabolic pathway influences another, and glucuronidation is effectively both a detoxication route and is also responsible for increasing toxicity.

The coplanarity of benzo[*a*]pyrene and many carcinogenic polycyclic hydrocarbons is of interest. It has been suggested that this flat structure allows intercalation of the hydrocarbon within the DNA molecule, thereby facilitating reaction of the intermediate with the nucleic acid.

It is clear that the effects of induction or inhibition of the metabolism will be complex, due to the large number of possible metabolic pathways through which benzo[*a*]pyrene may be metabolized. For instance, the microsomal enzyme inducer 5,6-benzoflavone inhibits the carcinogenicity of benzo[*a*]pyrene to mouse lung and skin, whereas inhibitors such as SKF 525A may increase the tumour production from certain polycyclic hydrocarbons.

Dimethylnitrosamine

Many nitrosamines are potent carcinogens, and dimethylnitrosamine is one of the most intensely studied of this group of compounds. It is a hepatotoxin, a mutagen and a carcinogen and also has an immunosuppressive effect. Single doses cause kidney tumours, whereas low-level chronic exposure results in liver tumours. It may also cause tumours in the stomach, oesophagus and central nervous system. Although dimethylnitrosamine is evenly distributed throughout the mammalian body, acute single doses cause centrilobular hepatic necrosis indicating that metabolism is a factor in this toxicity. The major route of metabolism accounting for about 67% *in vivo* is demethylation to monomethylnitrosamine (figure 7.3). This reaction is catalysed by cytochromes P-450 but may involve more than one step and the second may not be catalysed by cytochromes P-450. There are two isoenzymes of cytochrome P-450 involved in the first oxidation step, a low-affinity and a high-affinity form. The first intermediate, which has been postulated to be a free radical, may either lead to the production of nitrite, methylamine and formaldehyde or *N*-hydroxymethylnitrosamine as shown in figure 7.3. Like many *N*-hydroxymethyl compounds this intermediate rearranges to yield the monomethyl nitrosamine, which then rearranges to give methyldiazohydroxide, then methyldiazonium ion and finally **methyl carbonium ion**. The latter metabolite is a highly reactive **alkylating agent** which will methylate nucleophilic sites on nucleic acids and proteins. Other routes of metabolism have also been postulated.

The acute toxicity is reduced in animals pretreated with the microsomal enzyme inducers 3-methylcholanthrene and phenobarbital, and these pretreatments also protect animals from the carcinogenic effects. Phenobarbital and 3-methylcholanthrene reduce demethylation. However, inhibition of metabolism by certain compounds also protects animals against the toxic and carcinogenic

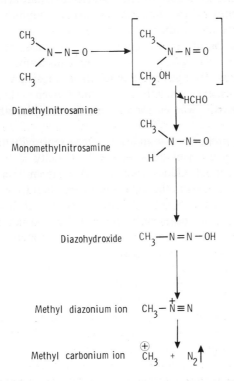

FIGURE 7.3. Metabolism of dimethylnitrosamine to the reactive carbonium ion intermediate responsible for methylation of nucleic acids and thought to be the ultimate carcinogen.

effects of dimethylnitrosamine. These data are consistent with there being other routes of metabolism, some of which may be detoxication pathways, which may explain the protective effects of some microsomal enzyme inducing agents.

Dimethylnitrosamine is a potent methylating agent, and the degree of **methylation of DNA** *in vivo* in various tissues correlates with the susceptibility of those tissues to tumour induction. Thus, metabolism is also important for the carcinogenicity. The major site of DNA methylation is the N^7 position of guanine. However, there is also methylation on the O^6 position of guanine, and this correlates better than N^7 methylation with both the carcinogenicity and mutagenicity of dimethylnitrosamine. This O^6 methylation produces mispairing of guanine with thymidine causing a GC to AT base transition. The ability of the cell to remove this error before cell division is a critical factor in the susceptibility of a particular tissue to the development of tumours.

The methylation of DNA, however, is greater in the liver than in any other organ, including the kidney, after a single dose of dimethylnitrosamine: yet such single doses rarely cause hepatic tumours, but do induce tumours in the kidney. Therefore, it seems that the liver may initially have greater resistance to the carcinogenicity of dimethylnitrosamine, due to protective mechanisms, but that these may be compromised by repeated doses of the carcinogen, thereby allowing tumour induction in the liver after chronic exposure.

Treatment of neonatal animals, or those subjected to partial hepatectomy, with single doses of dimethylnitrosamine, leads to hepatic tumours, as in these cases the liver cells are actively dividing and are more susceptible. It has been found that the liver cells in culture are susceptible at particular stages in the cell cycle.

The metabolic activity of various tissues is also a determinant of susceptibility to tumour production. The gastrointestinal tract is resistant to dimethylnitrosamine carcinogenesis, even when the compound is given orally, as the metabolic activity of this organ is low as far as activation of dimethylnitrosamine is concerned.

The **diet** of the animal may also influence organ sensitivity to this carcinogen. For instance, protein-deficient diets reduce the toxicity of the compound but increase the incidence of kidney tumours. A concomitant increase in the methylation of nucleic acids in the kidney is observed under these conditions. It was concluded from this and other evidence that a protein-deficient diet decreased metabolism of dimethylnitrosamine in the liver, but did not decrease it in the kidney, and so this and other organs were exposed to a higher concentration of unchanged carcinogen in the protein-deficient animals. Dimethylnitrosamine also interferes with the immune system causing a depression of humoral immunity. There is evidence that this is due to a metabolite of dimethylnitrosamine, but does not involve the metabolic pathway(s) responsible for the carcinogenicity or hepatotoxicity.

Vinyl chloride

Vinyl chloride, or vinyl chloride monomer (VCM) as it is commonly known, is the starting point in the manufacture of the ubiquitous plastic polyvinyl chloride (pvc). Vinyl chloride is a gas, normally stored as a liquid under pressure which when heated polymerizes to yield pvc. This plastic was introduced a number of years ago, and there have been many workers exposed or potentially exposed to vinyl chloride during the course of their working lives. However, safety standards in factories and working practices have not always been as rigorous as they are today and were perhaps not always observed. In some cases workers were required to enter reaction vessels periodically to clean them, despite the fact that they still contained substantial amounts of vinyl chloride. This was sufficient for some of the workers to be overcome by solvent narcosis. Thus, the acute poisoning caused elation followed by lethargy, loss of consciousness (at 70 000 p.p.m.), hearing and vision loss (at 16 000 p.p.m.) and vertigo (at 10 000 p.p.m.).

Chronic exposure to vinyl chloride results in 'vinyl chloride disease' which comprises **Raynaud's phenomenon**, skin changes (akin to scleroderma), changes to the bones of the hands (acro-osteolysis) and liver damage in some cases. The bone changes, mainly to the distal phalanges, are due to aseptic necrosis following ischaemic damage due to degeneration and occlusion of small blood vessels and capillaries. The liver may become enlarged, perhaps due to enzyme induction, and also fibrotic although liver function tests are normal. It has been suggested that the vinyl chloride syndrome has an immunological basis, as circulating immune complexes are deposited in vascular epithelium. Also, raised plasma immunoglobulins and complement activation are features. This may result from reactive metabolites forming antigens by reacting with proteins.

Vinyl chloride is also suspected of having effects on the reproductive systems of both males and females. The most severe toxic effect, which probably resulted from repeated high level exposure and was confined to the cleaners of the reaction vessels, was not immediately apparent. This was a type of liver tumour known as **haemangiosarcoma**, which was very rare and was observed only in epidemiological studies of workers in this industry. This type of liver tumour has now also been produced in experimental animals. Fortunately, the hygiene and safety standards applied to working with vinyl chloride are now stricter. However, these changes occurred with the benefit of hindsight and with more foresight the tragedy might have been avoided. Thus, vinyl chloride is now known to be both mutagenic and carcinogenic in experimental animals. Haemangiosarcoma is a tumour of the reticuloendothelial cells (sinusoidal cells), not hepatocytes, and gives rise to tumours of the hepatic vasculature.

Vinyl chloride is readily absorbed through skin and is rapidly eliminated, either unchanged or as metabolites. The metabolism is catalysed by cytochromes P-450, and vinyl chloride induces its own metabolism but also destroys cytochrome P-450. The tumour incidence in rats correlates with the amount metabolized rather than the exposure concentration (table 7.1). Indeed, it seems that the metabolism is saturable, and hence the incidence of tumours reaches a maximum (table 7.1). This indicates that the dose response for tumour incidence is not linear but reaches a maximum at around 20% incidence. However, it means that extrapolation from the maximum tumour incidence to determine a safe concentration is scientifically unsound without a knowledge of the dose–response curve. This example also shows the importance of metabolic data, and illustrates

Table 7.1. Correlation between exposure concentration of vinyl chloride (VC), metabolism and induction of hepatic angiosarcoma in rats.

Exposure concentration (p.p.m of VC)	μgVC/l of air	μgVC metabolized	% liver angiosarcoma
50	128	739	2
250	640	2435	7
500	1280	3413	12
2500	6400	5030	22
6000	15360	5003	22
1 0000	25600	5521	15

Data from Gehring *et al.* (1978) *Toxicol. Appl. Pharmac.*, **44**, 581.

the need for incorporation of such data into risk assessment models.

Mono-oxygenase-mediated metabolism gives rise to an epoxide intermediate and to **chloroacetaldehyde** (figure 7.4). The latter may in fact be derived from the epoxide. Other metabolites detected are *N*-acetyl-*S*-(hydroxyethyl)cysteine, carboxyethyl cysteine, monochloroacetic acid and thiodiglycolic acid.

Both the epoxide and chloroacetaldehyde react with glutathione and may cause depletion of the tripeptide. The epoxide also reacts with DNA and RNA and both metabolites react with protein. The oxirane ring formed by oxidation is highly reactive and is almost certainly the reactive intermediate. Both the epoxide intermediate and chloroacetaldehyde are mutagenic and the epoxide is known to bind to DNA at the N-7 position of deoxyguanosine and also to RNA to give $1,N^6$-ethenoadenosine and $3,N^4$-ethenocytidine.

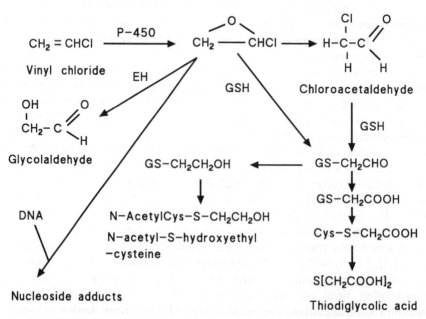

FIGURE 7.4. The metabolism of vinyl chloride. EH: epoxide hydrolase.

The epoxide metabolite of vinyl chloride may be a substrate for epoxide hydrolase which would yield glycolaldehyde. It has been shown that epoxide hydrolase blocks the reaction of vinyl chloride with adenosine, whereas alcohol dehydrogenase blocks binding to protein. In conclusion, it seems that the DNA binding is due to the epoxide, whereas the protein binding is probably due to chloroacetaldehyde. *In vitro* addition of glutathione blocks the binding to protein, but not the binding to DNA. Using isolated rat hepatocytes it was shown that radiolabelled vinyl chloride was metabolized and became bound to protein and DNA, and the binding to DNA could be blocked with epoxide hydrolase. These experiments showed that the epoxide and other metabolites were stable enough to leave the hepatocyte. An epoxide hydrolase inhibitor did not affect binding to protein, and glutathione transferase does not seem to be very important in the metabolism of the epoxide. The epoxide can rearrange to give chloroacetaldehyde which can then bind to protein and glutathione.

DNA adducts isolated from rat liver after chronic exposure to vinyl chloride and adducts with calf liver DNA added to *in vitro* preparations in which chloroethylene oxide or chloroacetaldehyde were formed, were found to be the same. Thus, both etheno-deoxyadenine and etheno-deoxycytidine were detected (figure 7.5). These adducts follow alkylation at N-1 of deoxyadenine and N-3 of deoxycytidine. Similar adducts were detected with RNA. In both cases an imidazole ring is formed by the addition to the nucleotide which is co-planar and shields hydrogen bonding sites. Etheno-deoxycytidine becomes similar to deoxyadenine in configuration and this would result in etheno-deoxycytidine simulating deoxyadenine in base pairing. The imidazole ring resembles an alkyl group and blocks base-pairing sites. Overall the changes might be expected to lead to dA-dT to dC-dG transversions from etheno-deoxyadenine and dC-dG to dA-dT transversions from etheno-deoxycytidine. These findings are all consistent with vinyl chloride, chloroacetaldehyde and chloroethylene oxide causing base-pair substitutions.

The carcinogenicity of vinyl chloride was eventually predictable from the animal studies, but the occupational histories and case histories of the workers were vital in tracking down the cause of the haemangiosarcoma. The other toxic effects may be due, at least in part, to metabolism to toxic, reactive metabolites which react with proteins. The lessons from this example are that safety standards need to be stringent in factories, and that animal studies are important in

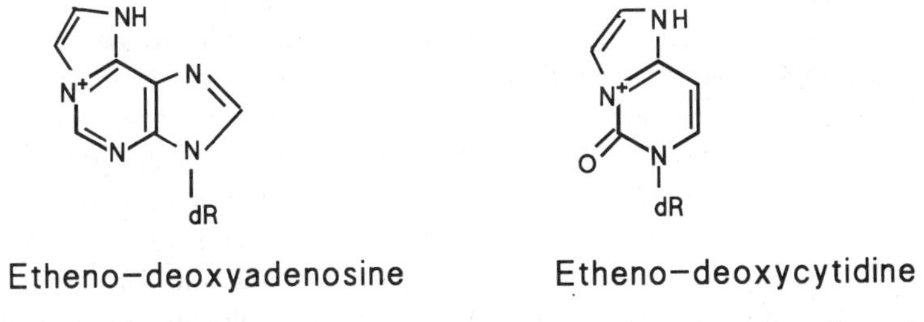

Etheno—deoxyadenosine   Etheno—deoxycytidine

FIGURE 7.5. The structures of the nucleoside adducts etheno-deoxyadenosine and etheno-deoxycytidine. dR = deoxyribose.

l toxicity and highlighting the type of toxic effect that might be
should be known before human exposure occurs. As a
this type of industrial problem legislation is now in force in most
countries which deals specifically with industrial chemicals. For
U.K. all chemicals produced in quantities of greater than 1 tonne
have . go toxicity testing, while strict occupational hygiene limits, known
variously as occupational exposure limits (U.K.) or threshold limit values (TLV)
(U.S.A.) for industrial chemicals are enforced.

Tissue lesions: liver necrosis

Carbon tetrachloride

The hepatotoxicity of carbon tetrachloride has probably been more extensively
studied than that of any other hepatotoxin and there is now a wealth of data
available. Its toxicity has been studied both from the biochemical and
pathological view points and therefore the data available provides particular
insight into mechanisms of toxicity.

Carbon tetrachloride is a simple molecule which, when administered to a
variety of species, causes centrilobular hepatic necrosis and fatty liver. It is a very
lipid-soluble compound and is consequently well distributed throughout the
body, but despite this its major toxic effect is on the liver, irrespective of the route
of administration. It should be noted that it does have other toxic effects, and
there are species and sex differences in toxicity. Chronic administration or
exposure causes liver cirrhosis, liver tumours and also kidney damage. The reason
for the liver being the major target is that the toxicity of carbon tetrachloride is
dependent on metabolic activation by the cytochromes P-450 system. Therefore,
the liver becomes the target as it contains the greatest concentration of
cytochromes P-450, especially in the centrilobular region which is where the
damage is greatest. Low doses of carbon tetrachloride cause only fatty liver and
destruction of cytochromes P-450. Interestingly, a low dose of carbon
tetrachloride protects the animals against the hepatotoxicity of a subsequent
larger dose due to this destruction of cytochromes P-450. The destruction of
cytochromes P-450 occurs particularly in the centrilobular and mid-zonal areas
of the liver. It is also selective for a particular isoenzyme, CYP 1A2 in the rat,
whereas other isoenzymes such as P-450 IA1 are unaffected.

Although carbon tetrachloride was originally thought to be resistant to
metabolic attack, it is now clear that it is metabolized by cytochromes P-450. As
the microsomal enzymes are involved in the metabolic activation, pretreatment
with enzyme inducers and inhibitors, respectively increases and decreases the
toxicity. However, in this instance cytochrome P-450 is functioning in the
reductive mode, catalysing the addition of an electron which then allows
homolytic cleavage and the loss of a chloride ion and the formation of the
trichloromethyl radical (figure 7.6). The resulting trichloromethyl radical may
then undergo one of several reactions. It may abstract a hydrogen atom from a

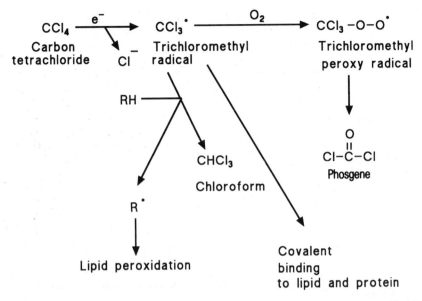

FIGURE 7.6. The microsomal enzyme mediated metabolic activation of carbon tetrachloride to the trichloromethyl radical. This radical may either react with oxygen or abstract a hydrogen atom from a suitable donor (R) to yield a secondary radical, or react covalently with lipid or protein. If R is a polyunsaturated lipid, (R) a lipid radical (R·) is formed which can undergo peroxidation (see figure 6.9).

suitable donor such as the methylene bridges on polyunsaturated fatty acids or a thiol group. This will produce chloroform which is a known metabolite of carbon tetrachloride (figure 7.6) both *in vitro* in the absence of oxygen and *in vivo*. The other product will be a lipid radical or thiol radical depending on the source of the hydrogen atom (figure 7.7; figure 6.10).

The trichloromethyl radical is reactive, but it has been suggested it is insufficiently reactive to account for the toxicity, and the **trichloromethyl peroxy radical** (figure 7.6) has been postulated as an alternative. Whichever is the reactive species, it will damage the immediate vicinity of the cytochromes P-450, including the enzyme itself and the endoplasmic reticulum. Thus, the trichloromethyl free radical covalently binds to microsomal lipid and protein, and reacts directly with membrane phospholipids and cholesterol. The trichloromethyl peroxy radical may produce **phosgene** and electrophilic chlorine. However, various studies have suggested that although the initiating event may be the formation of the trichloromethyl radical, this is not the major cause of damage. The covalent binding to protein occurs in the absence of oxygen, but destruction of cytochromes P-450 and other enzymes of the endoplasmic reticulum requires the presence of oxygen. Rather, the production of a reactive radical metabolite is the start of a *cascade* of events such as lipid peroxidation, thought to be the cause of the damage (figures 6.9 and 7.7). The lipid peroxidation process yields products which may spread the cellular damage beyond the endoplasmic reticulum. Thus, formation of lipid peroxides results in the breakdown of unsaturated lipids to give

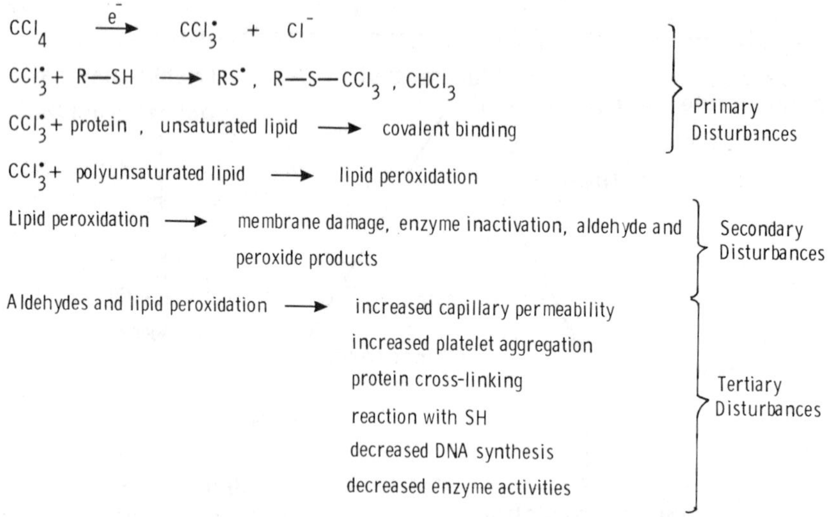

$$CCl_4 \xrightarrow{\ e^- \ } CCl_3^{\bullet} + Cl^-$$

$$CCl_3^{\bullet} + R—SH \longrightarrow RS^{\bullet}, \ R—S—CCl_3 \ , \ CHCl_3$$

$$CCl_3^{\bullet} + protein \ , \ unsaturated \ lipid \longrightarrow covalent \ binding$$

$$CCl_3^{\bullet} + polyunsaturated \ lipid \longrightarrow lipid \ peroxidation$$

Primary Disturbances

Lipid peroxidation $\longrightarrow$ membrane damage, enzyme inactivation, aldehyde and peroxide products

Secondary Disturbances

Aldehydes and lipid peroxidation $\longrightarrow$

increased capillary permeability

increased platelet aggregation

protein cross-linking

reaction with SH

decreased DNA synthesis

decreased enzyme activities

Tertiary Disturbances

FIGURE 7.7. The sequence of cellular events following the metabolism of carbon tetrachloride to a reactive free radical.

carbonyl compounds such as **4-hydroxynonenal** and other hydroxyalkenals. Such compounds are known to have many biochemical effects, such as inhibition of protein synthesis and inhibition of the enzyme glucose-6-phosphatase. 4-Hydroxynonenal will react with both microsomal protein and thiol groups and may inhibit enzymes by the latter process. There is, however, no overall depletion of glutathione, although there is a reduction in certain protein thiol groups which may be related to this. 4-Hydroxynonenal will also increase plasma membrane permeability *in vitro*. All of the effects of this reactive aldehyde are also caused by carbon tetrachloride itself. Perhaps the most significant effect of this product of lipid peroxidation is the reduction in calcium sequestration by the endoplasmic reticulum which leads to a rise in cytosolic free calcium. The calcium pump seems to be very sensitive to lipid peroxidation and substances which interact with thiol groups.

The first events occurring after a toxic dose of carbon tetrachloride can be observed or detected biochemically around the endoplasmic reticulum. Within 1 min of dosing, carbon tetrachloride is covalently bound to microsomal lipid and protein in the ratio of 11:3. **Conjugated dienes**, indicators of lipid peroxidation, can be detected in lipids within 5 min. These changes in lipids may reflect a transient stage in the alteration of the polyunsaturated fatty acids in the endoplasmic reticulum, but they are not dose related. Within 30 min of dosing protein synthesis is depressed reflecting changes in the ribosomes and rough endoplasmic reticulum, and loss of ribosomes can be detected by electron microscopy. Also, the cytochromes P-450 content of the cell and its activity are decreased. Other indicators of damage to the endoplasmic reticulum are inhibition of the enzyme glucose-6-phosphatase and of calcium sequestration.

Between 1 and 3 h after dosing with carbon tetrachloride triglycerides accumulate in hepatocytes, detectable as fat droplets, and there is continued loss of enzyme activity in the endoplasmic reticulum. Cellular calcium accumulates and the rough endoplasmic reticulum becomes vacuolated and sheds ribosomes. The smooth endoplasmic reticulum shows signs of membrane damage and eventually contracts into clumps. At later time-points lysosomal damage may become apparent when the centrilobular cells are damaged, cells may begin to show intracellular structural modifications and eventually the plasma membrane ruptures. It is now generally accepted that although metabolic activation to a reactive radical may be the first event, the production of lipid radicals and subsequent chain reactions leading to reactive products of lipid peroxidation are the important mediators of cellular toxicity. There may be many targets for these reactive products, but interference with calcium homeostasis seems to be one of the most important. Other targets will be membrane lipids, which may fragment during peroxidation, and thiol-containing enzymes. Clearly, there are many biochemical and structural effects occuring during carbon tetrachloride hepatotoxicity, and a single critical event which leads to cellular necrosis may not be easily identified, although the crucial events probably include increased permeability and eventual disruption of the plasma membrane. Fatty liver may have a simpler basis in the inhibition of protein synthesis which is known to result in the decreased production of the lipoprotein complex responsible for the transport of lipids out of the hepatocyte (see Chapter 6).

Paracetamol

Paracetamol (acetaminophen) is a widely used analgesic and antipyretic drug which is relatively safe when taken at prescribed therapeutic doses. However, it has become increasingly common for overdoses of paracetamol to be taken for suicidal intent. In the U.K., for example, around 200 deaths a year result from overdoses of paracetamol. When taken in overdoses, paracetamol causes primarily centrilobular hepatic necrosis, but this may also be accompanied by renal damage and failure.

By measurement of the blood level of paracetamol in overdose cases it is possible to estimate the likely outcome of the poisoning, and hence determine the type of treatment. Measurement of the blood level of paracetamol and its various

Table 7.2. Mean plasma concentration and half-life of unchanged paracetamol in patients with and without paracetamol-induced liver damage.

Patients	Plasma paracetamol half-life (h)	Plasma paracetamol concentrations (μg/ml)	
		4h after ingestion	12h after ingestion
No liver damage (18)	$2 \cdot 9 \pm 0 \cdot 3$	163 ± 20	$29 \cdot 5 \pm 6$
Liver damage (23)	$7 \cdot 2 \pm 0 \cdot 7$	296 ± 26	124 ± 22

Numbers of patients in parentheses.
Data of Prescott and Wright (1973) *Br. J. Pharmac.*, **39**, 602.

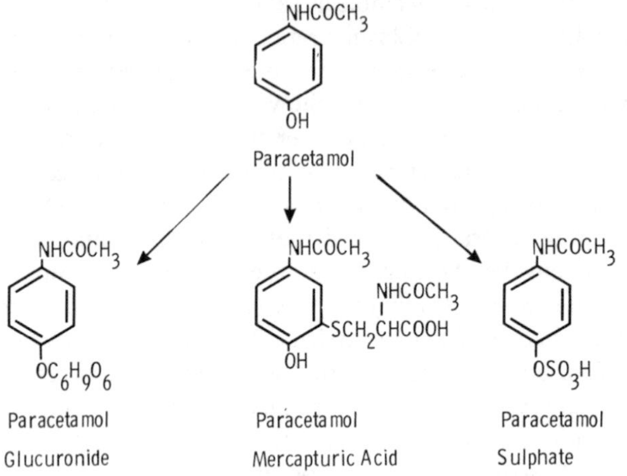

FIGURE 7.8. The major metabolites of paracetamol.

metabolites at various times after the overdose showed that the half-life was increased several-fold (table 7.2), and the patients who sustained liver damage had an impaired ability to metabolize paracetamol to conjugates (figure 7.8).

Paracetamol causes centrilobular hepatic necrosis in various species, although there are substantial species differences in susceptibility (figure 7.9). Experimental animal species susceptible to paracetamol have been used to study the mechanism underlying the hepatotoxicity. It was found that the degree of hepatic necrosis caused by paracetamol was markedly increased by pretreatment of animals with microsomal enzyme inducers and, conversely, reduced in animals given inhibitors of these enzymes (table 7.3). Similarly, humans who are exposed to enzyme-

Table 7.3.   The effect of various pretreatments on the severity of hepatic necrosis from and covalent binding of paracetamol to mouse tissue protein.

Treatment	Dose of Paracetamol (mg/kg)	Severity of liver necrosis[3]	Covalent binding of ^{3}H-Paracetamol (nmol/mg protein)[4]	
			Liver	Muscle
None	375	1–2+	$1\cdot02\pm0\cdot17$	$0\cdot02\pm0\cdot02$
[1]Piperonyl butoxide	375	0	$0\cdot33\pm0\cdot05$	$0\cdot01\pm0\cdot01$
[1]Cobaltous chloride	375	0	$0\cdot39\pm0\cdot11$	$0\cdot01\pm0\cdot03$
α-[1]Naphthylisothiocyanate	375	0	$0\cdot11\pm0\cdot06$	$0\cdot01\pm0\cdot02$
[2]Phenobarbital	375	2–4+	$1\cdot6\pm0\cdot1$	$0\cdot02\pm0\cdot02$

[1]Microsomal enzyme inhibitor; [2]microsomal enzyme inducer.
[3]Data of Mitchell *et al.* (1973) *J. Pharmac. Exp. Ther.*, **187**, 185; [4]Data from Jollow *et al.* (1973) *J. Pharmac. Exp. Ther.*, **187**, 175, and from Mitchell and Jollow (1975) *Gastroenterology*, **68**, 392. Severity of necrosis: $1+ <6\%$ necrotic hepatocytes; $2+ >6\% <25\%$ necrotic hepatocytes; $3+ >25\% <50\%$ necrotic hepatocytes; $4+ >50\%$ necrotic hepatocytes.

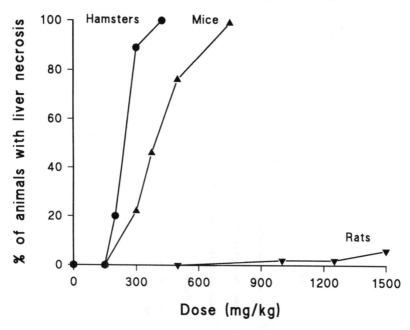

FIGURE 7.9. Species differences in the dose–response relationship for paracetamol-induced liver damage. Data from Potter *et al.*, *Pharmacology* (1974) **12**, 129 and Mitchell *et al.*, *J. Pharmacol. Exp. Ther.* (1973) **187**, 185.

inducing drugs are more susceptible to the hepatotoxic effects of paracetamol, and resistance to these effects has been described in a patient taking the drug **cimetidine** which is an inhibitor of the microsomal enzymes.

In studies in experimental animals there was a lack of correlation between tissue levels of unchanged paracetamol and the necrosis. Furthermore, when radiolabelled paracetamol was administered several important findings were noted:

(a) The radiolabel was bound to liver protein covalently and to a much greater extent than in other tissues such as the muscle (table 7.3).

(b) Autoradiography revealed that the binding was located primarily in the centrilobular areas of the liver which also suffered the damage.

(c) The covalent binding to liver protein was inversely related to the concentration of unchanged paracetamol in the liver.

(d) The covalent binding was increased and decreased by inducers and inhibitors of the microsomal enzymes respectively (table 7.3).

(e) The extent of binding was dose related and increased markedly above the threshold dose for hepatotoxicity (figure 7.10).

(f) The covalent binding to liver protein was time dependent, but that to muscle protein was not.

These findings suggested that a reactive metabolite of paracetamol was

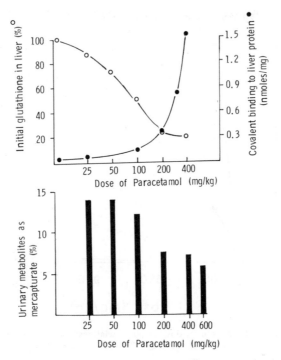

FIGURE 7.10. Relationship between hepatic glutathione, covalent binding of radio-labelled paracetamol to hepatic protein and urinary excretion of paracetamol mercapturic acid after different doses of paracetamol. Adapted from Mitchell *et al.* (1975). In *Handbook of Experimental Pharmacology*, Vol. 28, Part 3, *Concepts in Biochemical Pharmacology*, edited by J. R. Gilette and J. R. Mitchell (Berlin: Springer-Verlag).

responsible for the hepatotoxicity and covalent binding rather than the parent drug.

More recent immunohistochemical studies using anti-paracetamol antibodies have shown that covalent binding of a paracetamol metabolite occurs in the damaged centrilobular regions of human liver after overdoses.

The metabolic profile is straightforward and, as indicated by the urinary metabolites (figure 7.8), mainly involves conjugation with glucuronic acid and sulphate, with a small amount excreted as a mercapturic acid or *N*-acetylcysteine conjugate (about 5% of the dose in humans). None of these metabolites is chemically reactive and therefore not likely to react with hepatic protein. However, the excretion of the mercapturic acid was found to be dose dependent and decreased in experimental animals after heptotoxic doses (figure 7.10). There was also a corresponding depletion of hepatic glutathione to around 20% or less of the normal level (figure 7.10). Thus, there was an apparent relationship between the metabolism, covalent binding to liver protein and hepatic glutathione which resulted in the scheme proposed in figure 7.11. After overdoses of paracetamol the normally adequate detoxication of the reactive metabolite of paracetamol by glutathione was overwhelmed and hepatic glutathione levels were depleted allowing the reactive metabolite to damage the liver cell. Synthesis of new glutathione is inadequate to cope with the rate of depletion. There is thus a marked dose threshold for toxicity which occurs when the hepatic glutathione is depleted by 80% or more of control levels. Studies *in vitro* have established that the reactive intermediate is produced by cytochromes P-450, requires NADPH and molecular oxygen and is inhibited by carbon monoxide. Using ^{18}O it was established that as no ^{18}O was incorporated, an epoxide is not the reactive intermediate. The reactive intermediate is believed to be **N-acetyl-p-benzoquinone imine (NAPQI)**, although the preceding step(s) are not yet clear, *N*-hydroxyparacetamol is not the precursor. However 3-hydroxyparacetamol has also been shown to be a reactive metabolite. Using human liver microsomes it has been found that cytochromes P-450 IIE1 and P-450 IA2 are responsible for the production of a reactive metabolite which binds to microsomal protein; 70% of the covalent binding in the liver is to cysteine residues in hepatic proteins through the 3-position on the benzene ring. Furthermore this binding does not seem to be random and is to specific proteins in both mouse and human liver *in vivo*, and in microsomes and isolated hepatocytes *in vitro*. Some of the binding seems to be similar in both mice and humans. Thus both species show binding of a paracetamol metabolite to various proteins, but especially to a 58 kDa protein. The function of these proteins and the others arylated is currently unknown, but the finding that a particular protein is the target in both species suggests that it may be important in the hepatotoxicity. The role of covalent binding is still unclear as studies *in vitro* in isolated hepatocytes and *in vivo* have shown that certain agents can protect against the cytotoxicity of paracetamol or reduce the

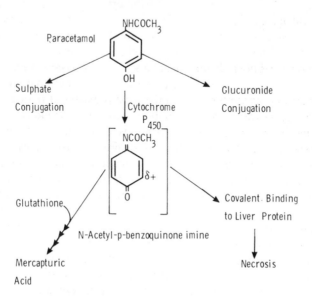

FIGURE 7.11. Proposed metabolic activation of paracetamol to a toxic, reactive intermediate *N*-acetyl-*p*-benzoquinone imine (NAPQI). This can react with glutathione (GSH) to form a conjugate or with tissue proteins. Alternatively NAPQI can be reduced back to paracetamol by glutathione, forming oxidized glutathione (GSSG).

cytotoxicity, despite that fact that substantial covalent binding has occurred. Although the role, if any, of covalent binding to protein in the development of the necrosis is not yet clear, there has been no demonstration of hepatotoxicity without binding.

However, the reactive metabolite will cause other changes as well as binding to protein. Thus, NAPQI will react both chemically and enzymatically with glutathione to form a conjugate, and will also oxidize it to GSSG and in turn be reduced back to paracetamol. This cyclical process may explain the extensive depletion of glutathione which occurs. NADPH will also reduce NAPQI and in turn be oxidized to NADP, although reduction via glutathione is probably preferential. NADPH oxidation may also result from GSSG reduction via GSH peroxidase (figure 6.10).

Analogues of paracetamol which are unable to undergo covalent binding to protein are still hepatotoxic and can undergo a redox reaction with glutathione. However, oxidative stress has not been demonstrated *in vivo* and there are differences between *in vivo* data and that obtained in isolated hepatocytes.

The depletion of glutathione and NADPH will allow the oxidation of protein sulphydryl groups, which is thought to be an important step in the toxicity. Thus, glutathione and protein sulphydryl groups, such as those on Ca^{2+} transporting proteins, are important for the maintenance of intracellular calcium homeostasis. Thus, paracetamol and NAPQI cause an increase in cytosolic calcium and paracetamol inhibits the Na^+/K^+ ATPase pump in isolated hepatocytes.

As well as the formation of NAPQI there are various other possible metabolic pathways, including deacetylation and radical formation, which may or may not play a role in the hepatotoxicity. The importance of and interrelationships between covalent binding to particular hepatic proteins, cyclical oxidation and reduction of glutathione, oxidation of protein thiol groups and the intracellular calcium level are currently unclear. However, these events are not mutually exclusive and so it is possible that all are a series of necessary events occurring at particular stages in the development of paracetamol hepatotoxicity.

The dramatic species differences in the hepatotoxicity of paracetamol (figure 7.9) are due to the differences in metabolic activation, as indicated by *in vitro* studies using liver microsomal preparations and hepatocytes from various species. The activity of these microsomal preparations, using covalent binding to protein as an end-point, approximately correlated with the *in vivo* toxicity. The sensitivity of the rat to paracetamol is increased by pretreatment with phenobarbital, as it increases the activity of the cytochromes P-450 system responsible for the metabolic activation to NAPQI, and therefore more is metabolized by this pathway. This is also the case in humans, whereas in the hamster phenobarbital pretreatment decreases the hepatotoxicity as a result of an increase in glucuronic acid conjugation, and hence detoxication, due to induction of glucuronosyl tranferase. Pretreatment of hamsters with a polycyclic hydrocarbon such as 3-methylcholanthrene, however, increases the toxicity. Another factor which will increase the toxicity is a reduction in detoxication pathways such as by depletion of glutathione with compounds such as diethyl maleate (figure 4.60), and inhibition of sulphate and glucuronic acid conjugation by treatment with salicylamide. The latter treatment diverts more paracetamol through the toxic pathway.

Studies in human subjects have revealed that there is considerable variation between individuals in their rate of metabolism of paracetamol via the toxic, oxidative pathway. Thus, at one end of the frequency distribution some individuals, approximately 5% of the particular population studied, metabolized paracetamol at a rate similar to hamsters and mice. Other individuals at the opposite end of the frequency distribution have rates of oxidation about one-quarter of the highest rates, and these individuals are probably more akin to rats, the resistant species.

Although the investigation of the mechanism underlying paracetamol hepatotoxicity has been of intrinsic toxicological interest, there has also been a particularly significant benefit that has arisen from this work. This is the development of an **antidote** which is now successfully used for the treatment of paracetamol overdose. The antidote now most commonly used is *N*-**acetylcysteine**, although methionine is also used in some cases as it can be given orally. There are various mechanisms by which *N*-acetylcysteine may act:

(a) It promotes the synthesis of glutathione utilized in the conjugation of the reactive metabolite, NAPQI.
(b) It stimulates the synthesis of glutathione used in the protection of protein thiols.
(c) It may relieve the saturation of sulphate conjugation which occurs during paracetamol overdose.
(d) It may itself be involved in the reduction of NAPQI.

Although originally thought to be effective *only* when administered up to 10–12 h after an overdose, significantly *N*-acetylcysteine may be effective when given 15 h or more after an overdose. Clearly the majority of the metabolic activation and covalent binding to protein will have taken place by this time, and therefore it is unlikely that it will be functioning by reacting with the NAPQI. Therefore it has been proposed that *N*-acetylcysteine may also protect the liver cells against the subsequent changes which occur after the reaction of the reactive metabolite with cellular constituents. **Methionine**, however, is not effective when given at such later times, possibly because synthesis of glutathione from methionine is inhibited or compromised by paracetamol toxicity. This is possibly due to inhibition of the thiol-containing enzymes which are involved in the synthetic pathway.

Bromobenzene

Bromobenzene is a toxic industrial solvent which causes centrilobular hepatic necrosis in experimental animals. It may also cause renal damage and bronchiolar necrosis. The study of the mechanism underlying the hepatotoxicity of bromobenzene has been of particular importance in leading to a greater understanding of the role of glutathione and metabolic activation in toxicity.

The metabolism of bromobenzene is complex, but the pathway thought to be responsible for the hepatotoxicity is that leading to 4-bromophenol, postulated to proceed via the 3,4-oxide (figure 7.12), a pathway catalysed by the microsomal mono-oxygenase system. An alternative metabolic pathway is the formation of 2-bromophenol, via the 2,3-oxide. The use of strains of mice with different degrees of microsomal mono-oxygenase activity towards epoxidation at the 2,3- and 3,4-positions showed a correlation between the activity for 3,4-epoxidation and centrilobular necrosis. It was also observed that radiolabelled bromobenzene became covalently bound to liver protein in the necrotic tissue of the centrilobular area. One of the major urinary metabolites of bromobenzene is a mercapturic acid in which *N*-acetylcysteine is conjugated on the 4-position (figure 7.12). This results from an initial conjugation of the 3,4-oxide, a reactive metabolite, with glutathione, followed by enzymatic removal of the glutamyl and glycinyl residues. It was observed that pretreatment of rats with the microsomal enzyme inducer phenobarbital, resulted in greater hepatotoxicity, increased excretion of the glutathione conjugate from the 3,4-oxide and greater covalent binding to liver

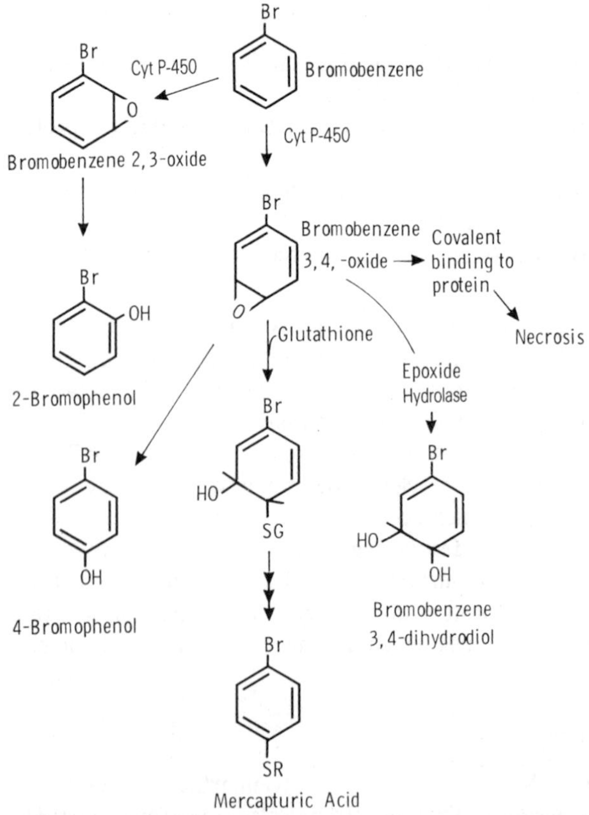

FIGURE 7.12. Metabolism of bromobenzene. The bromobenzene 2,3-oxide and 3,4-oxide may undergo chemical rearrangement to the 2- and 4-bromophenol respectively. Bromobenzene 3,4-oxide may also be conjugated with glutathione and in its absence react with tissue proteins. An alternative, detoxication pathway is hydration to the 3,4-dihydrodiol via epoxide hydrolase.

protein. Inhibitors of the microsomal enzymes decreased hepatotoxicity and covalent binding to liver protein.

Investigation of the level of glutathione in the liver of animals revealed that bromobenzene caused a dose-dependent depletion of the tripeptide which coincided with a decrease in the excretion of the mercapturic acid and an increase in covalent binding (figure 7.13). These observations culminated in the hypothesis that glutathione protects the hepatocyte against the reactive metabolite, bromobenzene 3,4-oxide by conjugating with it, chemically, enzymatically or both. After an hepatotoxic dose, however, there is sufficient of the reactive metabolite to deplete the available glutathione from the liver. The reactive metabolite is therefore not detoxified by conjugation with glutathione, and is able to react with cellular macromolecules such as protein. Studies *in vitro* revealed that the addition of glutathione to the incubation mixture reduced the amount of covalent binding to protein, and the production of hydroxylated metabolites, yet the overall amount of metabolism remained the same. More recently it has been shown *in vitro* that these reactive epoxides react at different rates with different proteins. Thus the 3,4-oxide reacts with histidine groups on microsomal protein more avidly than the 2,3-oxide which preferentially reacts with cysteine residues

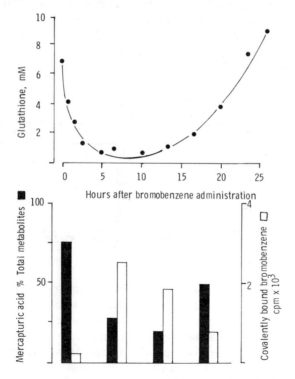

FIGURE 7.13. Relationship between hepatic glutathione, covalent binding of radio-labelled bromobenzene to hepatic protein and urinary excretion of bromophenyl mercapturic acid in rats. Animals were given bromobenzene (10 mmol/kg, i.p.) and then radiolabelled bromobenzene at various times thereafter. Adapted from Gillette (1973) *5th Int. Congr. Pharmacology*, Vol. 2 (Basel: Karger).

on haemoglobin. However, although the 2,3-oxide appears to be chemically more stable than the 3,4-oxide, the latter has been detected in the blood *in vivo* and in the incubation medium of hepatocytes *in vitro* after exposure to bromobenzene. This data means that the stability of the bromobenzene 3,4-oxide would allow it to leave the endoplasmic reticulum where it is formed and react with macromolecules in various parts of the cell. In confirmation of this are data which indicate that the conjugation with glutathione is catalysed by a cytosolic glutathione transferase. This apparent stability perhaps seems at odds with the concept of chemically reactive metabolites, but there may be an *optimal* reactivity for toxic metabolites; not too great to result in indiscriminate reactions with macromolecules, yet sufficient to react with and modify certain proteins.

As well as detoxication via reaction with glutathione, the reactive 3,4-epoxide can be removed by hydration to form the dihydrodiol, a reaction which is catalysed by epoxide hydrolase (also known as epoxide hydratase). This enzyme is induced by pretreatment of animals with the polycyclic hydrocarbon 3-methylcholanthrene, as can be seen from the increased excretion of 4-bromophenyldihydrodiol (table 7.4). This induction of a detoxication pathway offers a partial explanation for the decreased hepatotoxicity of bromobenzene observed in such animals. A further explanation, also apparent from the urinary metabolites, is the induction of the form of cytochrome P-450 which catalyses the formation of the 2,3-epoxide. This potentially reactive metabolite readily rearranges to 2-bromophenol, and hence there is increased excretion of 2-bromophenol in these pretreated animals (table 7.4).

Table 7.4.   Effect of 3-methylcholanthrene (3-MC) pretreatment on the urinary metabolites of bromobenzene.

Metabolites	% Total urinary metabolites	
	Untreated	3-MC treated
4-Bromophenylmercapturic acid	72	31
4-Bromophenol	14	20
4-Bromocatechol	6	10
4-Bromophenyldihydrodiol	3	17
2-Bromophenol	4	21

Data from Zampaglione *et al.* (1973) *J. Pharmac. Exp. Ther.*, **187**, 218.
Dose of bromobenzene 10 mmol/kg to rats.

The metabolites 2- and 4-bromophenol can also be metabolized by a further oxidation pathway to yield catechols and quinones, some of which are cytotoxic and potentially hepatotoxic. Thus, *in vitro* studies have indicated that bromoquinones and bromocatechols may be responsible for some of the covalent binding to protein and reaction with glutathione. However, administration of primary phenolic metabolites does not cause hepatotoxicity. At least seven glutathione conjugates have been identified as metabolites of bromobenzene and its primary phenolic metabolites.

The mechanism of hepatotoxicity is therefore currently unclear. It has been suggested that lipid peroxidation is responsible rather than covalent binding to protein. Arylation of other low molecular weight nucleophiles such as coenzyme A and pyridine nucleotides also occurs and may be involved in the toxicity. Bromobenzene is known to cause the inhibition or inactivation of enzymes containing SH groups. It also causes increased breakdown of phospholipids and inhibits enzymes involved in phospholipid synthesis. Arylation of sites on plasma membranes which are crucial for their function may also occur. However, it seems likely that the toxicity is a mixture of such events.

Bromobenzene is also **nephrotoxic,** possibly due to the production of a reactive metabolite and covalent binding to protein. It has been suggested that the reactive metabolite is formed in the liver and then transported to the kidney. 2-Bromophenol and 2-bromohydroquinone are both nephrotoxic. It seems that the ultimate toxic metabolite is a **diglutathionyl conjugate** formed from **2-bromohydroquinone** (figure 7.14) as this will cause the same lesions in the kidney as bromobenzene, 2-bromophenol and 2-bromohydroquinone when administered to animals.

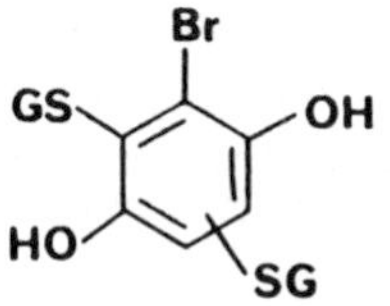

FIGURE 7.14. Diglutathionyl conjugate formed from 2-bromohydroquinone.

Thus, bromobenzene hepatotoxicity is probably the result of metabolic activation to a reactive metabolite which covalently binds to protein and other macromolecules and other cellular molecules. It may also stimulate lipid peroxidation and biochemical effects, such as the inhibition of SH-containing enzymes, may also play a part.

Isoniazid and iproniazid

Isoniazid and iproniazid are chemically similar drugs having different pharmacological effects; they may both cause liver damage after therapeutic doses are given. Isoniazid is still widely used for the treatment of tuberculosis, but iproniazid is now rarely used as an antidepressant.

The major routes of metabolism for isoniazid are acetylation to give acetylisoniazid, followed by hydrolysis to yield isonicotinic acid and acetylhydrazine (figure 7.15). The acetylation of isoniazid in human populations is genetically determined and therefore shows a bimodal distribution (see page 150). Thus, there are two acetylator phenotypes, termed rapid and slow acetylators, which may be distinguished by the amount of acetylisoniazid excreted or by the plasma half-life of isoniazid.

Mild **hepatic dysfunction,** detected as an elevation in serum transaminases, is now well recognized as an adverse effect of isoniazid and occurs in 10–20% of patients. Possibly as many as 1% of these cases progress to severe hepatic damage

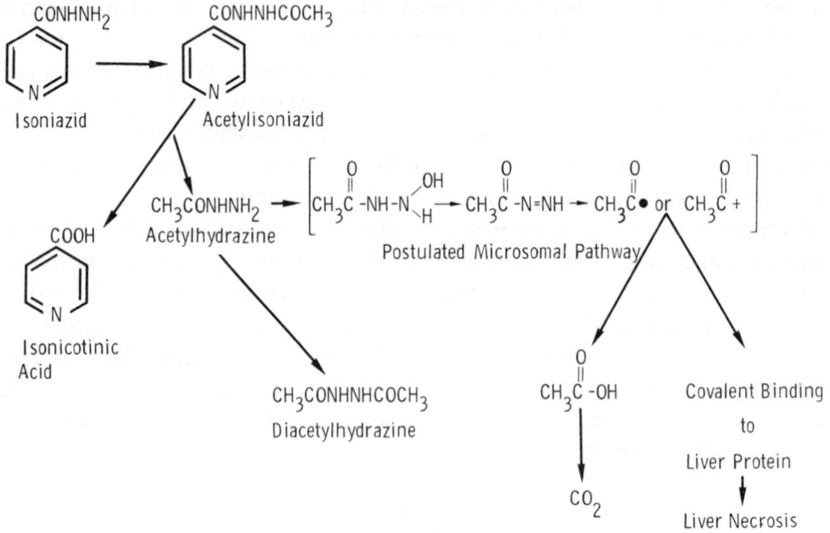

CONHNH$_2$
(Isoniazid) → CONHNHCOCH$_3$ (Acetylisoniazid)

Isoniazid

Acetylisoniazid

COOH
Isonicotinic Acid

CH$_3$CONHNH$_2$
Acetylhydrazine

$$\left[CH_3\overset{O}{\underset{\|}{C}}-NH-N\overset{OH}{\underset{H}{<}} \rightarrow CH_3\overset{O}{\underset{\|}{C}}-N=NH \rightarrow CH_3\overset{O}{\underset{\|}{C}}\bullet \text{ or } CH_3\overset{O}{\underset{\|}{C}}+ \right]$$

Postulated Microsomal Pathway

CH$_3$CONHNHCOCH$_3$
Diacetylhydrazine

CH$_3\overset{O}{\underset{\|}{C}}$-OH

CO$_2$

Covalent Binding
to
Liver Protein
↓
Liver Necrosis

FIGURE 7.15. Metabolism of isoniazid. The acetylhydrazine released by the hydrolysis of acetylisoniazid is further metabolized to a reactive intermediate thought to be responsible for the hepatotoxicity.

and it has been suggested that this latter, more severe form, of hepatotoxicity may have a different underlying mechanism. However, the greater incidence of hepatotoxicity reported in rapid acetylators has since been questioned. It seems that the incidence of the mild form of isoniazid hepatotoxicity is not related to the acetylator phenotype, but the incidence of the rarer, more severe form is more common in slow acetylators. However, initial suggestions that patients with the rapid acetylator phenotype (see page 150) were more susceptible to such damage, prompted a study of the relationship between metabolism and toxicity.

Animal studies revealed that both acetylisoniazid and acetylhydrazine were hepatotoxic, causing centrilobular necrosis in phenobarbital-pretreated animals. Conversely, pretreatment of animals with microsomal enzyme inhibitors reduced necrosis. Inhibition of the hydrolysis of acetylisoniazid (figure 7.15) with **bis-*p*-nitrophenylphosphate** (figure 5.32), an acyl amidase inhibitor, reduced its hepatotoxicity, but not that of acetylhydrazine. Further studies revealed that metabolism of acetylhydrazine by the microsomal mono-oxygenases was responsible for the hepatotoxicity, resulting in the covalent binding of the acetyl group to liver protein. This data indicated that **acetylhydrazine** was the toxic metabolite produced from acetylisoniazid, and therefore isoniazid. Acetylhydrazine required metabolic activation via the microsomal enzymes to a reactive acylating intermediate which would react with protein (table 7.5).

The proposed activation of acetylhydrazine involves *N*-hydroxylation, followed by loss of water to yield **acetyldiazine**, an intermediate which would fragment to yield acetyl **radical** or acetyl **carbonium ion** (figure 7.15). Glutathione was not depleted by hepatotoxic doses of acetylhydrazine, indicating that unlike bromobenzene or paracetamol toxicity, it does not have a direct protective role.

Table 7.5.   Effect of various pretreatments on the hepatotoxicity, covalent binding and metabolism to CO_2 of acetylisoniazid and acetylhydrazine.

	Acetylisoniazid (200 mg/kg)			Acetylhydrazine (20 mg/kg)		
Pretreatment	[4]Necrosis	[5]Covalent binding (nmol/mg protein)	[5] [14]CO_2 (% Dose)	[4]Necrosis (30 mg/kg)	[5]Covalent binding (nmol/mg protein)	[5] [14]CO_2 (% Dose)
None	0–+	0·20	10	0–+	0·15	29
[1]Phenobarbital	+ +	0·31	12	+ + +	0·19	35
Phenobarbital + [2]Cobalt chloride	0	0·15	4	+	0·09	22
Phenobarbital + [3]bis-*p*-phosphate-nitrophenyl	0	0·11	4	+ + +	0·23	37

[1]Microsomal enzyme inducer; [2]microsomal enzyme inhibitor; [3]acylamidase inhibitor.
[4]Data from Mitchell *et al.* (1976) *Ann. Intern. Med.*, **84**, 181.
[5]Data from Timbrell *et al.* (1980) *J. Pharmac. Exp. Ther.*, **213**, 364.
Severity of necrosis: 1+ <6% necrotic hepatocytes; 2+ >6% <25% necrotic hepatocytes; 3+ >25% <50% necrotic hepatocytes; 4+ >50% necrotic hepatocytes.

The role of acetylhydrazine in isoniazid hepatotoxicity is complex, as the production of the toxic intermediate involves several steps. Further study has indicated that acetylhydrazine is detoxified by further acetylation to diacetlhydrazine (figure 7.15) and that isoniazid may interact with acetylhydrazine metabolism. It is therefore clear that the relative rates of production, detoxication and activation of acetylhydrazine are very important determinants of the hepatotoxicity of isoniazid.

The acetylation of isoniazid and acetylhydrazine are both subject to genetic variability, so the production and detoxication pathways are both influenced by the acetylator phenotype. The hepatotoxicity of isoniazid is therefore dependent on a complex interrelationship between genetic factors and individual variation in the pharmacokinetics of isoniazid. Studies of the pharmacokinetics of acetylhydrazine in human subjects after a dose of isoniazid have revealed that although the peak plasma concentration of acetylhydrazine is higher in rapid acetylators, when the plasma concentration of acetylhydrazine is plotted against time, the area under the curve is greater in slow acetylators. Thus, the overall exposure to acetylhydrazine is greater in the slow acetylators. Another possible toxic metabolite is **hydrazine** itself which causes fatty liver and may cause liver necrosis in some animal species. This metabolite arises by metabolic hydrolysis of isoniazid, but the amount of isoniazid which is metabolized by this route is difficult to estimate as the other product of the hydrolysis, isonicotinic acid, can also arise from acetylisoniazid (figure 7.15). Hydrazine has been detected in the plasma of human subjects treated with isoniazid and was found to accumulate in slow, but not rapid, acetylators. The role of hydrazine in isoniazid hepatotoxicity, if any, remains to be clarified.

Iproniazid hepatotoxicity has a similar basis to that of isoniazid toxicity. Hydrolysis of iproniazid yields **isopropylhydrazine** (figure 7.16) which is extremely hepatotoxic in experimental animals. Isopropylhydrazine is metabolically activated by the microsomal enzymes to a reactive alkylating species which covalently binds to liver protein. One of the products of this metabolic route has been shown to be propane gas, the presence of which indicates that the isopropyl **radical** may be the reactive intermediate. This was further suggested by double labelling experiments, which showed that the whole isopropyl moiety is bound to protein without loss of hydrogen atoms (figure 7.16). Sulphydryl compounds reduce the covalent binding to protein *in vitro*, and S-isopropyl conjugates have been isolated from preparations *in vitro*, which confirms the role of the isopropyl group as an alkylating species. However, sulphydryl compounds are not depleted *in vivo* by hepatotoxic doses of isopropylhydrazine.

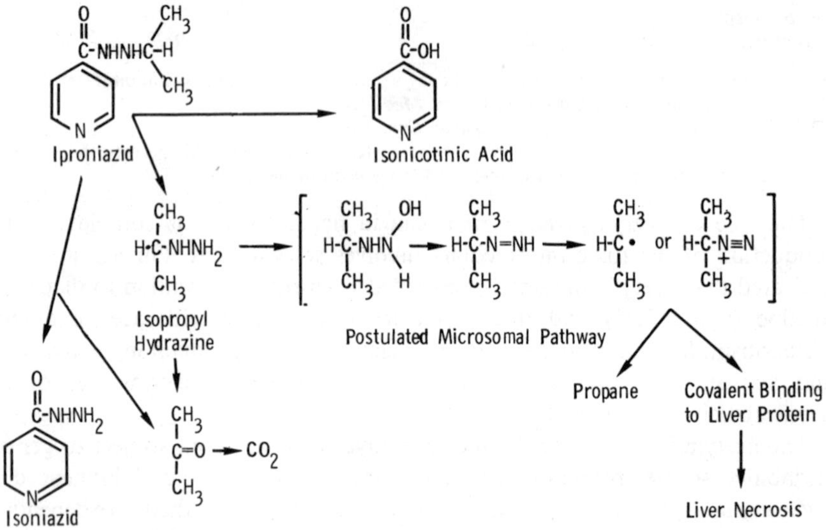

FIGURE 7.16. Metabolism of iproniazid. The isopropylhydrazine moiety released by hydrolysis is further metabolized to a reactive intermediate thought to be responsible for the hepatotoxicity.

Thus the hepatotoxicity of iproniazid and isoniazid may involve the alkylation or acylation of tissue proteins and other macromolecules in the liver. How these covalent interactions lead to the observed hepatocellular necrosis is at present not understood.

Tissue lesions: kidney damage

Chloroform

Unlike the halogenated hydrocarbons discussed below, chloroform is nephrotoxic following metabolic activation via the microsomal enzyme system. Chloroform is also hepatotoxic and again this involves cytochromes P-450-mediated activation, although in male mice the kidney is susceptible at doses which are not hepatotoxic. There are considerable species and strain differences in susceptibility to chloroform-induced nephrotoxicity, and males are generally more susceptible than females. Thus, in male mice chloroform causes primarily renal damage, while in female mice it causes hepatic necrosis. The renal damage caused by chloroform is necrosis of the proximal tubular epithelium which may be accompanied by fatty infiltration, swelling of tubular epithelium and casts in the tubular lumen. The damage can be detected biochemically by the presence of protein and glucose, a decrease in the excretion of organic ions and increased blood urea (BUN).

The mechanism is believed to involve metabolic activation in the kidney itself. Thus, when radiolabelled chloroform was given to mice, in the kidney the radiolabel was localized in the tubular cells which were necrotic. Certain microsomal enzyme inducers such as 3-methylcholanthrene decreased the nephrotoxicity but not hepatotoxicity of chloroform, and phenobarbital pretreatment had no effect on nephrotoxicity but increased hepatotoxicity. Pretreatment with polybrominated biphenyls, however, increased toxicity to both target organs and also increased mixed function oxidase activity in both. *In vitro* studies have shown that microsomal enzyme-mediated metabolism of chloroform to CO_2 occurs in renal tissue, and is associated with cytotoxicity and covalent binding of metabolites to microsomal protein. The sex difference is maintained *in vitro* with cortical slices from male mice showing a toxic response and metabolizing chloroform to a reactive metabolite which binds to protein, but tissue from female mice was neither susceptible nor did it metabolize chloroform. These findings are consistent with the lower level of cytochromes P-450 in the kidneys of female mice when compared with males. The finding that **deuterated chloroform, CDCl3,** was less nephrotoxic than the hydrogen analogue, indicated that cleavage of the C-H bond was involved in the metabolism, was a rate

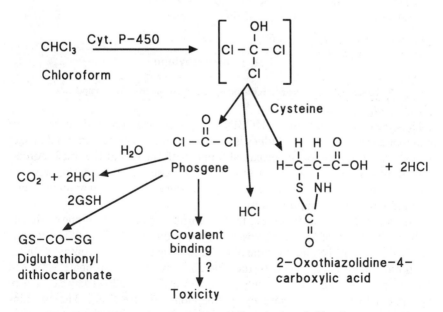

FIGURE 7.17. The proposed metabolic activation of chloroform.

limiting step and was necessary for the nephrotoxicity. Studies using rabbits indicated that phosgene was produced via a cytochrome P-450-mediated reaction. Addition of cysteine reduced covalent binding and yielded a conjugate, 2-oxothiazolidine-4-carboxylic acid, and in rabbit renal microsomes this was increased by prior induction with phenobarbital. This data, in conjunction with studies on the hepatotoxicity of chloroform, suggested that oxidative metabolism to phosgene was responsible for the nephrotoxicity in mice and rabbits (figure 7.17).

Haloalkanes and alkenes

A variety of halogenated alkanes and alkenes such as hexachlorobutadiene, **chlorotrifluoroethylene, tetrafluoroethylene** and trichloroethylene (figure 7.18) are nephrotoxic. Studies have shown that metabolic activation is necessary for toxicity, but this does not involve cytochromes P-450.

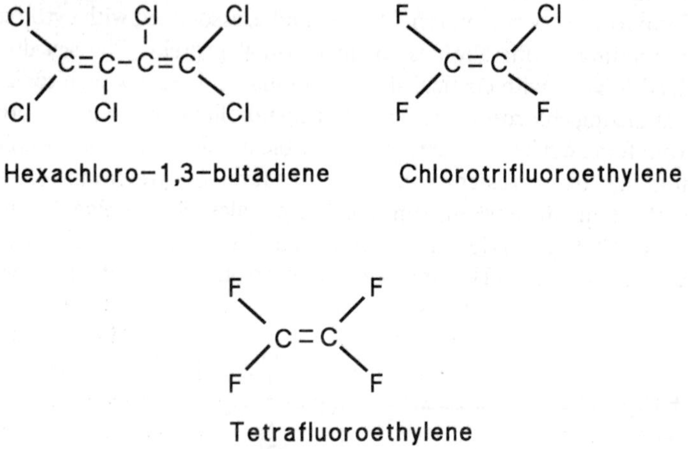

Hexachloro–1,3–butadiene   Chlorotrifluoroethylene

Tetrafluoroethylene

FIGURE 7.18. The structures of some nephrotoxic halogenated compounds.

Thus, **hexachlorobutadiene (HCBD)** is a potent nephrotoxin in a variety of mammalian species, and the kidney is the major target. The compound damages the pars recta portion of the proximal tubule with the loss of the brush border. The result is renal failure detected as glycosuria, proteinuria, loss of concentrating ability and reduction in the clearance of inulin, p-aminohippuric acid and tetraethylammonium ion.

The mechanism seems to involve first the formation of a glutathione conjugate in the liver, catalysed by the microsomal glutathione S-transferase. The hepatic metabolite is then secreted into the bile and thereby comes into contact with the gut flora. There is therefore further metabolism of the conjugate with first cleavage of the glutamyl moiety by γ-glutamyltransferase followed by cleavage of the cysteinyl-glycine conjugate by a dipeptidase (figure 7.19). The resulting cysteine conjugate can then be absorbed from the gastrointestinal tract. In the

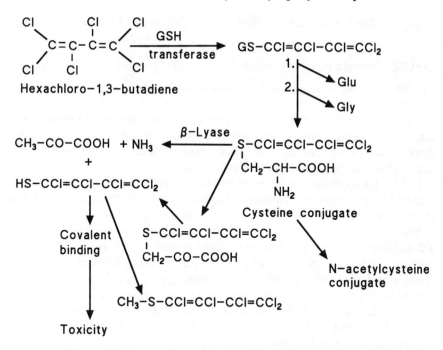

FIGURE 7.19. The metabolic activation of hexachlorobutadiene. The glutathione conjugate is degraded to a cysteine conjugate in two stages involving the action of γ-glutamyltransferase (1) and cysteinyl glycinase (2).

rat, in which the glutathione–HCBD conjugate is detectable in bile, cannulation of the bile duct prevents the nephrotoxicity of HCBD. Thus, **enterohepatic recirculation** may play an important role in the toxicity. The cysteine conjugate is then absorbed from the gastrointestinal tract and transported via the blood to the kidney where it can either undergo acetylation and be excreted as a mercapturic acid, or further metabolized by the enzyme cysteine conjugate β-lyase **(C–S lyase)** to a reactive thiol. The other products of this reaction are a keto acid and ammonia. β-Lyase is present in the liver and kidney. In the kidney it occurs in the cytosol, mitochondria and brush border membranes, especially in the pars recta section. The reactive alkylating fragment binds to protein, DNA and glutathione. The sensitivity of the kidney may be due to its concentrating ability and the active uptake processes. Thus, the nephrotoxicity and covalent binding of metabolites to renal protein can be reduced by treating animals with the organic acid probenecid which competes with the cysteine conjugate for the active uptake system. The cytotoxicity of the glutathione conjugate of HCBD *in vitro* is reduced by inhibitors of γ-glutamyltransferase and β-lyase indicating that both these enzymes are essential for the toxicity.

Trichloroethylene is metabolized similarly and gives rise to dichlorovinyl cysteine. It has been found that *S*-(1,2-**dichlorovinyl)-L-cysteine (DCVC)** and *S*-(2-**chloroethyl)-DL-cysteine (CEC)** (figure 7.20) are both nephrotoxic when administered to animals causing renal proximal tubular necrosis. *S*-(2-chloro-

$$CHCl= CCl-S-CH_2-CH-COOH$$
$$NH_2$$

S-(1,2-dichlorovinyl)-cysteine

$$CHClCH_2-S-CH_2-CH-COOH$$
$$NH_2$$

S-(2-chloroethyl)-cysteine

FIGURE 7.20. The structures of *S*-(1,2-dichlorovinyl)-L-cysteine and *S*-(2-chloroethyl)-DL-cysteine.

ethyl)-DL-cysteine does not require β-lyase activation in order to be nephrotoxic, but can rearrange, possibly to a reactive episulphonium ion, by nucleophilic displacement of the chlorine atom. These compounds decrease the activity of the renal tubular anion and cation transport system.

If the β-lyase enzyme, or the renal basolateral membrane transport system, or γ-glutamyltransferase or cysteinylglycinase are inhibited, the nephrotoxicity of DCVC can be reduced, indicating that each of these processes is involved.

The toxic thiol produced by the action of β-lyase on DCVC is **1,2-dichlorovinylthiol**. This may be further metabolized or may rearrange to give reactive products such as thioacetyl chloride or chlorothioketene. These are reactive and unstable alkylating or acylating agents, and hydrogen sulphide, which is toxic, has also been shown to be formed. The conjugate of hexachlorobutadiene, DCVC and CEC all inhibit mitochondrial respiration *in vitro* and this may be the ultimate target in the kidney. Damage to DNA also occurs and this could underlie the toxicity.

Acetylation of DCVC to yield the *N*-acetylcysteine conjugate does not give nephrotoxic metabolites, as these conjugates are not substrates for β-lyase. Deacetylation of the *N*-acetylcysteine conjugates, however, may occur and give rise to the nephrotoxic cysteine conjugate. Similarly analogues such as *S*-(1,2-dichlorovinyl)-DL-α-methylcysteine which cannot undergo β-elimination are not nephrotoxic.

Thus, these similar examples illustrate that **glutathione conjugation** may not always lead to a reduction in toxicity. Furthermore, they illustrate that several factors are responsible for the kidney being the target organ: active uptake processes, the presence of β-lyase and γ-glutamyltransferase and possibly the importance of mitochondria in the brush border membranes which are the ultimate target.

Antibiotics

Aminoglycosides

These drugs are an important cause of renal failure in humans. Chemically they are carbohydrates with glycosidic linkages to side-chains containing various numbers of amino groups (figure 7.21). **Gentamycin** for example has five amino groups. They are therefore cationic, and there is some correlation between the number of ionizable amino groups and the nephrotoxicity. Aminoglycosides are excreted by glomerular filtration which is followed by active reuptake via a high-capacity, low-affinity transport system. The aminoglycosides remain in the kidney for longer than in other exposed tissues such as the liver. Thus, with gentamycin the time for the tissue concentration to be reduced by half is five to six days. This suggests that gentamycin is binding to tissue components and will accumulate with repeated dosing. Gentamycin achieves a concentration in the renal cortex of two to five times that in the blood. It is believed that the cationic aminoglycosides bind to anionic phospholipids, especially phosphoinositides, on the brush border of the proximal tubule. The bound aminoglycoside is then engulfed by endocytosis and stored in secondary lysosomes. The gentamycin thus achieves a high concentration in the proximal tubular cells, and it is these cells which are damaged and suffer necrosis. The binding of the aminoglycoside leads to overall impairment of the phosphatidyl inositol cascade and to inhibition of phospholipid breakdown due to effects on phospholipase action. The binding depends on the number and position of charged groups on the aminoglycoside. There is altered membrane phospholipid composition, and the activity of transport systems such as Na^+/K^+ ATPase, adenylate cyclase and Mg^{2+}, K^+ and Ca^{2+} transport are impaired. Oxidative phosphorylation in the mitochondria is decreased, and this will also adversely affect the transport systems which are so vital to the kidney. The accumulation of gentamycin in lysosomes gives rise to myeloid inclusion bodies by impairment of phospholipid degradation. The lysosomes subsequently rupture and release hydrolytic enzymes which contribute to the cell damage. Lipid peroxidation occurs and hydroxyl radicals have been implicated although these may be secondary rather than primary events.

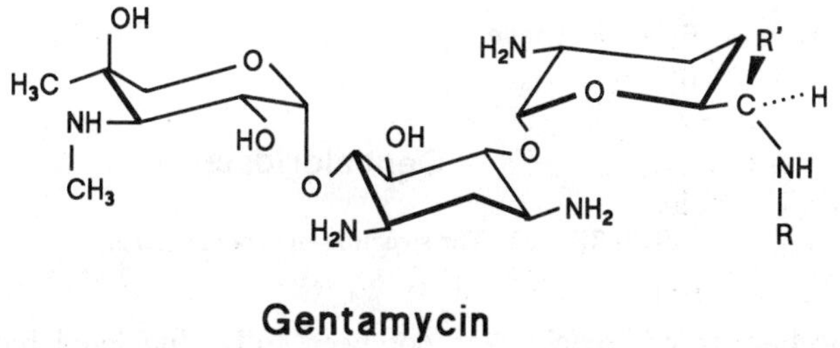

Gentamycin

FIGURE 7.21. The structure of the aminoglycoside antibiotic gentamycin. The drug may be used as a mixture and R and R′ = H or CH_3 depending on the particular component.

Alterations of glomerular structure have also been reported as another feature of the nephrotoxicity. Thus, a reduction in the glomerular filtration coefficient and the number and size of fenestrations in the glomerular endothelium have been observed.

Some aminoglycosides also cause damage to the ear, in particular to the cochlear apparatus. **Gentamycin**, for example, damages the hair cells in the cochlear, a measure of the toxicity which can be detected by scanning electron microscopy. There is a clear dose–response relationship between the number of cells destroyed and the increasing dose of gentamycin. The ototoxicity can also be detected in the intact animal by the Preyer reflex test which determines the threshold sound level for particular frequencies which an animal responds to by movement of the external ear.

Cephalosporins

These drugs, such as **cephaloridine** (figure 7.22), are used therapeutically for the treatment of bacterial infections, but may cause kidney damage in human patients. Thus, acute doses cause proximal tubular necrosis, selectively damaging the S2 segment of the tubule (figure 6.8). Cephaloridine accumulates in the kidney, especially the cortex, to a greater extent than in other organs. It is an organic anion, but has a cationic (pyridyl) group and it is actively secreted by the renal tubular cationic transport system into the tubular lumen. However, there is also an anionic transport system for re-uptake into the tubular cell which is more active and the overall result is active accumulation. Again binding to intracellular receptors may be involved. The movement of cephaloridine via facilitated diffusion out of the tubular cell into the lumen also seems to be restricted, and hence the intracellular concentration remains high. Blocking active transport with **probenecid** decreases both the nephrotoxicity and the concentration of cephaloridine.

$$R_1-CH_2 - CO-NH-CH_2-CH_2 \quad CH_2$$

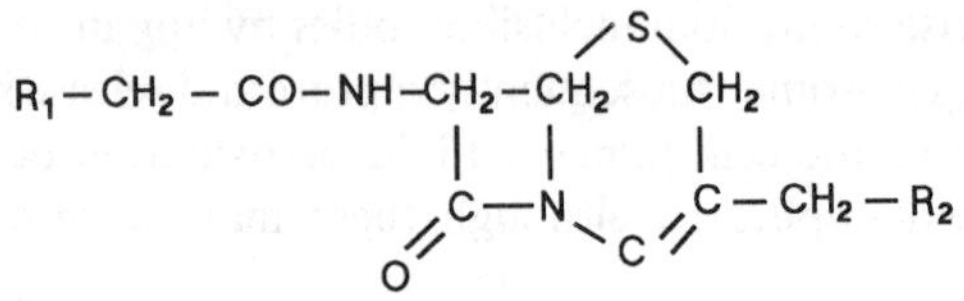

R_1 = Thiophene

R_2 = Pyridine

Cephaloridine

FIGURE 7.22. The structure of cephaloridine.

The mechanism of toxicity has not been fully elucidated but may involve metabolic activation. A reactive intermediate has been suggested as inhibition of cytochromes P-450 decreases the toxicity and inducers of the mono-oxygenases increase toxicity. However, these treatments also affect the renal concentration of cephaloridine in a manner consistent with the effect on toxicity. Glutathione is depleted seemingly by oxidation rather than conjugation, and prior depletion of glutathione increases the toxicity. NADPH is also depleted and consequently GSSG cannot be reduced back to GSH. Vitamin E- and selenium-deficient animals are more susceptible to the nephrotoxicity, and so it has been suggested that superoxide anion may be formed and that lipid peroxidation could be responsible for the toxicity.

Again, therefore, the susceptibility is the result of accumulation of the xenobiotic to reach concentrations not achieved in other tissues. This is then followed by metabolic activation.

Tissue lesions: lung damage

4-Ipomeanol

4-Ipomeanol (figure 7.23) is a pulmonary toxin produced by the mould *Fusarium solani*, which grows on sweet potatoes. The pure compound produces lung damage in a number of species when given intraperitoneally. This lung damage is manifested as oedema, congestion and haemorrhage. These are probably secondary or tertiary pathological changes resulting from the primary lesion, which is necrosis of the non-ciliated bronchiolar cells, also known as the **Clara cells**. Toxicologically, 4-ipomeanol is of particular interest as it is a specific lung toxin which selectively damages one cell type, the Clara cell in the smaller bronchioles, although after high doses other cell types and those of the larger airways may also be damaged.

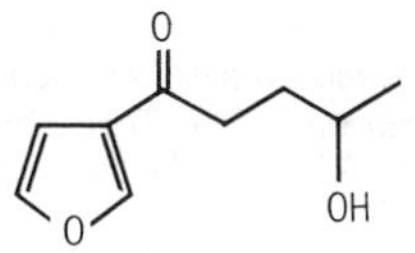

FIGURE 7.23. Structure of 4-ipomeanol.

The elucidation of the mechanism has revealed that this specificity is due to a requirement for metabolic activation for which the Clara cell is particularly suited. Using radiolabelled 4-ipomeanol it was found that the compound was localized particularly in the lungs (when expressed as nmol/g wet weight of tissue), and was covalently bound to lung protein. This binding was five times that seen in the liver. Furthermore, autoradiography revealed that the radiolabelled 4-ipomeanol was bound to the Clara cells, which were necrotic, whereas ciliated and other cells lining the airways did not show necrosis or covalently bound radioactivity.

Studies *in vitro* showed that 4-ipomeanol was metabolically activated by the microsomal mono-oxygenases to an alkylating species which would covalently interact with protein. Although the level of cytochromes P-450 in the lung is lower than that in the liver, the relative rate of alkylation of protein in the lung is greater than that in the liver. Thus, the V_{max} for the covalent binding to protein is higher for the lung microsomal enzyme than for the liver, whereas the apparent K_m for the lung enzyme is about one tenth that for the liver microsomal enzyme preparation. The lung enzyme, therefore, has a greater affinity for the ipomeanol and activates it more rapidly. Glutathione inhibited the binding to protein *in vitro*.

Studies *in vivo* indicated that the cause of death was probably pulmonary oedema, as the time-course for lethality and oedemagenesis were similar (figure 7.24). Also, the pulmonary oedema and lethality showed a very similar dose–response and the covalent binding to pulmonary protein was similarly dose-dependent (figure 7.25).

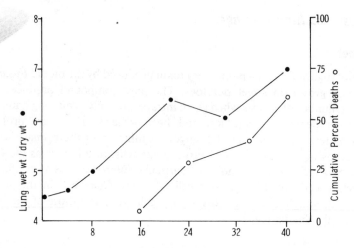

FIGURE 7.24. Time-course of lung oedemagenesis and lethality in animals given 4-ipomeanol. Adapted from Boyd *et al.* (1978) *J. Pharmac. Exp. Ther.*, **207**, 687.

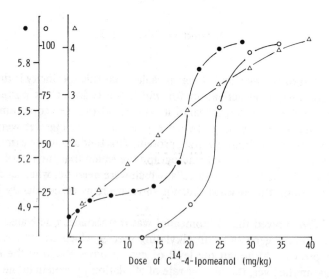

FIGURE 7.25. Dose-response curves for lethality (%, ○), oedemagenesis (wet weight : dry weight ratios, ●) and pulmonary covalent binding (nmol/mg protein, △) of radiolabelled 4-ipomeanol given i.p. to the rat. Adapted from Boyd *et al.* (1978) *J. Pharmac. Exp. Ther.*, **207**, 687.

Inhibitors of the microsomal enzymes decreased the covalent binding and the Clara cell necrosis and increased the LD_{50} even though the blood and pulmonary levels of ipomeanol were increased. The microsomal enzyme inducer phenobarbital decreased binding to both liver and lung protein and correspondingly increased the LD_{50}. In contrast, 3-methylcholanthrene induction increased binding in the liver and potentiated liver necrosis, but again decreased lung binding and lung damage. The LD_{50} was again increased.

Diethyl maleate treatment, which depletes glutathione, markedly increased covalent binding to protein and decreased the LD_{50}. Analogues of 4-ipomeanol in which the furan ring was replaced by a methyl or phenyl group were very much less toxic and did not covalently bind to lung or liver protein to any great extent.

There is good correlation between the toxicity as measured by lethality or LD_{50}, oedemagenesis and covalent binding of radiolabelled ipomeanol to protein (figure 7.25). The reason for the pulmonary selectivity is not yet clear, but liver necrosis may occur if hepatic metabolic activation is increased by 3-methylcholanthrene treatment. The particular susceptibility of the Clara cell in the lung could reflect a deficiency in protective mechanisms as well as the level of microsomal enzyme activity. It seems likely that the metabolic activation takes place *in situ* rather than the reactive metabolite being transported from the liver.

In conclusion, the pulmonary toxicity of 4-ipomeanol seems to be due to metabolic activation of the furan ring, probably by epoxidation, catalysed by the cytochromes P-450 mono-oxygenases found in the Clara cell. The reactive intermediate binds to protein and possibly other macromolecules, initiating cellular necrosis followed by oedema.

Paraquat

Paraquat is a bipyridylium herbicide which is widely sold and used, but unfortunately has been involved in both accidental and intentional cases of poisoning. This usually occurs after oral ingestion rather than percutaneous absorption, although it will damage the skin and cause inflammation as it is a strong irritant. After absorption of a toxic dose there will be abdominal pain, vomiting and diarrhoea. The main target organs are the lungs and kidneys, although there may also be cardiac and liver toxicity. Such multi-organ toxicity will occur with relatively large doses. Thus, the toxic effects are dose related and so there may be minimal damage which is reversible. Larger doses cause alveolar oedema followed by pulmonary fibrosis if the patient survives beyond a few days. Pulmonary fibrosis and respiratory failure may develop up to six weeks after ingestion. Within 24 h of a toxic dose renal tubular function is affected, and so excretion is reduced and this will exacerbate the toxicity.

Mechanism of toxicity

The main target organ, and that usually involved with lethality, is the lung. The damage observed is destruction of the Type I and II alveolar epithelial cells. The reason for the susceptibility of the lung is because paraquat is selectively taken up into the Type I and II cells by an active transport system. The lung : plasma ratio of paraquat in experimental animals dosed with the compound may reach values of 6 or 7 : 1 despite a fall in the plasma level after dosing. Paraquat thus reaches a higher concentration in the Type I and II alveolar cells than most other tissues, but it is not bound to tissue components or metabolized. The uptake system shows saturation kinetics, with an apparent K_m of 70 μM and V_{max} of 300 nmol/g lung/h. These kinetic constants are similar in the rat and human, and therefore

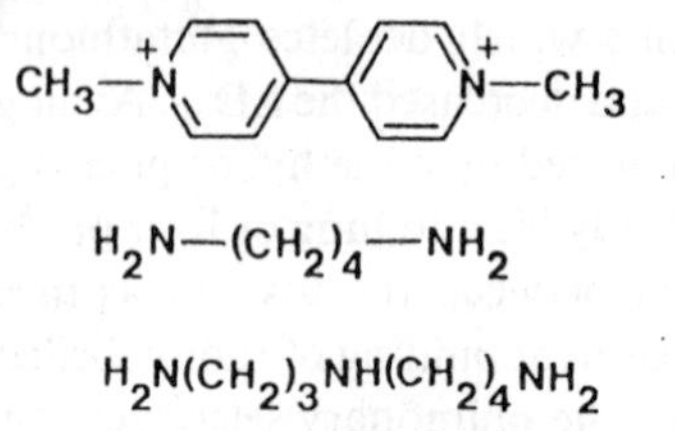

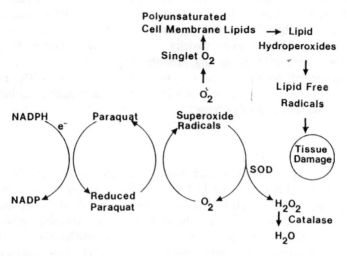

FIGURE 7.26. The structures of the herbicide paraquat (A) and the polyamines putrescine (B) and spermine (C). From Timbrell, J. A., *Introduction to Toxicology*, Taylor and Francis, London, 1989.

FIGURE 7.27. The proposed mechanism of toxicity of paraquat. SOD is the enzyme superoxide dismutase. From Timbrell, J. A., *Introduction to Toxicology*, Taylor and Francis, London, 1989. (Adapted from *Poisoning-Diagnosis and Treatment*, eds. J. A. Vale and T. J. Meredith, Update Books, London, 1981.)

the rat is a good model species. The system which takes up paraquat into the lung normally transports polyamines such as **putrescine** and **spermine** (figure 7.26). However, the size and shape of the paraquat molecule is similar to these amines. The preferred structure has two cationic nitrogen atoms separated by at least four methylene groups (6.6 Å), and it seems that the quaternary nitrogen atoms in paraquat have a separation which is sufficiently similar to allow it to be a substrate for this active transport system.

The mechanism of toxicity involves an initial reaction between paraquat and an electron donor such as NADPH (figure 7.27). Paraquat readily accepts an electron and forms a stable radical cation. However, under aerobic conditions this electron is transferred to oxygen giving rise to superoxide. As there is a ready supply of oxygen in the lung tissue, this process can then be repeated and a redox cycle is set up. This cycle causes two effects, both of which may be responsible for the toxicity:

(a) it produces superoxide which causes a variety of toxic effects

(b) it depletes NADPH and so may compromise the alveolar cell and reduce its ability to carry out essential biochemical and physiological functions.

The superoxide produced can be detoxified by the action of the enzyme **superoxide dismutase** which produces hydrogen peroxide:

$$O_2^{\cdot -} + O_2^{\cdot -} \rightarrow 2H^+ + H_2O_2$$

The hydrogen peroxide is then removed by catalase:

$$2H_2O_2 \rightarrow 2H_2O + O_2$$

However, the superoxide dismutase may be overwhelmed by the amount of superoxide being produced after a toxic dose of paraquat. The superoxide may then cause lipid peroxidation via the production of **hydroxyl radicals**. These may be produced from hydrogen peroxide in the presence of transition metal ions (see Chapter 6). The resulting lipid peroxides, if not detoxified, may give rise to lipid radicals and membrane damage. However, although there is experimental evidence that paraquat causes lipid peroxidation, there is little direct evidence from *in vivo* studies. This may reflect the fact that only a small proportion of lung cells are affected. Lipid peroxidation will oxidize glutathione (GSH) and the reduction of this oxidized glutathione (GSSG), catalysed by glutathione reductase, will also utilize NADPH and hence contribute to the depletion of this nucleotide (figure 7.27). The hexose monophosphate shunt will be stimulated to produce more NADPH, but this will be used for further reduction of paraquat. The lung is therefore a target for the toxicity of paraquat for two reasons:

(i) accumulation of the compound to toxic levels in certain lung cells
(ii) presence of high concentrations of oxygen.

The result is redox cycling which produces active oxygen species which can deplete NADPH and glutathione and potentially cause peroxidation of membrane lipids.

Treatment

There is no antidote to paraquat poisoning; therefore prevention of absorption of paraquat from the gastrointestinal tract using **Fullers Earth** as adsorbent and gastric lavage are usually employed. Experimental studies have shown that repeatedly giving an adsorbent and using gastric lavage can reduce the toxicity in rats and this is used in patients. **Haemoperfusion** can also be utilized to reduce the blood concentration once absorption has started to take place. However, once the paraquat has accumulated in the lungs, little can be done to alter the course of the toxicity.

Neurotoxicity

Isoniazid

As well as being implicated as a hepatotoxin, the antitubercular drug isoniazid may also cause peripheral neuropathy with chronic use. In practice this can be avoided by the concomitant administration of vitamin B_6 (pyridoxine).

In experimental animals, however, chronic dosing with isoniazid causes degeneration of the peripheral nerves. The biochemical basis for this involves interference with **vitamin B_6** metabolism. Isoniazid reacts with pyridoxal phosphate to form a hydrazone (figure 7.28) which is a very potent inhibitor of pyridoxal phosphate kinase. The hydrazone has a much greater affinity for the enzyme ($100–1000 \times$) than the normal substrate, pyridoxal. The result of this is a depletion of tissue pyridoxal phosphate. This cofactor is of importance particularly in nervous tissue for reactions involving decarboxylation and transamination. The decarboxylation reactions are principally affected however, with the result that transamination reactions assume a greater importance.

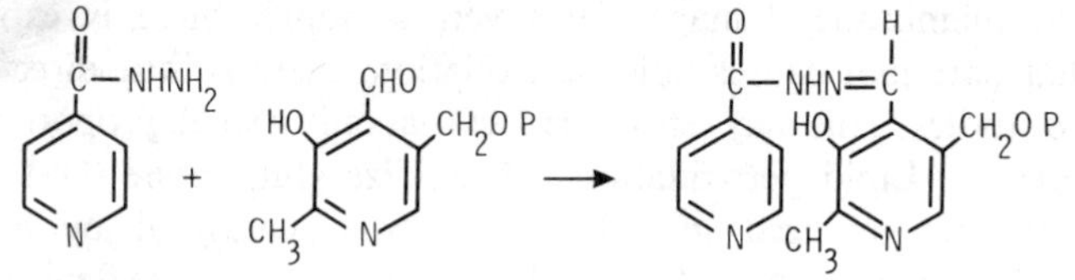

FIGURE 7.28. Reaction of isoniazid with pyridoxal phosphate to form a hydrazone.

In man peripheral neuropathy due to isoniazid is influenced by the acetylator phenotype (see page 149), being predominantly found in slow acetylators. This is probably due to the higher plasma level of isoniazid in this phenotype. In this case, therefore, acetylation is a detoxication reaction, removing the isoniazid and rendering it unreactive towards pyridoxal phosphate.

6-Hydroxydopamine

6-Hydroxydopamine is a selectively neurotoxic compound which damages the sympathetic nerve endings. It can be seen from Figure 7.29 that 6-hydroxydopamine is structurally very similar to dopamine and noradrenaline, and because of this it is actively taken up into the synaptic system along with other catecholamines. Once localized in the synapse the 6-hydroxydopamine destroys the nerve terminal. A single small dose of 6-hydroxydopamine destroys all the nerve terminals and possibly the nerve cells as well.

The mechanism of the destruction of the nerve terminals is thought to involve oxidation of 6-hydroxydopamine to a *p*-quinone, the production of a **free radical** or of superoxide anion. It seems that a reactive intermediate is produced which reacts covalently with the nerve terminal and permanently inactivates it.

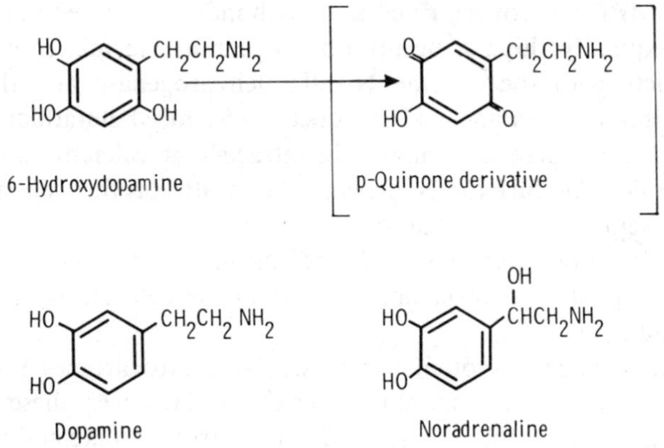

FIGURE 7.29. Structures of dopamine and noradrenaline, the analogue 6-hydroxydopamine and a possible oxidation product.

Factors which influence the disposition of catecholamines will affect the toxicity. For instance, compounds which inhibit the uptake of noradrenaline reduce the destruction of adrenergic nerve terminals but not of dopaminergic ones. Interference with the oxidative metabolism of catecholamines also influences the toxicity of 6-hydroxydopamine.

This example of selective neurotoxicity particularly illustrates the potential importance of a similarity in chemical structure between an endogenous compound and a foreign compound in the distribution, localization and type of toxic effect caused.

1-Methyl-4-phenyl-1,2,3,6-tetrahydropyridine (MPTP)

MPTP is a contaminant of a meperidine analogue which was synthesized illicitly and used by drug addicts. Some of the drug addicts using this drug suffered a syndrome similar to Parkinson's disease. It was found that the contaminant MPTP was responsible for the symptoms. When monkeys were exposed to MPTP they showed similar symptoms to the human victims. Furthermore, it was found that MPTP caused destruction of dopamine cell bodies in the substantia nigra of the brain. Although this example is similar to 6-hydroxydopamine, it also illustrates other important factors which determine the target organ and particular cell type which is affected by the toxicity.

MPTP is not directly neurotoxic but must first be metabolically activated to MPP+ in a two-step reaction (figure 7.30). As MPP+ is charged, and therefore should not cross the blood–brain barrier, it is presumed to be formed in the brain itself. MPTP is metabolized by **monoamine oxidase B** to MPDP+ in astrocytes and is then oxidized to MPP+ (figure 7.30). Although the MPP+ is formed outside the neurone, it is then taken up by the dopamine reuptake system into dopaminergic neurones where it accumulates and may associate with neuromelanin. Presumably, however, as it is a charged molecule, it cannot cross the blood–brain barrier and be removed from the central nervous system.

Furthermore, MPP⁺ is concentrated in mitochondria via an energy-dependent carrier. Consequently, high concentrations occur in the mitochondria. Here MPP⁺ interacts with the enzyme NADH dehydrogenase and this inhibits oxidative phosphorylation and ATP production. Reduced glutathione levels are also decreased, and there are changes in intracellular calcium concentration. These biochemical perturbations presage the death of the cell and so the dopaminergic neurones are destroyed.

It has been suggested that superoxide radicals may be formed and also other radical species, but the involvement of these in the cytotoxicity of MPTP is currently speculative.

MPTP is metabolized by other routes involving cytochromes P-450, FAD-dependent mono-oxygenases and aldehyde oxidase. However, these seem to be detoxication pathways as they divert MPTP away from uptake and metabolism in the brain. However, MPTP may inhibit its own metabolism by cytochromes P-450 and thereby reduce one means of detoxication.

This example illustrates the importance of structure and physicochemical properties in toxicology. MPTP is sufficiently lipophilic to cross the **blood–brain barrier** and gain access to the astrocytes. The structure of the metabolite is

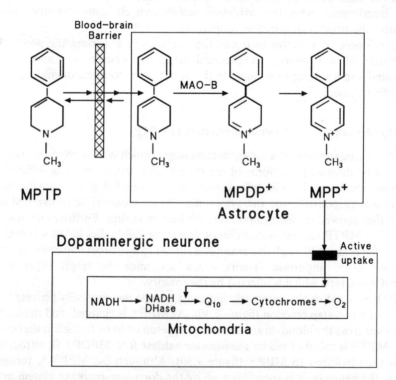

FIGURE 7.30. The metabolic activation and proposed mechanism of toxicity of MPTP in the central nervous system. Thus MPP⁺ is actively taken into the neurone where it inhibits mitochondrial electron transport between NADH dehydrogenase (NADH DHase) and coenzyme Q (Q₁₀). MPTP:1-Methyl-4-phenyl-1,2,3,6-tetrahydropyridine.

important for uptake via the **dopamine system**, hence localizing the compound to a particular type of neurone. Again, uptake into mitochondria is presumably a function of structure as a specific energy-dependent carrier is involved.

Exaggerated and unwanted pharmacological effects

The exaggerated or unwanted pharmacological responses are the most common toxic effects of drugs observed clinically, as opposed to direct toxic effects on tissues. However, this type of response may also be observed with other compounds such as the toxic cholinesterase inhibitors used as pesticides and nerve gases.

Organophosphorus compounds

There are many different cholinesterase inhibitors which find use, particularly as insecticides but also as nerve gases for use in chemical warfare. Organophosphorus insecticides are the most widely used and the most frequently involved in fatal human poisonings. They may be absorbed through the skin and there have been accidental poisoning cases arising from such exposure. Accidental contamination of food with insecticides such as parathion has led to a significant number of deaths. There are two types of toxic effects: **inhibition of cholinesterases** and **delayed neuropathy.**

Cholinesterase inhibition

The inhibition of cholinesterases results in a number of physiological effects. The enzyme acetylcholinesterase, found in various tissues including the plasma, is responsible for hydrolysing acetylcholine. This effectively terminates the action of the neurotransmitter at the synaptic nerve endings which occur in the central nervous system and other tissues such as glands and smooth muscle. The inhibition leads to accumulation of acetylcholine, and therefore the signs of poisoning resemble excessive stimulation of cholinergic nerves. The toxicity becomes apparent when there is about 50% inhibition of acetylcholinesterase, and at 10–20% of normal activity death occurs.

The toxic effects can be divided into three types as the accumulation of acetylcholine leads to symptoms which mimic the muscarinic, nicotinic and CNS actions of acetylcholine. **Muscarinic receptors** for acetylcholine are found in smooth muscles, the heart and exocrine glands. Therefore the signs and symptoms are tightness of the chest, wheezing due to bronchoconstriction, bradycardia and constriction of the pupils (miosis). Salivation, lacrimation and sweating are all increased, and peristalsis is increased leading to nausea, vomiting and diarrhoea.

Nicotinic signs and symptoms result from the accumulation of acetylcholine at motor nerve endings in skeletal muscle and autonomic ganglia. Thus, there is fatigue, involuntary twitching and muscular weakness which may affect the muscles of respiration. Hypertension and hyperglycaemia may also reflect the action of acetylcholine at sympathetic ganglia.

Accumulation of acetylcholine in the CNS leads to a variety of signs and symptoms including tension, anxiety, ataxia, convulsions, coma and depression of the respiratory and circulatory centres.

The cause of death is usually respiratory failure due partly to neuromuscular paralysis, central depression and bronchoconstriction.

The onset of symptoms depends on the particular organophosphorus compound, but is usually relatively rapid, occurring within a few minutes to a few hours and the symptoms may last for several days. This depends on the metabolism and distribution of the particular compound and factors such as lipophilicity. Some of the organophosphorus insecticides such as **malathion** for example (figure 5.10), are metabolized in mammals mainly by hydrolysis to polar

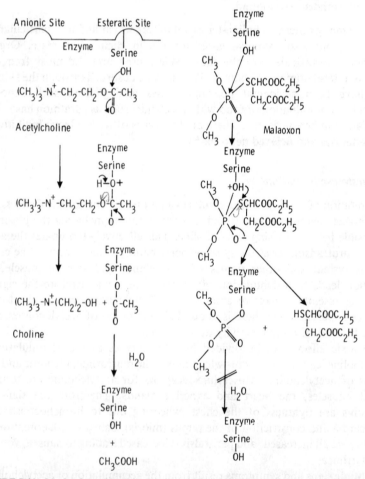

FIGURE 7.31. Mechanism of hydrolysis of acetylcholine by acetylcholinesterase and blockade of the enzyme by malaoxon. With malaoxon as substrate, the final step, regeneration of the enzyme by hydrolysis, is blocked (≠) leading to inactivated enzyme.

metabolites which are readily excreted, whereas in the insect oxidative metabolism occurs which produces the cholinesterase inhibitor. Metabolic differences between the target and non-target species are exploited to maximize the selective toxicity. Consequently, malathion has a low toxicity to mammals such as the rat in which the LD_{50} is about $10 \, g/kg$.

Mechanism of inhibition of cholinesterases

Inhibition of the cholinesterase enzymes depends on blockade of the active site of the enzyme, specifically the site which binds the ester portion of acetylcholine (figure 7.31). The organophosphorus compound is thus a pseudosubstrate. However, in the case of some compounds such as the **phosphorothionates** (parathion and malathion for example), metabolism is necessary to produce the inhibitor. In both cases metabolism by the microsomal mono-oxygenase enzymes occurs in which the sulphur atom attached to the phosphorus is replaced by an oxygen (figure 5.10).

With malaoxon the P=O group binds to a serine hydroxyl group at the esteratic site of the cholinesterase enzyme in an analogous manner to the C=O group in the normal substrate acetylcholine (figures 7.31 and 7.32). With acetylcholine the enzyme–substrate complex breaks down to leave acetylated enzyme, which is then rapidly hydrolysed to regenerate the serine hydroxyl group and hence the functional enzyme. With organophosphorus compounds such as malathion (figure 7.31), the bound organophosphorus compound also undergoes cleavage to release the corresponding thiol or alcohol, leaving phosphorylated enzyme. However, unlike the acetylated enzyme intermediate produced with acetylcholine, the phosphorylated enzyme is only hydrolysed very slowly, if at all. Consequently, the active site of the enzyme is effectively blocked. If the phosphorylated enzyme undergoes a change known as ageing in which one of the groups attached to the phosphorus atom is lost, then the inhibition may become

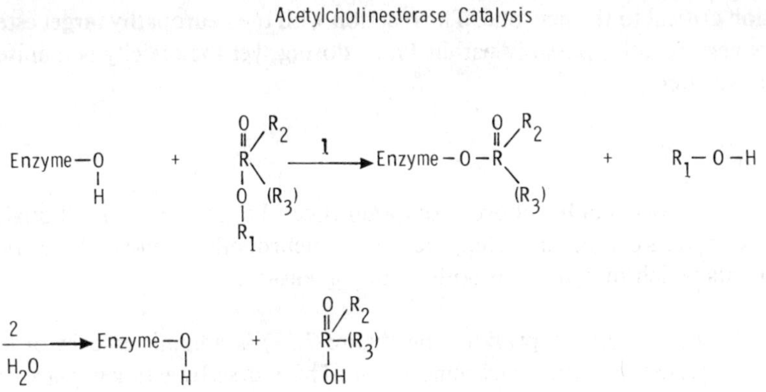

Acetylcholinesterase Catalysis

FIGURE 7.32. General scheme for acetylcholinesterase action. R may = C or P. If R = P, then (R_3) is present. The group OR_1 may be replaced by SR_1, giving R_1SH on hydrolysis (reaction 1). IF R = P, reaction 2 is very slow, giving inactivated enzyme. The rate of hydrolysis or reactivation depends on the nature of R_2 and R_3.

more permanent. However, the toxicity will ultimately depend on the affinity of the enzyme for the inhibitor and the rate of hydrolysis of the phosphorylated intermediate. These will in turn depend on the nature of the substituent groups on the phosphorus atom (figure 7.32). Furthermore, the rates of metabolism to the active metabolite and via other routes will also be determinants of the toxicity.

Although a particular organophosphorus compound may be rapidly metabolized and therefore not accumulate in the animal, chronic dosing may cause a cumulative toxic effect due to the slow rate of reversal of the inhibition. Thus the rate of regeneration may be slow with a half-life of the order of 10–30 days. Thus, in some cases resynthesis of the enzyme may be the limiting factor. Thus repeated, small non-toxic doses can cause eventual toxicity when sufficient of the cholinesterase is inhibited. Different cholinesterases have different half-lives. The acetylcholinesterases present in red blood cells, which is similar to that in the nervous tissue, remains inhibited after **di-isopropylfluorophosphate** for the lifetime of the red blood cell.

Delayed neuropathy

Certain organophosphorus compounds, such as **tri-orthocresyl phosphate**, cause this toxic effect. The symptoms, which may result from a single dose, may not be apparent for 10–14 days afterwards. The result is degeneration of peripheral nerves in the distal parts of the lower limbs which may spread to the upper limbs. Pathologically it is observed that the nerves undergo 'dying back' with axonal degeneration followed by myelin degeneration. The effect does not seem to be dependent on cholinesterase inhibition as tri-orthocresyl phosphate is not a potent cholinesterase inhibitor but causes delayed neuropathy. It seems that there is a covalent interaction with a membrane-bound protein, known as neuropathy target esterase, and the organophosphorus compound which may disturb metabolism in the neurone. This is followed by an ageing process which involves loss of a group from the phosphorylated protein. This protein seems to have a function critical to the neurone. The reaction with the neuropathy target esterase occurs very rapidly, possibly within 1 h of dosing, yet the toxicity is manifested only days later.

Treatment

Poisoning by organophosphorus compounds can be treated, and although the acute symptoms can be alleviated, the delayed neuropathy cannot. There are two treatments which may be used both involving **antidotes**:

(i) The compound **pralidoxime** (figure 7.33) is administered in order to regenerate the acetylcholinesterase. This it does by acting in place of the serine hydroxyl group in the enzyme and forming a complex with the organophosphorus moiety (figure 7.33). Pralidoxime must be administered quickly after the poisoning as if the phosphorylated enzyme is allowed to age then it will no longer be an effective antidote.

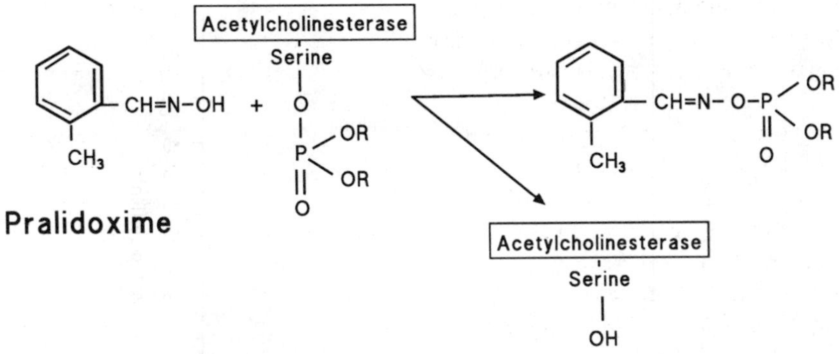

FIGURE 7.33. The mechanism of reaction of the antidote pralidoxime with phosphorylated acetylcholinesterase. The original acetylcholinesterase is thereby regenerated.

The reactivation rate will depend on the particular organophosphorus compound.

(ii) The physiological effects of the accumulation of acetylcholine can be antagonized by the administration of **atropine** and the symptoms alleviated.

If both atropine and pralidoxime are given together there is a much greater effect than if either are given alone. Thus, when used after experimental parathion poisoning, each treatment alone only increased the LD_{50} about twice, whereas in combination the LD_{50} was increased 128 times.

Digitalis glycosides

Some of the toxic effects of the digitalis glycosides have been known since digitalis was first described by William Withering in 1785. Overdoses cause vomiting, diarrhoea, visual disturbances, hypotension, slow pulse and ventricular tachycardia, eventually leading to ventricular fibrillation, delirium and convulsions. Toxic effects such as vomiting are not infrequent in the clinical use of the drug, as it has a narrow margin of safety, or a low therapeutic ratio. The dose is therefore critical for any given patient, and monitoring the plasma level of the drug may be necessary to avoid toxicity.

There is wide individual variation in the response to digitalis, which has a long half-life and which therefore may accumulate when certain dose regimens are used. This accumulation, the low therapeutic index (see page 21) and the individual variation in response are responsible for the toxicity encountered in the normal clinical use of the drug. It can be seen from figure 7.34 that **digitoxin** (one of the digitalis glycosides) will be toxic, causing vomiting, in three out of 100 patients at the ED_{50} dose. The dose therefore has to be individualized to try to avoid this.

The toxicity of the digitalis glycosides, such as cardiac arrhythmias, is the result of an exaggerated pharmacological effect. The mechanism involves alterations in impulse formation and/or conduction due to an extension of the alteration of

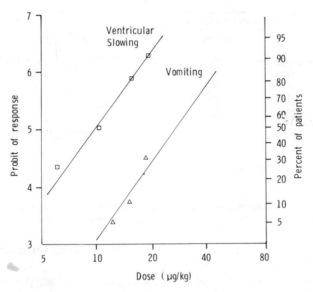

FIGURE 7.34. Comparison of the pharmacological and toxic effects of digitoxin in man. The pharmacological effect was a 40–50% decrease in heart rate; the toxic effect, nausea and vomiting, occurred after a single oral dose. Adapted from Marsh (1951) *Outline of Fundamental Pharmacology* (Springfield, Ill.: Charles C. Thomas).

$Na^+/K^+ - ATPase$ activity. Digitalis glycosides and K^+ compete for this enzyme and the glycosides may cause a loss of intracellular K^+ and an increase in Na^+. Thus toxic doses cause an efflux of potassium ions from the heart tissue, and therefore an infusion of K^+ may be used as an antidote to the toxicity. The manifestations of toxicity are gastrointestinal disturbances giving rise to nausea, vomiting and diarrhoea, central effects, neurologic and ophthalmologic disorders and cardiac arrhythmias. The pattern of toxic effects is variable but may occur after only a small change in the dose or plasma level.

Diphenylhydantoin

As with the digitalis glycosides, the toxic effects of the anticonvulsant drug diphenylhydantoin result from elevated plasma levels of the drug. This can simply be due to inappropriate dosage, but other factors may also be involved in the development of toxicity.

The toxic effects observed are nystagmus, ataxia, drowsiness and sometimes more serious effects on the CNS. These toxic effects are clearly dose-related and correlate well with the plasma levels of the unchanged drug. High plasma levels may be the result of defective metabolism as well as of excessive dosage (see page 154). Diphenylhydantoin is metabolized by hydroxylation of the aromatic ring (figure 7.35) and this is the major route of metabolism. However, a genetic trait has been described in man in which there appears to be a deficiency in this metabolic route. The consequences of this are decreased metabolism and the

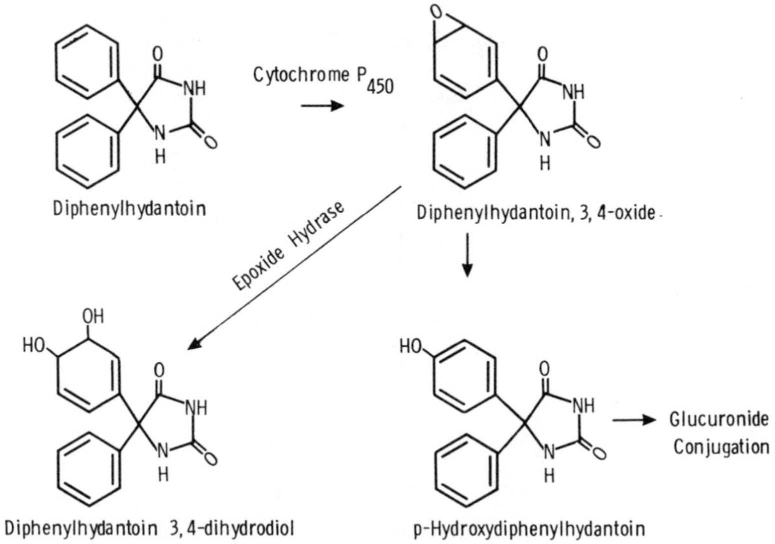

FIGURE 7.35. Metabolism of diphenylhydantoin.

appearance of toxicity after therapeutic doses, due to the elevated plasma levels of the unchanged drug.

Another cause of toxicity with diphenylhydantoin can be the result of interactions with other drugs such as **isoniazid**. This drug is a non-competitive inhibitor of the aromatic hydroxylation of diphenylhydantoin. Consequently, the elimination of diphenylhydantoin is impaired in the presence of isoniazid and the plasma level is greater than anticipated for a normal therapeutic dose. Furthermore, it was found there was a significant correlation between the rate of acetylation of isoniazid and the appearance of adverse effects of diphenylhydantoin to the central nervous system. This greater incidence in toxicity in slow acetylators corresponded with a greater increase in the plasma level of diphenylhydantoin when the drugs were taken in combination. This was due to higher plasma levels of isoniazid in slow acetylators causing a greater degree of inhibition of diphenylhydantoin metabolism.

Succinylcholine

Succinylcholine is a neuromuscular blocking agent which is used clinically to cause muscle relaxation. Its duration of action is short due to rapid metabolism – hydrolysis by cholinesterases (pseudocholinesterase or acylcholine acyl hydrolase) in the plasma and liver to yield inactive products (figure 7.36). Thus, the pharmacological action is terminated by the metabolism. However, in some patients the effect is excessive, with prolonged muscle relaxation and apnoea lasting as long as two hours compared with the normal duration of a few minutes.

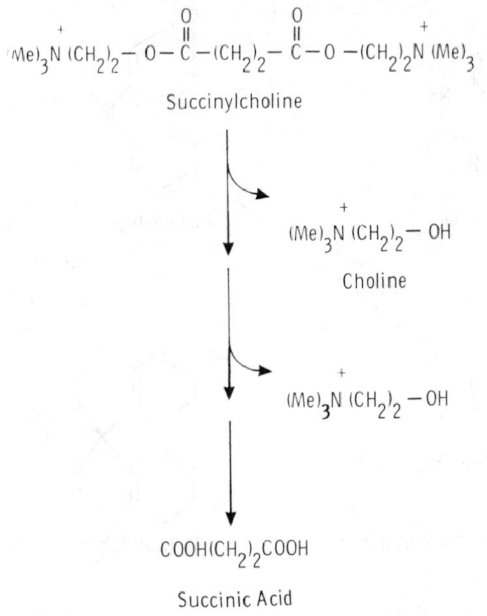

$$\overset{+}{(Me)_3}N \ (CH_2)_2 - \ O - \overset{\overset{O}{\|}}{C} - (CH_2)_2 - \overset{\overset{O}{\|}}{C} - O - (CH_2)_2 \overset{+}{N} \ (Me)_3$$

Succinylcholine

$$\overset{+}{(Me)_3}N \ (CH_2)_2 - OH$$

Choline

$$\overset{+}{(Me)_3}N \ (CH_2)_2 - OH$$

COOH(CH$_2$)$_2$COOH

Succinic Acid

FIGURE 7.36. Metabolism of succinylcholine by pseudocholinesterases.

This occasional toxicity is due to a deficiency in the hydrolysis of succinylcholine; therefore the parent drug circulates unchanged for longer periods of time and consequently the pharmacological effect is prolonged. This lack of hydrolysis is due to the presence of an atypical pseudocholinesterase. This aberrant enzyme hydrolyses various substrates, including succinylcholine, at greatly reduced rates and its affinity for both substrates and inhibitors is markedly different from that of the normal enzyme. The occurrence of this abnormal **pseudocholinesterase** is genetically determined and is controlled by two alleles at a single genetic locus. Homozygotes for the abnormal gene produce only the abnormal enzyme, heterozygotes produce a mixture of normal and abnormal enzymes. Homozygotes for the normal gene produce the normal pseudocholinesterase, the heterozygotes show an intermediate response. The abnormality shows a gene frequency of about 2% in certain groups (British, Greeks, Portuguese, North Africans) is rare in others (Australian Aborigines, Filipinos) and absent from some (Japanese, South American Indians).

There are other substrates for the enzyme such as diacetylmorphine and substance P and other variants have been described some of which may lack enzymic activity.

Physiological effects

Aspirin

Aspirin (acetylsalicylic acid) and other salicylates are still a common cause of human poisoning, both therapeutic and suicidal, and account for a significant number of deaths each year. Although the toxicity has a biochemical basis, some of the effects caused are clearly physiological and consequently it has been used as an example in this category.

Although aspirin is well absorbed from the gastrointestinal tract, after an overdose the plasma levels may rise for as long as 24 h because if there is a large mass of tablets present in the stomach this reduces their dissolution rate. Aspirin undergoes metabolism, mainly hydrolysis to **salicylic acid**, then hydroxylation to gentisic acid and conjugation with glucuronic acid and glycine (figure 7.37). The formation of the glycine and glucuronic acid conjugates is saturable, and therefore the half-life for elimination increases with increasing doses and the steady-state blood level therefore increases disproportionately with increasing dose. When this saturation of conjugation occurs there is more salicylic acid present in the blood, and therefore renal excretion of salicylic acid, which is sensitive to changes in the pH of urine, becomes more important.

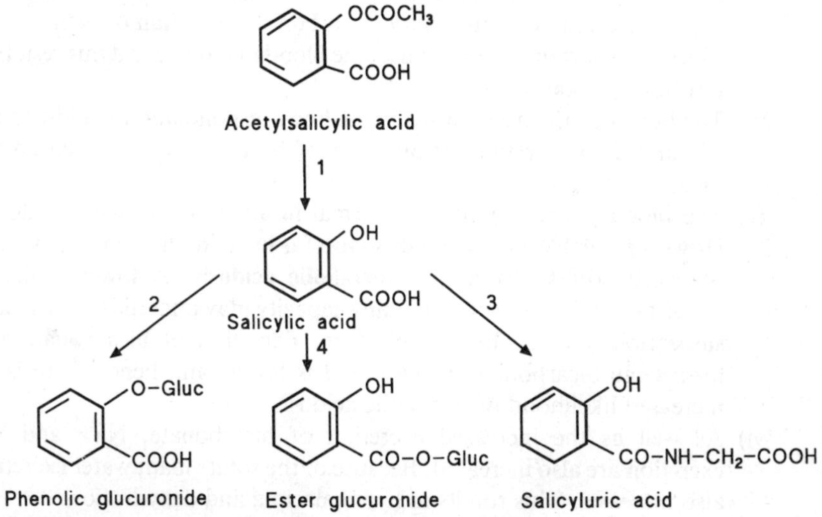

FIGURE 7.37. The metabolism of aspirin (acetylsalicylic acid). Step 1 (hydrolysis) yields the major metabolite, salicyclic acid, which is conjugated with glucuronic acid (pathways 2 and 4) or glycine (pathway 3). All these three pathways (2,3 and 4) are saturable.

The therapeutic blood level is greater than 150 mg/l but symptoms of toxicity occur at blood levels of around 300 mg/l. Therefore, knowledge of the blood level is important, particularly when aspirin is given in repeated doses. In children therapeutic overdosage is responsible for the majority of fatalities from aspirin. When an overdose is suspected measurement of the plasma level on two or more occasions will allow an estimate to be made of the severity of the overdose and whether the plasma level has reached its maximum. For interpretation of the blood level the **Done nomogram** can be used. An overdose of 50 300 mg tablets in adults will give rise to moderate to severe toxicity and a blood level of 500–750 mg/l at 12 h. The blood level must be interpreted with caution however because:

(a) the presence of metabolic acidosis will complicate the interpretation because this will alter the distribution of salicylic acid (see below)

(b) with therapeutic overdose after repeated dosing, tissue levels may be higher than expected from the blood level.

Salicylic acid is mainly responsible for the toxic effects of aspirin, and it has a number of metabolic and physiological effects some of which are interrelated:

(i) The initial effect of salicylate is stimulation of the depth of respiration due to increased production of carbon dioxide and greater use of oxygen. This is due to the **uncoupling of oxidative phosphorylation** in the mitochondria which increases the rate of oxidation of substrates.

(ii) As salicylate enters the central nervous system it stimulates the respiratory centre to give further increased depth and also rate of respiration. This causes the blood level of carbon dioxide to fall as there is greater alveolar permeability to carbon dioxide than oxygen.

(iii) The loss of carbon dioxide causes the blood pH to rise and this results in **respiratory alkalosis.**

(iv) The homeostatic mechanisms in the body respond and the kidneys are stimulated to excrete large amounts of bicarbonate so the blood pH falls.

(v) The blood pH may return to normal in adults with mild overdoses. However, the blood pH can drop too far in children or more severely-poisoned adults resulting in **metabolic acidosis**. A lower buffering capacity or plasma-protein binding capacity may underlie the increased susceptibility to acidosis in children. Excretion of bicarbonate also means the bicarbonate in the blood is lower, and hence there is an increased likelihood of metabolic acidosis.

(vi) As well as the increased excretion of bicarbonate, Na^+ and K^+ excretion are also increased. Because of the solute load, water excretion also increases. This results in hypokalaemia and dehydration.

(vii) As well as these effects salicylate also affects Krebs' cycle, carbohydrate metabolism, lipid metabolism, protein and amino acid metabolism.

(viii) Krebs' cycle is inhibited at the points where α-ketoglutarate dehydrogenase and succinate dehydrogenase operate. This causes an increase in organic acids and an accumulation of glutamate.

(ix) ATP is depleted due to the uncoupling of oxidative phosphorylation, and hence fat metabolism and glycolysis are stimulated.

(x) Increased glycolysis leads to an increase in lactic and pyruvic acids. This contributes to the metabolic acidosis.

(xi) Salicylate also inhibits aminotransferases leading to increased amino acid levels in blood and aminoaciduria. This also increases the solute load and contributes to the dehydration.

(xii) The uncoupling of oxidative phosphorylation stops ATP being produced by this process and so the energy is dissipated as heat. The patient therefore suffers hyperpyrexia and sweating. Again this produces dehydration.

(xiii) Increased glycolysis to compensate for the loss of ATP leads to first hyper- and then hypoglycaemia especially in the brain. Mobilization of hepatic glycogen stores occurs through glucose-6-phosphatase activation via epinephrine release, possibly due to an effect on the central nervous system. This causes an initial hyperglycaemia, but then when the glucose is completely utilized, hypoglycaemia occurs.

(xiv) The metabolic acidosis, if it occurs, allows salicylate to distribute to the central nervous system and other tissues more readily due to an increase in the proportion of the non-ionized form (figure 7.38). Hence the metabolic acidosis will exacerbate the toxicity to the central nervous system. This results in irritability, tremor, tinnitus, hallucinations and delirium.

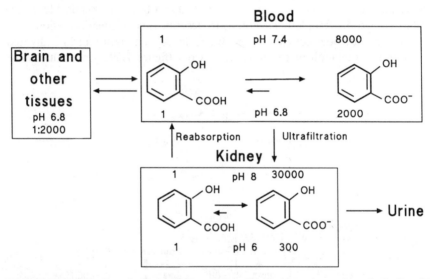

FIGURE 7.38. The effect of pH on the dissociation, distribution and excretion of salicylic acid. The numbers represent the proportions of ionized and non-ionized salicylic acid.

The specific treatment for salicylate poisoning, apart from gastric lavage and aspiration to remove the drug from the stomach, is based on a knowledge of the biochemical mechanisms underlying the toxicity. There is no antidote but treatment may be successful. Thus treatment involves:

(a) correction of the metabolic acidosis with intravenous bicarbonate. This will also increase urine flow and cause it to become more alkaline (**alkaline diuresis**) and therefore facilitate excretion of salicylic acid and its conjugated metabolites. As the blood pH rises the ionization of the salicylic acid increases causing a change in the equilibrium and distribution of salicylate which diffuses out of the central nervous system (figure 7.38).

(b) correction of solute and fluid loss and hypoglycaemia with intravenous K+ and dextrose and oral fluids.

(c) sponging of the patient with tepid water to reduce the hyperpyrexia.

Haemoperfusion and **haemodialysis** may be used in very severely poisoned patients.

Biochemical effects: lethal synthesis and incorporation

Fluoroacetate

Monofluoroacetic acid (fluoroacetate, figure 7.39) is a compound found naturally in certain South African plants, and which causes severe toxicity in animals eating such plants. The compound has also been used as a rodenticide. The toxicity of fluoroacetate was one of the first to be studied at a basic biochemical level, and Peters coined the term lethal synthesis to describe this biochemical lesion.

Fluoroacetate does not cause direct tissue damage and is not intrinsically toxic but requires metabolism to fluoroacetyl CoA (figure 7.39). Other fluorinated

$$\text{F}-\text{CH}_2-\text{COOH} \xrightarrow{\text{CoASH}} \text{F}-\text{CH}_2-\text{COSCoA}$$

Fluoroacetate · · · · · · · · · · · Fluoroacetyl CoA

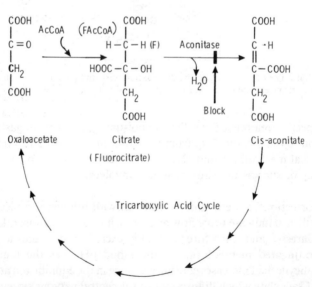

FIGURE 7.39. Metabolism of fluoroacetic acid to fluoroacetyl CoA (FAcCoA) and mechanism underlying blockade of the tricarboxylic acid cycle. Fluorocitrate cannot be dehydrated to *cis*-aconitate by aconitase and therefore blocks the cycle at this point.

compounds which are metabolized to fluoroacetyl CoA therefore produce the same toxic effects. For instance, compounds such as fluoroethanol and fluorofatty acids with even numbers of carbon atoms may undergo β-oxidation to yield fluoroacetyl CoA.

Fluoroacetyl CoA is incorporated into the tricarboxylic acid cycle (TCA cycle) in an analogous manner to acetyl CoA, combining with oxaloacetate to give fluorocitrate (figure 7.39). However, **fluorocitrate** inhibits the next enzyme of the TCA cycle, aconitase, and there is a build-up of both fluorocitrate and citrate. The TCA cycle is blocked and the mitochondrial energy supply is disrupted. The inhibition arises from the fact that the aconitase is able to bind fluorocitrate but cannot carry out the dehydration to *cis*-aconitate as the carbon–fluorine bond is stronger than the carbon–hydrogen bond (figure 7.39). Fluorocitrate is therefore a pseudosubstrate. As well as inhibiting cellular respiration, inhibition of the TCA cycle will also reduce the supply of 2-oxoglutarate. This may decrease the removal of ammonia via formation of glutamic acid and glutamine, and this might account for the convulsions seen in some species after exposure to fluoroacetate. The toxicity is manifested as a malfunction of the CNS and heart, giving rise to nausea, apprehension, convulsions and defects of cardiac rhythm, leading to ventricular fibrillation. Fluoroacetate and fluorocitrate do not appear to inhibit other enzymes involved in intermediary metabolism, and the di- and tri-fluoroacetic acids are not similarly incorporated and therefore do not produce the same toxic effects.

Galactosamine

D-Galactosamine is an amino sugar (figure 7.40a) normally found *in vivo* only in acetylated form in certain structural polysaccharides. Administration of single doses of this compound to certain species results in a dose-dependent hepatic damage resembling viral hepatitis, with focal necrosis and periportal inflammation. As well as a reversible hepatitis after acute doses, galactosamine may also cause chronic progressive hepatitis, cirrhosis and liver tumours. As all the liver cells are affected the damage tends to be diffuse throughout the lobule. A number of biochemical events have been found to occur. RNA and protein

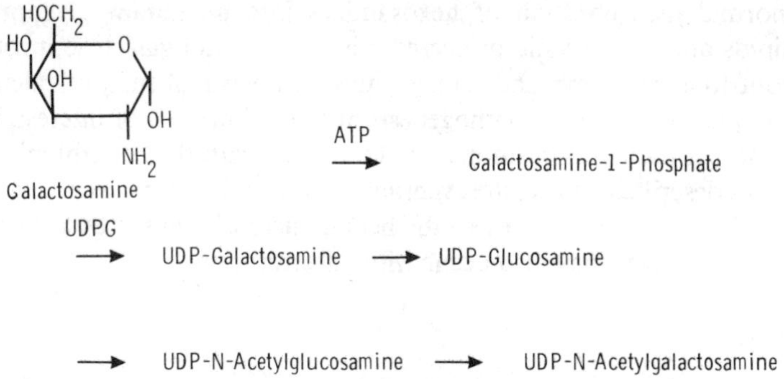

FIGURE 7.40a. Metabolism of galactosamine.

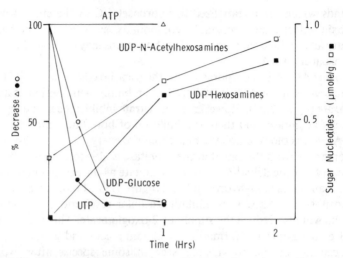

FIGURE 7.40b. The effect of galactosamine on biochemical parameters. The graph shows the effect on ATP (△), UTP (●), UDP-glucose (○), UDP-*N*-acetylhexosamines (□) and UDP-hexosamines (3401). Adapted from Decker and Kepler (1974) *Rev. Physiol. Biochem. Pharmac.*, **71**, 78.

synthesis are inhibited, membrane damage can be observed 2 h after dosing. There is also an increase in intracellular calcium which may be the crucial event which leads to cell death (see Chapter 6). Also, levels of uridine triphosphate (UTP) and uridine diphosphate glucose (UDP-glucose) fall dramatically within the first hour after the administration of galactosamine. There is also a concomitant rise in UDP-hexosamines and UDP-*N*-acetylhexosamines (figure 7.40b).

The rapid and extensive depletion of UTP is thought to be the basic cause of the toxicity. Galactosamine combines with UDP to form UDP-galactosamine (figure 7.40a), which sequesters UDP, and therefore does not allow cycling of uridine nucleotides. Consequently, synthesis of RNA and hence protein are depressed and sugar and mucopolysaccharide metabolism are disrupted. The latter effects may possibly explain the membrane damage which occurs. The toxic effects of galactosamine can be alleviated by the administration of UTP or its precursors.

Abnormal incorporation of hexosamines into **membrane glycoproteins** or glycolipids and a concomitant decrease in glucose and galactose incorporation may contribute to membrane damage, and the abnormal entry of calcium in the cell may play a role in the pathogenesis of the lesion. It is of interest, however, that although the mechanism may be compared with that of ethionine hepatotoxicity as described below, the eventual lesions are different. With ethionine the lesion is fatty liver in contrast to the hepatic necrosis caused by galactosamine, although fatty liver may also occur with galactosamine.

Ethionine

Ethionine is a hepatotoxic analogue of the amino acid methionine (figure 7.41). Ethionine is an antimetabolite which has similar chemical and physical properties

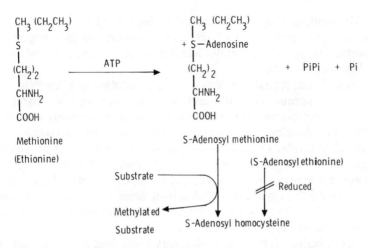

FIGURE 7.41. The role of methionine in methylation reactions and the mechanisms underlying ethionine hepatotoxicity. After the substrate is methylated the *S*-adenosyl homocysteine remaining is broken down into homocysteine and adenine, both of which are re-utilized. When *S*-adenosyl ethionine is formed, however, this recycling is reduced (‡) and a shortage of adenine and hence ATP develops.

to the naturally occurring amino acid. After acute doses ethionine causes fatty liver but prolonged administration results in liver cirrhosis and hepatic carcinoma. Some of the toxic effects may be reversed by the administration of methionine. The effects may be produced in a variety of species, athough there are differences in response. The rat also shows a sex difference in susceptibility, the female animal showing the toxic response rather than the male.

After a single dose of ethionine, **triglycerides** accumulate in the liver, the increase being detectable after four hours. After 24 hours the accumulation of triglycerides is maximal, being 15–20 times the normal level. Initially the fat droplets accumulate on the endoplasmic reticulum in periportal hepatocytes and then in more central areas of the liver. Some species develop hepatic necrosis as well as fatty liver, and nuclear changes and disruption of the endoplasmic reticulum may also be observed.

Chronic administration causes proliferation of bile duct cells leading to hepatocyte atrophy, fibrous tissue surrounding proliferated bile ducts and eventually cirrhosis and hepatocellular carcinoma.

The major biochemical changes observed are a striking depletion of ATP, impaired protein synthesis, defective incorporation of amino acids, and the appearance of RNA and proteins containing the ethyl rather than the methyl group. The plasma levels of triglycerides, cholesterol, lipoprotein and phospholipid are all decreased.

The mechanism underlying this toxicity is thought to involve a deficiency of ATP. Methionine acts as a methyl donor *in vivo*, in the form of *S*-adenosyl methionine, and ethionine forms the corresponding *S*-adenosyl ethionine. However, the latter analogue is relatively inert as far as recycling the adenosyl moiety is concerned, and this is effectively trapped as *S*-adenosyl ethionine (figure

7.41). The resulting lack of ATP leads to inhibition of protein synthesis and a deficiency in the production of the apolipoprotein complex responsible for transporting triglycerides out of the liver. Consequently, there is an accumulation of triglycerides.

The reduction in protein synthesis obviously has other ramifications, such as a deficiency in hepatic enzymes and a consequent general disruption of intermediary metabolism. Methylation reactions are presumably also affected.

S-Adenosyl ethionine carries out ethylation reactions or ethyl transfer and this is presumably involved in the carcinogenesis. Administration of ethionine to animals leads to the production of ethylated bases such as ethyl guanine. This may account for the observed inhibition of RNA polymerase and consequently of RNA synthesis. Incorporation of abnormal bases into nucleic acids and the production of impaired RNA may also lead to the inhibition of protein synthesis and misreading of the genetic code.

The depletion of ATP is the preliminary event leading to the pathological changes and the ultrastructural abnormalities of the nucleus and cytoplasmic organelles. Administration of ATP or precursors reverses all of these changes. The exact mechanism underlying the carcinogenesis is less clear, but presumably involves inhibition of RNA synthesis or the production of abnormal ethylated nucleic acids and hence disruption of transcription, translation or possibly replication. It is of interest that ethionine is not mutagenic in the Ames test, with or without rat liver homogenate. However, Williams and Weisburger (see Bibliography) suggested that ethionine may be carcinogenic after metabolism to vinyl homocysteine (in which vinyl replaces ethyl) which is highly mutagenic.

Biochemical effects: interaction with specific protein receptors

Carbon monoxide

Despite the fact that carbon monoxide is no longer present in the gas used for domestic cooking and heating, it is still a major cause of poisoning. Indeed, it is one of the most important single agents involved in accidental and intentional poisoning resulting in the deaths of several hundred people in Britain each year. There are in fact many sources of carbon monoxide: improperly burnt fuel in domestic fires, car exhausts, coal gas, furnace gas, cigarette smoke and burning plastic. Thus traffic policemen, firemen and those trapped in fires and certain types of factory workers may all be potentially at risk. Because carbon monoxide is not irritant and has no smell, a dangerous concentration may be reached before the victim is aware of anything untoward. The pure gas is also used in laboratories and it may be formed endogenously from the solvent **methylene chloride** by metabolism (figure 7.42). In the latter case people using the solvent as a paint stripper in the home or for degreasing machinery in industry, and who inhale sufficient quantities, may be overcome some time after the exposure due to the slow release of methylene chloride from adipose tissue and metabolism to carbon monoxide. There is also a small amount of carbon monoxide produced normally

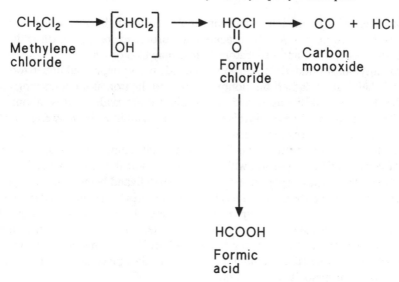

FIGURE 7.42. Metabolism of methylene chloride to carbon monoxide.

from metabolism of the protoporphyrin ring of haemoglobin. As a result of the pioneering studies of Haldane in the 19th century and subsequent case studies, the symptoms are well documented. The concentration in the blood is normally measured as percent carboxyhaemoglobin, as carbon monoxide binds to haemoglobin (see below).

A level of 60% carboxyhaemoglobin will normally be fatal even for a few minutes, and if not fatal may cause permanent damage; it is rare to find a blood level of greater than 80% carboxyhaemoglobin. At a level of 20% carboxy-haemoglobin there may be no obvious symptoms but the ability to perform tasks can be impaired. When the level reaches 20–30% the victim may have a headache, raised pulse rate, a dulling of the senses and a sense of weariness. At levels of 30–40% the symptoms will be the same but more pronounced, the blood pressure will be low and exertion may lead to faintness. At 40–60% carboxyhaemoglobin there will be weakness and incoordination, mental confusion and a failure of memory. At concentrations of 60% carboxyhaemoglobin and above, the victim will be unconscious and will suffer convulsions. There are many other clinical features: nausea, vomiting, pink skin, mental confusion, agitation, hearing loss, hyperpyrexia, hyperventilation, decrease in light sensitivity, arrhythmias, renal failure and acidosis.

The spectrum of pathological effects includes peripheral neuropathy, brain damage, myocardial ischaemia and infarction, muscle necrosis and pulmonary oedema. However, the main target organs are the brain and heart. This is because these organs have a relative inability to sustain an oxygen debt and they utilize aerobic metabolic pathways extensively. The brain damage may be due to several mechanisms including metabolic acidosis, hypotension, metabolic inhibition and decreased blood flow and oxygen availability. The progressive hypotension which is observed may be an important contributor to the ischaemia which occurs. Neural damage may follow both acute and chronic exposure. Death is due to brain tissue hypoxia, and respiratory failure may also occur.

The mechanism underlying carbon monoxide poisoning is well understood at the biochemical level. Carbon monoxide binds to the iron atom in haemoglobin at the same binding site as oxygen, but it binds more avidly, indeed about 240 times more strongly. The product is carboxyhaemoglobin which may contain one or more carbon monoxide molecules.

The haemoglobin molecule has four polypeptide chains, 2α and 2β, and each has a porphyrinic haem group with one iron atom at the centre which has five bonds utilized in the porphyrin structure, the sixth ligand being free for oxygen, (figure 7.43). Thus, haemoglobin has four binding sites for oxygen, and therefore carbon monoxide can take up some of these. Loss of oxygen from the haemoglobin molecule causes a change in the tertiary structure of the whole molecule, a phenomenon known as **co-operativity**. Loss of molecules of oxygen from a fully oxygenated haemoglobin complex takes place in four steps, each with a different dissociation constant:

$$Hb_4O_8 \rightarrow Hb_4O_6 + O_2$$
$$K_1$$

where dissociation constant, $K_1 = \dfrac{[Hb_4O_6][O_2]}{[Hb_4O_8]}$

$$K_2 = \dfrac{[Hb_4O_4][O_2]}{[Hb_4O_6]}$$

$$K_3 = \dfrac{[Hb_4O_2][O_2]}{[Hb_4O_4]}$$

$$K_4 = \dfrac{[Hb_4][O_2]}{[Hb_4O_2]}$$

(The smaller the dissociation constant, the more tightly the oxygen is bound.) When all four oxygen molecules are bound, each is equivalent. Loss of the first oxygen occurs when the partial pressure of the ambient oxygen falls. This loss of oxygen triggers a co-operativity change that facilitates the loss of the second oxygen molecule; thus $K_2 > K_1$. Similarly loss of the second oxygen facilitates loss of the third oxygen ($K_3 > K_2$). Loss of the fourth oxygen does not normally occur. Loss of the first oxygen requires a change in the partial pressure of O_2 of 60 mm Hg, the second requires a drop of 15 mm Hg, and the third of 10 mm Hg. With a level of 50% carboxyhaemoglobin, a fall in the partial pressure of O_2 of about 90 mm Hg is required for the loss of the first oxygen. This is due to the fact that the most common species will be $Hb_4 (O_2)_2(CO)_2$, so the co-operativity normally expected is confined to one oxygen. Thus, binding of carbon

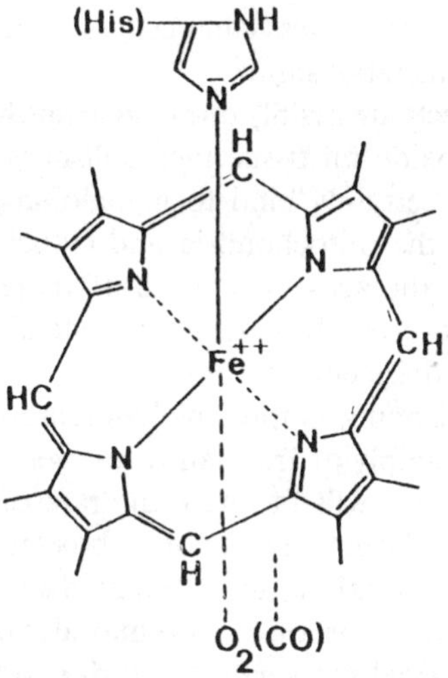

FIGURE 7.43. The haem moiety of the haemoglobin molecule showing the binding of the oxygen molecule to the iron atom. As shown in the diagram, carbon monoxide binds at the same site. His = side chain of the amino acid histidine. From Timbrell, J. A., *Introduction to Toxicology*. Taylor and Francis, London, 1989.

monoxide causes a shift in the oxygen dissociation curve resulting in more avid binding of oxygen, and so it is very much less readily given up in the tissues. Also, the binding of one or more carbon monoxide molecules to haemoglobin results, just as does the binding of oxygen, in an **allosteric change** in the remaining oxygen binding sites. The affinity of the remaining haem groups for oxygen is increased and so the dissociation curve is also distorted. The consequence of this is that the tissue anoxia is very much greater than would be expected from a simple loss of oxygen carrying capacity by replacement of some of the oxygen molecules on the haemoglobin molecule.

The binding of carbon monoxide to haemoglobin is described by the Haldane equation:

$$\frac{[COHb]}{[HbO_2]} = M \frac{[P_{CO}]}{[P_{O_2}]}$$

where $M = 220$ for human blood.

Therefore when $P_{CO} = 1/220 \times P_{O_2}$ the haemoglobin in the blood will be 50% saturated with carbon monoxide. Since air contains 21% oxygen, approximately 0·1% carbon monoxide will give this level of saturation. Hence, carbon monoxide is potentially very poisonous at low concentrations. The rate at which the arterial blood concentration of carbon monoxide reaches an equilibrium with the alveolar concentration will depend on other factors such as exercise and the efficiency of the lungs. Other factors will also affect the course of the poisoning. Thus, physiological changes in blood flow through organs, such as shunting, may occur. Peripheral vasodilation due to hypoxia may exceed the cardiac output and hence fainting and unconsciousness can occur. Lactic acidaemia will also result from impaired aerobic metabolism.

Whether the toxic effects are mainly due to **anaemic hypoxia** or to the **histotoxic effects** of carbon monoxide on tissue metabolism is a source of controversy. Carbon monoxide will certainly bind to myoglobin and cytochromes such as cytochrome oxidase in the mitochondria and cytochrome P-450 in the endoplasmic reticulum, and the activity of both of these enzymes is decreased by carbon monoxide exposure. However, the general tissue hypoxia will also decrease the activity of these enzymes.

Treatment for the poisoning simply involves removal of the source of carbon monoxide and either a supply of fresh air or oxygen. Use of hyperbaric oxygen (2·5 atmospheres pressure) will facilitate the rate of dissociation of carboxyhaemoglobin. Thus, the plasma half-life of carboxyhaemoglobin can be reduced from 250 min in a patient breathing air to 23 min in a patient breathing hyperbaric oxygen (figure 7.44). Adding carbon dioxide may also be useful as this reduces the half-life to 12 min at normal pressure. This is due to the stimulation by carbon dioxide of alveolar ventilation and acidaemia which increases the dissociation of carboxyhaemoglobin.

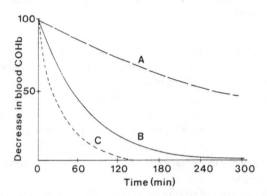

FIGURE 7.44. The dissociation of carboxyhaemoglobin in the blood of a patient poisoned with carbon monoxide. The graph show the effects of breathing air (A), oxygen (B) or oxygen at increased pressure (2.5 atmospheres) (C) on the rate of dissociation. From Timbrell, J.A., *Introduction to Toxicology*, Taylor and Francis, London, 1989. (Data from *Poisoning-Diagnosis and Treatment*, Eds. J.A. Vale and T.J. Meredith, Update Books, London, 1981.)

Cyanide

Poisoning with cyanide may occur in a variety of ways: accidental or intentional poisoning with cyanide salts, which are used in industry or in laboratories; as a result of exposure to hydrogen cyanide in fires when polyurethane foam burns; from **sodium nitroprusside** which is used therapeutically as a muscle relaxant and produces cyanide as an intermediate product; and from the natural product amygdalin which is found in apricot stones for example. **Amygdalin** is a glycoside which releases glucose, benzaldehyde and cyanide when degraded by the enzyme β-glucosidase in the gut flora of the gastrointestinal tract. There are many other natural sources of cyanide such as *Cassava* tubers for example. There are various reports of human victims poisoned with cyanide, the most famous of which was the case of the influential Russian monk, **Rasputin**. From such case studies the approximate lethal dose and toxic blood level is known. Thus, a dose of around 250–325 mg of potassium or sodium cyanide is lethal in humans, and a lethal case would have a blood level of around 1 μg/ml. A blood level of greater than 0·2 μg/ml is toxic. The symptoms of poisoning include headache, salivation, nausea, anxiety, confusion, vertigo, convulsions, paralysis, unconsciousness, coma, cardiac arrhythmias, hypotension and respiratory failure. Both venous blood and arterial blood remain oxygenated, and hence the victim may appear pink.

The pathological effects seem to involve damage to the grey matter, and possibly the white matter, in the brain.

The mechanism underlying the toxicity of the cyanide ion is relatively straightforward and involves reversible binding to and inhibition of the cytochrome a–cytochrome a_3 complex (cytochrome oxidase) in the mitochondria. This is the terminal complex in the electron transport chain which transfers electrons to oxygen, reducing it to water (figure 7.45). Cyanide therefore blocks electron transport and oxygen is not utilized, so the peripheral oxygen tension rises and the unloading gradient for oxyhaemoglobin is decreased. The toxic effect is known as **histotoxic hypoxia**. Cyanide also directly stimulates chemoreceptors causing hyperpnoea. Cardiac irregularities often occur, but death is due to respiratory arrest resulting from a central effect as the nerve cells of the respiratory centre are particularly sensitive to hypoxia. The susceptibility of the brain to pathological damage may reflect the lower concentration of cytochrome oxidase in white matter.

Delayed **ischaemic anoxia** may also result from cyanide poisoning as a result of circulatory effects.

There are several antidotal treatments, but the blood level of cyanide should be determined if possible as treatment may be hazardous in some cases. Cyanide is metabolized in the body, indeed up to 50% of the cyanide in the circulation may be metabolized in 1 h. This metabolic pathway involves the enzyme **rhodanese** and **thiosulphate** ion which produces thiocyanate (figure 7.46). However, the crucial part of the treatment is to reduce the level of cyanide in the blood as soon

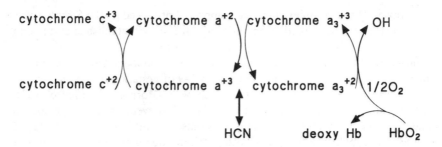

FIGURE 7.45. The site of action of cyanide in the electron transport chain. Hb: haemoglobin; HbO$_2$: oxyhaemoglobin.

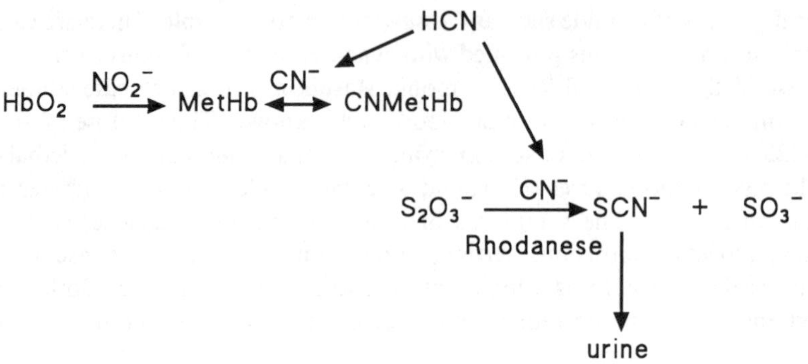

FIGURE 7.46. The metabolism and detoxication of cyanide. HbO$_2$: oxyhaemoglobin; MetHb: methaemoglobin; CNMetHb: cyanomethaemoglobin.

as possible and allow the cyanide to dissociate from the cytochrome oxidase. This is achieved by several means. **Methaemoglobin** will bind cyanide more avidly than cytochrome oxidase and therefore competes for the available cyanide. Converting some of the haemoglobin in the blood to methaemoglobin will therefore decrease the circulating cyanide level and reduce the inhibition of the mitochondrial enzyme. Inhalation of **amyl nitrite** which is volatile, may be used initially rapidly to oxidize haemoglobin to methaemoglobin, and this is followed by sodium nitrite given intravenously to continue the production of methaemoglobin. When methaemoglobin binds cyanide, cyanomethaemoglobin is formed which cannot carry oxygen but will release its cyanide. Therefore the administration of thiosulphate will facilitate the conversion of cyanide to thiocyanate and dissociation of cyanomethaemoglobin. The thiocyanate is excreted in urine. An alternative treatment, which is now usually used when possible, is the administration of **cobalt edetate** which is a chelating agent which binds cyanide and is excreted into the urine.

Teratogenesis

Actinomycin D

Actinomycin D is a complex chemical compound produced by the *Streptomyces* species of fungus and is used as an antibiotic. It is a well established and potent teratogen and is also suspected of being carcinogenic.

The teratogenic potency of actinomycin D shows a marked dependence on the time of administration, being active on days 7–9 of gestation in the rat, when a high proportion of surviving foetuses show malformations (figure 6.13). Administration of the compound at earlier times, however (figure 7.47), results in a high foetal death and resorption rate. This falls to about 10% on the 13th day of gestation. The malformations produced in the rat are numerous, including cleft palate and lip, spina bifida, ecto and dextrocardia, anencephaly and disorganization of the optic nerve. Abnormalities of virtually every organ system

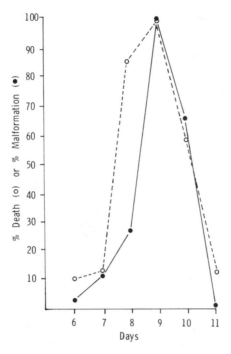

FIGURE 7.47. Embryolethality and tetatogenicity of actinomycin D. This graph shows the relationship between the time of dosing and susceptibility to malformations (●) or death (○). Data from Wilson (1965) *Ann. N.Y. Acad. Sci.*, **123**, 119.

may be seen at some time. Actinomycin D is particularly embryolethal, unlike other teratogens such as thalidomide (see below). This embryolethality shows a striking parallel with the incidence of malformations (figure 7.48). Although this has been observed to a certain extent with some other teratogens, in other cases embryolethality and malformations seem to be independent variables. It is well established that actinomycin D inhibits DNA-directed RNA synthesis by binding to guanosyl residues in the DNA molecule. This disrupts the transcription of genetic information and thereby interferes with the production of essential proteins. DNA synthesis may also be inhibited, being reduced by 30–40% *in utero*. It is clear that in the initial stages of embryogenesis, synthesis of RNA for protein production is vital and it is not surprising that inhibition of this process may be lethal.

It has been shown that radiolabelled actinomycin D is bound to the RNA of embryos on days 9, 10 and 11 of gestation. Using incorporation of tritiated uridine as a marker for RNA synthesis, it was shown that only on gestational days 9 and 10 was there significant depression of incorporation in certain embryonic cell groups. On days 7, 8, and 11 no significant depression of uridine incorporation was observed. The depression of RNA synthesis as measured by uridine incorporation therefore correlates approximately with teratogenicity. However, actinomycin D is also cytotoxic and this can be demonstrated as cell damage in embryos on days 8, 9 and 10 but not at earlier times or on day 11.

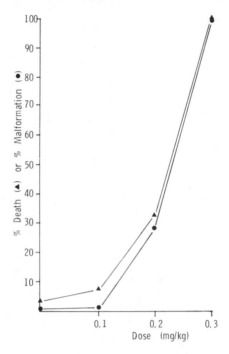

FIGURE 7.48. Embryolethality and teratogenicity of actinomycin D. This graph shows the dose–response relationship for these two toxic effects. Data from Wilson (1965) *Ann. N.Y. Acad. Sci.*, **123**, 119.

Therefore, the teratogenesis shows a correlation with cytotoxicity as well as with inhibition of RNA synthesis. Certainly the concentration of actinomycin D in the embryo after administration on day 11 was too low for effective inhibition of RNA synthesis.

Although it is clear that actinomycin D is cytotoxic and inhibits RNA synthesis, the role of these effects in teratogenesis is not yet clear. RNA synthesis is a vital process in embryogenesis, preceding all differentiation, chemical and morphological, and therefore inhibition of it would be expected to disturb growth and differentiation. This may account for the production of malformations after administration on gestational day 7 in the rat, a period not normally sensitive in this species for the production of malformations. Excessive embryonic cell death may also be teratogenic.

Diphenylhydantoin

Diphenylhydantoin is an anticonvulsant drug in common use which is suspected of being teratogenic in humans. There is at present insufficient data available to specify the exact type of malformation caused in humans, but in a few cases craniofacial anomalies, growth retardation and mental deficiency have been documented. Heart defects and cleft palate have also been described in some cases in humans. In experimental animals, however, diphenylhydantoin is clearly teratogenic and this has been shown repeatedly in mice and rats. The defects most commonly seen are orofacial and skeletal, the orofacial defect usually described being cleft palate. Rhesus monkeys, however, are much more resistant to the teratogenicity, only showing minor urinary tract abnormalities, skeletal defects and occasional abortions after high doses.

Although the teratogenicity of diphenylhydantoin in humans has not been demonstrated so clearly as that of thalidomide, the levels of drug to which experimental animals were exposed were not excessively high. The plasma levels of the unbound drug in the maternal plasma of experimental animals after teratogenic doses were only 2–3 times higher than those found in man after therapeutic doses. The teratogenic effects in experimental animals were found to occur near the maternal toxic dose.

The types of defect produced in mice by diphenylhydantoin at various times of gestation are shown in table 7.6. There is good correlation between the timing of these defects and the known pattern of organogenesis in the mouse.

Table 7.6. Timing of teratogenic effect of diphenylhydantoin in the mouse

| Malformation | \multicolumn{6}{c}{Treatment on gestational day} |
	6–8	9–10	11–12	13–14	15–16	17–19
Orofacial	0	24	63	57	4	0
Eye defects	0	19	0	0	0	0
Limb defects	0	12	0	0	0	0
CNS defects	0	28	0	0	0	0
Skeletal defects	0	44	52	57	3	0
Kidney defects	0	25	17	7	4	0

Dose of diphenylhydantoin: 150 mg/kg.
Figures are percentages of all surviving foetuses displaying the malformation.
Data from Harbison, R. D. and Becker, B. A. (1969) *Teratology*, **2**, 305.

It can be seen that the greatest number of malformations of any sort occur on gestational days 11–12, with these being mainly orofacial and skeletal.

Diphenylhydantoin-induced malformations show a clear dose–response, as can be seen from figure 7.49 for pregnant mice treated on the 11th, 12th and 13th days of gestation. No significant increase in foetal deaths was observed below 75 mg/kg, but above this dose level more than 60% embryolethality was observed *in utero*, indicating a very steep dose–response curve. The mechanism underlying diphenylhydantoin teratogenicity is not fully understood, but enough is known to make it an interesting example.

In both rats and humans, diphenylhydantoin is known to undergo aromatic hydroxylation, presumably catalysed by the microsomal cytochromes P-450 system. However, pretreatment of pregnant mice with inducers or inhibitors of microsomal drug oxidation decreases or increases the teratogenicity respectively. This paradoxical effect may simply be due to increases or decreases in the removal of the drug from the maternal circulation by metabolism and excretion following pretreatment.

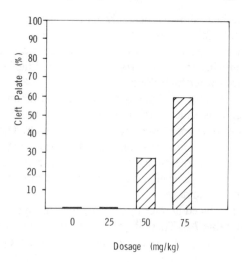

FIGURE 7.49. Dose–response relationship for diphenylhydantoin teratogenesis. The incidence of cleft palate in the surviving embryos is plotted against the dose of diphenylhydantoin given to pregnant mice on days 11, 12 and 13 of gestation. Adapted from Harbison and Becker (1969) *Teratology*, 2, 305.

It has recently been proposed that metabolic activation of diphenylhydantoin may be responsible for the teratogenicity. After the administration of radioactively labelled diphenylhydantoin to pregnant mice, radioactive drug or a metabolite was found to be covalently bound to protein in the embryo. It was shown that both the teratogenicity and embryolethality of diphenylhydantoin could be increased by using an inhibitor of epoxide hydrolase (see page 106), **trichloropropene oxide**. Similarly, the covalent binding of radiolabelled diphenylhydantoin to protein was also increased by this treatment.

Metabolism by epoxide hydrolase is effectively a detoxication pathway for reactive epoxides such as that proposed as an intermediate in diphenylhydantoin metabolism (figure 7.35), and inhibition of this enzyme therefore blocks the detoxication. It was postulated that the epoxide of diphenylhydantoin was the reactive and teratogenic intermediate produced by metabolism. The maternal plasma concentrations of drug in the mice used in this study were similar to those measured in humans after therapeutic doses of diphenylhydantoin had been given.

Although changes in the covalent binding of labelled diphenylhydantoin to foetal protein correlated with changes in teratogenicity, this does not prove a direct relationship. Indeed, other evidence suggests that diphenylhydantoin itself is the teratogenic agent. *In vitro*, using **micromass cell culture**, the drug was directly inhibitory and toxicity was increased in the presence of various cytochromes P-450 inhibitors. Studies in mice *in vivo* also showed a correlation between increased maternal plasma levels of the unchanged drug and increased teratogenicity. Thus, there is conflicting evidence at present, and whether the putative reactive intermediate of diphenylhydantoin which binds to protein is the ultimate teratogen, or whether it is the parent drug, awaits clarification.

Thalidomide

Despite the great interest in and notoriety of thalidomide as a teratogen, the underlying mechanisms of its teratogenicity are still not clear. It is, however, of particular interest in being a well established human teratogen.

Thalidomide was an effective sedative sometimes used by pregnant women for the relief of morning sickness, and the drug seemed remarkably non-toxic. However, it eventually became apparent that its use by pregnant women was associated with characteristic deformities in the offspring. These deformities were **phocomelia** (shortening of the limbs), and malformations of the face and internal organs have also occurred. It was the sudden appearance of cases of phocomelia which alerted the medical world, as this had been a hitherto rare congenital abnormality.

It became clear that in virtually every case of phocomelia, the mother of the malformed child had definitely taken thalidomide between the 3rd and 8th week of gestation. In some cases only a few doses had been taken during the critical period. Analysis of the epidemiological and clinical data suggested that thalidomide was almost invariably effective if taken on a few days or perhaps just one day between the 20th and 35th gestational days. It has been possible to pinpoint the critical periods for each abnormality. Lack of the external ear and paralysis of the cranial nerve occur on exposure during the 21st and 22nd days of gestation. Phocomelia, mainly of the arms, occurs after exposure on days 24–27 of gestation with the legs affected 1–2 days later. The sensitive period ends on days 34–36 with production of hypoplastic thumbs and anorectal stenosis.

Initially these malformations were not readily reproducible in rats or other experimental animals. It was later discovered that limb malformations could be produced in certain strains of white rabbits if they were exposed during the 8th–16th days of gestation. Several strains of monkeys were found to be susceptible and gave similar malformations to those seen in humans. It was eventually found that malformations could be produced in rats but only if they were exposed to the drug on the 12th day of gestation. At teratogenic doses thalidomide has no embryolethal effect, doses several times larger being necessary to cause foetal death.

The mechanism underlying thalidomide teratogenesis has never been completely elucidated, despite considerable research. However, the evidence suggests that a metabolite is responsible rather than the parent drug. The metabolism of thalidomide is complex, involving both enzymic and non-enzymic pathways. A number of metabolites arise by hydrolysis of the amide bond in the piperidine ring (figure 7.50) of which about 12 have been identified *in vivo*, and aromatic ring hydroxylation may also take place. The hydrolytic opening of the piperidine ring yields phthalyl derivatives of glutamine and glutamic acid. **Phthalylglutamic acid** may also be decarboxylated to the monocarboxylic acid derivative. Other monocarboxylic and dicarboxylic acid derivatives are also produced, some by opening of the phthalimide ring.

None of these metabolites was found to be teratogenic in the rabbit, but it has been subsequently found that they were unable to penetrate the embryo. Studies *in vivo* revealed that after the administration of the parent drug the major products detectable in the foetus were the monocarboxylic acid derivatives. It therefore seems that the parent drug may cross the placenta and enter the foetus and there be metabolized to the teratogenic metabolite.

Subsequently the phthalylglutamic acid (figure 7.50), a dicarboxylic acid, was shown to be teratogenic in mice when given on days 7–9 of pregnancy. However, other work has indicated that the phthalimide ring is of importance in the teratogenicity. Other studies have indicated that thalidomide acylates aliphatic amines such as putrescine (figure 7.26) and spermidine, histones, RNA and DNA and that it may also affect ribosomal integrity.

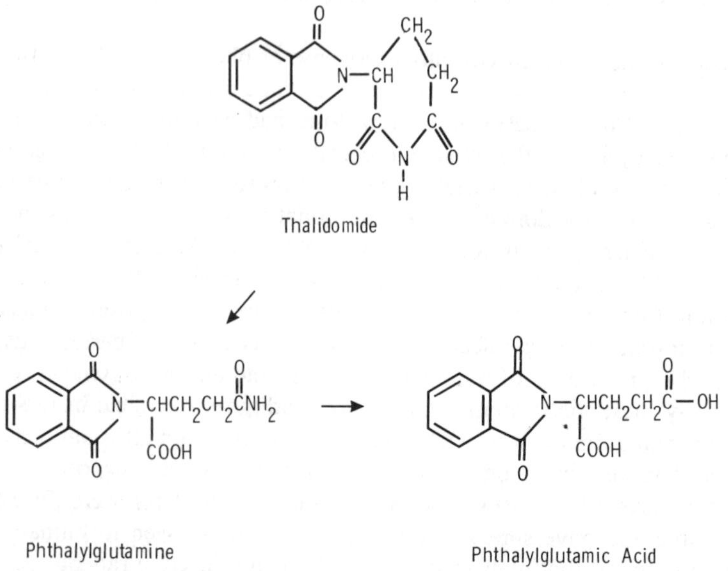

Thalidomide

Phthalylglutamine

Phthalylglutamic Acid

FIGURE 7.50. Structure of thalidomide and two of its hydrolysis products.

Evidence is now accumulating that a reactive metabolite(s), generated perhaps by a minor form of cytochrome P-450, is involved in the teratogenicity. This has been gained mainly using various *in vitro* systems such as tumour cells, human lymphocytes, limb bud and embryo culture and a variety of techniques. The identity of this metabolite is unknown, although some evidence suggests an epoxide is involved. Thus hepatic microsomes from pregnant rabbits and hepatic preparations (S–9) from rabbit, monkey and human foetuses all produced a metabolite which was toxic to human lymphocytes *in vitro*. Liver microsomes from pregnant rats did not. This species difference is in agreement with the difference in susceptibility to thalidomide teratogenesis.

Furthermore, inhibition of epoxide hydrolase increased and addition of the purified enzyme decreased the toxicity of thalidomide to human lymphocytes *in vitro*. This data suggested that an epoxide might be involved in the toxicity and this is supported by earlier metabolic studies *in vivo* which have shown the presence of phenolic metabolites in the urine of rabbits but not rats treated with the drug. However, other experimental data using an *in vitro* system involving tumour cells was inconsistent with this data and suggested that an epoxide intermediate was not involved. The elucidation of the ultimate teratogenic metabolite, whether cytochrome P-450-generated or not, awaits further research. It is also of interest that there is a difference in the toxicity of the isomers of thalidomide, with the $S-$ enantiomer being more embryotoxic than the $R+$ enantiomer.

Perhaps the particular lesson to be learnt from the tragedy of thalidomide is that a drug with low maternal and adult toxicity, which is similarly of low toxicity in experimental animals, may have high teratogenic activity. The human embryo seems to have been particularly sensitive to normal therapeutic doses in this case. However, it has resulted in new drugs now being more rigorously tested and also tested in pregnant animals.

Immunotoxicity

Halothane

Halothane is a very widely used anaesthetic drug which may cause **hepatic damage** in some patients. It seems there are two types of damage, however. One is a very rare, idiosyncratic, reaction resulting in serious liver damage with an incidence of about 1 in 35 000. The other form of hepatotoxicity is a mild liver dysfunction which is more common and occurs in as many as 20% of patients receiving the drug. The two different types probably involve different mechanisms.

The more common mild liver dysfunction is thought to be due to a direct toxic action of one of the halothane metabolites on the liver. It is manifested as raised serum transaminases (AST and ALT). Studies in experimental animals have indicated that under certain conditions, such as in phenobarbital-induced male rats exposed to halothane with a reduced concentration of oxygen, halothane is directly hepatotoxic. This is believed to be due to the reductive pathway of metabolism (figure 7.51) which produces a free radical metabolite. This metabolite will bind covalently to liver protein and can initiate lipid peroxidation, both of which have been detected in the experimental rat model. The other metabolites and fluoride ion have also been detected in experimental animals in support of this reductive pathway. Covalent binding to liver protein and the metabolic products of the reductive pathway have both been detected in a brain-dead human subject exposed to radiolabelled halothane during a transplant operation. The level of plasma fluoride correlated with the covalent binding. Thus, this reductive pathway which would be favoured in rats anaesthetized with halothane, and reduced oxygen concentration may be the cause of the direcdt toxic effect on the liver. The severe and rare hepatic damage has clearly different features from the mild hepatotoxicity. As well as centrilobular hepatic necrosis, patients suffer fever, rash, arthralgias and have serum tissue autoantibodies.

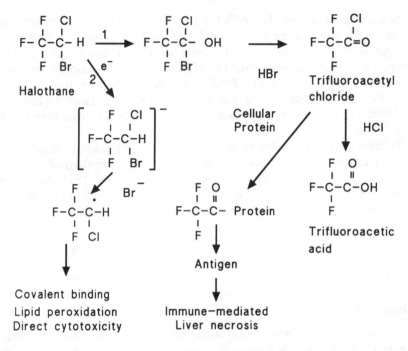

FIGURE 7.51. The metabolism of halothane and its proposed involvement in the toxicity. Pathway 1 (oxidative) and pathway 2 (reductive) are both catalysed by cytochrome P-450.

About 25% of patients have anti-microsomal antibodies. There are also several predisposing factors and those so far recognized are:

(a) multiple exposures, which seem to sensitize the patient to future exposures
(b) sex, females being more commonly affected than males in the ratio 1.8 : 1
(c) obesity, 68% of patients in one study were obese
(d) allergy, a previous history of allergy was found in one third of patients. After multiple exposures the incidence increases from 1 in 35 000 patients to 1 in 3700.

Halothane is believed to cause this severe hepatic damage via an immunological mechanism whereby antibodies to altered liver cell membrane components are generated by repeated exposure to a halothane-derived antigen. The immunological reaction which ensues is directed at the liver cells of the patient. These are destroyed, resulting in hepatic necrosis.

Initial studies using rabbits anaesthetized with halothane revealed that lymphocytes from patients with halothane-induced hepatic damage were sensitized against the liver homogenate from these rabbits. Furthermore, lymphocytes from these patients were cytotoxic towards the rabbit hepatocytes. Serum from patients suffering from halothane-induced hepatic damage contained antibodies of the IgG type, directed against an antigenic determinant on the surface of rabbit hepatocytes when these were derived from rabbits which had been anaesthetized with halothane. A subpopulation of lymphocytes (T-lymphocytes, killer lymphocytes) from normal human blood attacked rabbit hepatocytes after incubation with sera from patients, showing the antibodies were specific for cells altered by halothane. These studies suggested that patients had an immunological response to a halothane-derived antigen. Controlling the level of oxygen during anaesthesia indicated that the oxidative metabolic pathway was involved rather than the reductive pathway. Only sera or lymphocytes from those patients with the severe type of damage, not the more common hepatic damage, reacted to rabbit hepatocytes in this way, and only hepatocytes from rabbits exposed to halothane were targets. Thus, these and later studies distinguished between the mild and severe hepatic damage.

Further work has identified **trifluoroacetyl chloride** as the probable reactive metabolite which trifluoroacylates protein (figure 7.51). Removal of the trifluoroacyl moiety from the antigenic protein by chemical means abolished most of the interactions with antibodies from patients. Preparation of the suspected hapten, **trifluoroacyl-lysine**, inhibited the interaction between the antibody and altered cell protein. The major target proteins seem to be microsomal, that is derived from the smooth endoplasmic reticulum, and the majority of the antigenic activity is present in this fraction although it is also detectable on the surface of the hepatocyte. The antigens correspond to at least five polypeptide fractions in rat, rabbit and human liver and they appear to be similar although not identical in each species. These polypeptides have been characterized as having the following molecular weights: 100, 76, 59, 57 and 54 kDa, detected by immunoblotting with human serum from patients and rat, human or rabbit liver microsomal fraction. The 100- and 76-kDa antigens are the most common. The 59-kDa protein has been identified as a microsomal carboxyl-esterase and the 54-kDa protein may be cytochromes P-450. Each of these antigens which are recognized by patients' sera is trifluoroacylated. However, antibodies recognize epitopes which are a combination of the trifluoroacyl group and determinants of the polypeptide carrier. The trifluoroacyl group is essential and is recognized in the liver antigens by anti-trifluoroacyl antibodies. However, there are other determinants as trifluoroacyl-lysine, the putative hapten, would not totally inhibit the binding of antibody from patients' serum to the liver antigen. This phenomenon is known as hapten inhibition. Also, the sera from different patients varied in the recognition of the different antigens.

The production of the antigens was reduced in rats in which a deutero analogue of halothane was given and the metabolism of which would be reduced by an isotope effect. The antigens could be formed *in vitro*, but only under aerobic conditions and not under anaerobic conditions implicating the oxidative pathway rather than the reductive (figure 7.51).

It is not yet clear whether halothane-induced hepatitis is mediated by a humoral- or cell-mediated type of immune response, but it seems to involve both.

Thus, both specific circulating antibodies and sensitized T-lymphocytes are involved and it may be a **Type II hypersensitivity reaction** in which antibodies bind to tissue antigens. Certain cells such as killer lymphocytes have Fc receptors which bind to the Fc portion of the bound antibody (IgG). As a result the killer lymphocyte is activated and lyses the cell to which the antibody is bound. This is known as **antibody-dependent cell cytotoxicity (ADCC)**.

Trifluoroacetylchloride is very reactive and therefore if any escapes from the vicinity of the cytochromes P-450 in the smooth endoplasmic reticulum it might be expected to react with those proteins in highest concentration. The 59-kDa polypeptide, a microsomal carboxylesterase, constitutes 1·5% of the total microsomal protein and therefore it fulfils this criterion. The rate of degradation of the carrier molecule may also be an important factor in determining whether an antigen becomes an immunogen. The concentration of antigen which is presented to the immune system may similarly be a very important factor. Some of the antigenic polypeptides detected seem to be long lived. The resulting epitope density may depend on the number of lysine groups in the particular protein and this will in turn affect the immunogenicity of the antigen. Trifluoroacyl adducts have been detected on the outer surface of hepatocytes, but it is not yet clear how they get there.

The fact that the production of the trifluoroacetyl chloride is part of the major metabolic pathway and that the majority of patients produce trifluoroacylated proteins suggests that it is differences in the immune surveillance system or immune responsiveness which determine which patients will succumb to the immunotoxic effect.

Practolol

Practolol (figure 7.52) is an antihypertensive drug which had to be withdrawn from general use because of severe adverse effects which became apparent in 1974 about four years after the drug had been marketed. The toxicity of practolol was unexpected and when it occurred it was severe. Furthermore, it has never been reproduced in experimental animals, even with the benefit of hindsight.

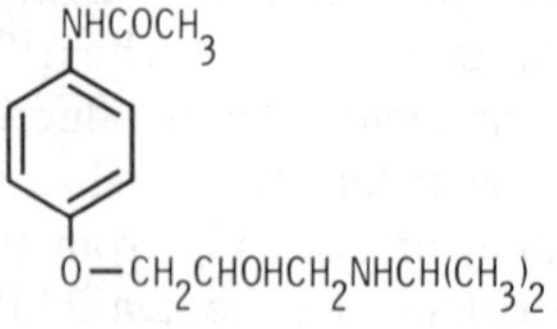

FIGURE 7.52. Structure of practolol.

The available evidence points to the involvement of an immunological mechanism, the basis of which is probably a metabolite-protein conjugate acting as an antigen. It seems unlikely that the pharmacological action of this drug, β-blockade, is involved, as other β-blocking drugs have not shown similar adverse effects. The syndrome produced by this drug features lesions to the eye, peritoneum and skin. Epidemiological studies have firmly established practolol as the causative agent in the development of this toxic effect, described as the **occulo-mucocutaneous syndrome**. This involved extensive and severe skin rashes, similar to psoriasis, and keratinization of the cornea and peritoneal membrane.

Studies *in vivo* showed that practolol was metabolized to only a limited extent in human subjects and experimental animals. Thus, 74–90% of a dose of practolol is excreted in the urine in humans, and of this, 80–90% is unchanged and less than 5% of the dose is deacetylated. No differences in metabolism were detected between patients with the practolol syndrome and those not suffering from it. However, in patients with the syndrome there were no antibodies detected to practolol itself. Hamsters treated with practolol showed a marked and persistent accumulation of the drug in the eye. Furthermore, in a hamster liver microsomal system *in vitro*, it was shown that practolol could be covalently bound to microsomal protein and also to added serum albumin. It was further shown that human sera from patients with the practolol syndrome contained antibodies which were specific for this practolol metabolite generated *in vitro*. Factors affecting the covalent binding of the practolol metabolite to protein *in vitro* also influenced the reaction between the patients' sera and the protein conjugate. The absence of NADPH, the presence of glutathione and the use of 3-methylcholanthrene-induced hamster liver microsomes, all reduced the binding and the immunological reaction. These results indicated a requirement for NADPH-dependent microsomal enzyme-mediated metabolism and binding to protein to produce a hapten recognizable by the antibodies in the patients' sera. However patients without the syndrome also had antibodies although these were present at a lower level.

The identity of the metabolite has not yet been established but it is neither the deacetylated nor 3-hydroxymetabolite found *in vivo*. However, deacetylation may be a prerequisite for the *N*-hydroxylation reaction which has been suggested. Although the exact antigenic determinants of the metabolite have not been elucidated, it is known that the complete isopropanolamine side chain is necessary.

Therefore it seems likely that practolol causes an adverse reaction which has an immunological basis due to the formation of an antigen between a practolol metabolite and a protein. The reason for the bizarre manifestations of the immunological reaction and the metabolite responsible have not yet been established.

Penicillin

Penicillin and its derivatives are very widely used antibiotics which cause more **allergic reactions** than any other class of drug. The incidence of allergic reactions to such drugs occurs in 1–10% of recipients. All four types of hypersensitivity reaction have been observed with penicillin. Thus high doses may cause haemolytic anaemia and immune complex disease, cell-mediated immunity may give rise to skin rashes and eruptions, and the most common reactions are urticaria, skin eruptions and arthralgia. Antipenicillin IgE antibodies have been detected consistent with an anaphylactic reaction. The anaphylactic reactions (Type 1; see Chapter 6) which occur in $0 \cdot 004$–$0 \cdot 015\%$ of patients may be life threatening.

Pencillin is a reactive molecule both *in vitro* and *in vivo* where it undergoes a slow transformation to a variety of products. Some of these products are more reactive than the parent drug and can bind covalently to protein, reacting with nucleophilic amino, hydroxyl, mercapto and histidine groups. Therefore a number of antigenic determinants may be formed from a single penicillin derivative, which may vary between individuals and cross-reactivity with other penicillins may also occur. Also IgM and IgG can act as **blocking antibodies** towards IgE, in which case the severe anaphylactic reactions do not occur. This makes it difficult to predict the outcome of penicillin allergy.

The mechanism of penicillin immunotoxicity relies on the formation of a covalent conjugate with soluble or cellular protein. There is a clear relationship between conjugation to proteins and cells and the ability to cause hypersensitivity in man. However, the immunogen(s) is still not known with certainty. There may be many different protein–drug conjugates with various breakdown products of penicillin, bound to varying extents. Thus the conjugate may have many drug molecules bound per protein molecule. The conjugate most commonly formed, a benzyl penicilloyl derivative, involves formation of an α-amide derivative with the ϵ-amino group of the amino acid lysine via the strained β-lactam ring in the penicillin molecule (figure 7.53). This may also occur after reaction of **penicillenic acid** with protein (figure 7.53). The thiazolidine ring may also break open and give

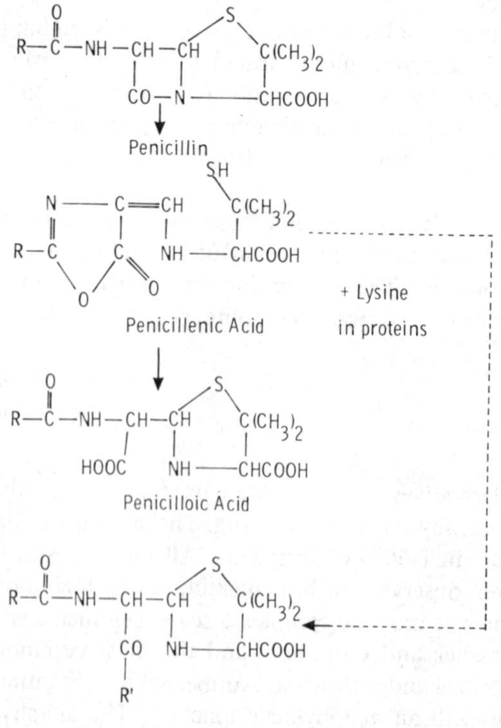

FIGURE 7.53. Structure of penicillin and derivatives arising by spontaneous transformation. For penicillin G, R = benzyl, R′ = protein. Adapted from De Weck (1962) *Int. Archs Allergy Appl. Immunol.*, **21**, 20.

rise to conjugates through the sulphur atom. However, it seems that the **penicilloyl derivative** is the major antigenic determinant and the disulphide-linked conjugate is the minor antigenic determinant. Thus, benzylpenicillenic acid is 40 times more reactive towards cellular proteins than benzylpenicillin itself and is highly immunogenic in the rat. This may be the ultimate immunogen. The importance of the benzyl penicilloyl amide conjugate was established using the technique of hapten inhibition where the penicillin derivative combines with the antibody and therefore inhibits the antibody–antigen reaction. This is normally visualized experimentally as a precipitation or agglutination of the antibody–antigen complex. Formation of penicilloic acid results in loss of immunogenicity. Not all of the drug–protein conjugates are immunogenic, however. For example, small amounts of penicillin are irreversibly bound to serum proteins such as albumin, but the penicilloyl groups do not seem to be accessible to antibodies.

Hydralazine

The drug hydralazine is a vasodilator used for the treatment of hypertension. In a significant proportion of individuals it causes a serious adverse effect, drug-induced **lupus erythematosus** (LE). This toxic effect, believed to have an immunological basis, involves a number of interesting features. Thus, there are a number of predisposing factors which make it possible to identify patients at risk, although the mechanism(s) underlying some of these predisposing factors are unknown. When the occurrence in the exposed population is examined in the light of some of these factors the incidence can be seen to be extremely high. The predisposing factors identified to date are:

(1) dose
(2) duration of therapy
(3) acetylator phenotype
(4) HLA type
(5) sex.

Thus, if the exposed population is divided into males and females and by the various doses administered, the incidence in susceptible populations can be appreciated (table 7.7).

Thus, the highest incidence recorded in one recent study was over 19% in the most susceptible population – females taking 100 mg of hydralazine twice daily.

If we examine each of these factors in turn.

DOSE
There is an increase in the incidence of hydralazine-induced LE in the exposed population with increasing dose as can be seen in table 7.7. However, patients who develop LE do not have a significantly different cumulative intake of hydralazine from those patients who do not develop the syndrome. This latter observation is consistent with the absence of a clear dose–response relationship in toxicity with an immunological basis.

Table 7.7.   Incidence of hydralazine-induced systemic lupus erythematosus.

Patients	Dose (mg)	Incidence over 3 years
All patients	all doses	6·7%
	25	0%
	50	5·4%
	100	10·4%
Men	All doses	2·8%
	25 or 50	0%
	100	4·9%
Women	All doses	11·6%
	25 or 50	5·5%
	100	19·4%

Data from Cameron, H. A. and Ramsay, L. E. (1984) *Brit. Med. J.*, **289**, 410.

DURATION OF THERAPY

The adverse effect develops slowly, typically over many months and there is a mean development time of around 18 months.

ACETYLATOR PHENOTYPE

The LE syndrome only develops in those patients with the slow acetylator phenotype. Metabolic studies have shown that the metabolism of hydralazine involves an acetylation step (figure 7.54) which is influenced by the acetylator phenotype.

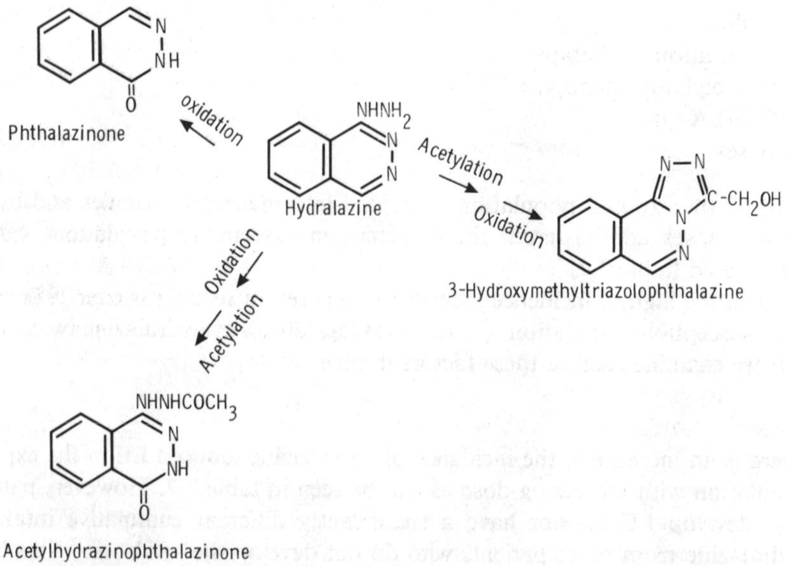

FIGURE 7.54. Some of the major routes of metabolism of hydralazine.

HLA TYPE

The tissue type seems to be a factor as there is a preponderance of the HLA-type DR4 in patients who develop the syndrome. Thus, in one study an incidence of 73% for this HLA type was observed in patients suffering the LE syndrome compared with an incidence of 33% in controls and 25% in patients not developing the LE syndrome.

SEX

Females seem to be more susceptible to hydralazine-induced LE than males. The ratio may be as high as 4:1.

Hydralazine-induced LE causes inflammation in various organs and tissues giving rise to a number of different symptoms. Thus patients suffer from arthralgia and myalgia, skin rashes and sometimes vasculitis. There may also be hepatomegaly, splenomegaly, anaemia and leucopenia. Two particular characteristics detectable in the blood are antinuclear antibodies and LE cells. **Antinuclear antibodies (ANA)** are directed against single-stranded DNA and de-oxyribonucleoprotein such as histones and occur in at least 27% of patients taking the drug, but not all of these develop the LE syndrome. Indeed, in one study over 3 years, 50% of patients had an ANA titre of 1:20 or more. Antibodies directed against other cellular and tissue constituents such as DNA and immunoglobulins are also present, and antibodies against synthetic hydralazine–protein conjugates have also been detected. There are thus various auto-antibodies present and if the auto-antigens are released by cellular breakdown, a **type III immune reaction** can occur where an immune complex is formed which is deposited in small blood vessels and joints, giving rise to many of the symptoms. The immunoglobulins IgG and IgE act as both auto-antibody and antigen and hence immune complexes form. Such complexes stimulate the complement system leading to inflammation, infiltration by polymorphs and macrophages and the release of lysosomal enzymes.

LE cells are neutrophil polymorphs which have phagocytosed the basophilic nuclear material of leucocytes which has been altered by interaction with anti-nuclear antibodies. The development of ANA requires a lower intake of hydralazine and occurs more quickly in slow acetylators than in rapid acetylators, and rapid acetylators have significantly lower titres of ANA than slow acetylators. There is also a significant correlation between the cumulative dose of hydralazine and the development of ANA, but as indicated above patients who develop LE do not have a significantly different cumulative intake of hydralazine from those patients who do not develop the syndrome.

The mechanism of hydralazine-induced LE is not currently understood but the evidence available indicates that it has an immunological basis. Hydralazine is a chemically reactive molecule and it is also metabolized to reactive metabolites, possibly **free radicals**, by the cytochromes P-450 system (figure 7.55), which bind covalently to protein. The production of the metabolite **phthalazinone** correlates with the binding *in vitro*. However, no antibodies against such conjugates with human microsomal protein were detected in the sera of patients with the LE syndrome. Hydralazine may also be a substrate for the **benzylamine oxidase** system found in vascular tissue, and for a **peroxidase-mediated** metabolic activation system which occurs in cells such as activated leucocytes. Thus a myeloperoxidase/H_2O_2/Cl^- system will metabolize hydralazine to phthalazinone and phthalazine, and this may also involve the production of reactive intermediates. This system has also been suggested to be involved in the activation of the drug **procainamide** which similarly causes a LE syndrome. Metabolic studies have shown that slow acetylators excrete more unchanged hydralazine and metabolize more via oxidative pathways (figure 7.54). Patients with the LE syndrome excrete more phthalazinone than control patients although this is not statistically significant. There is no difference between males and females in the nature or quantities of urinary metabolites of hydralazine detected however and so metabolic differences do not currently explain the sex difference in susceptibility. However, there is clearly scope for the formation of protein–drug conjugates which may be antigenic and hydralazine also reacts with DNA. Synthetic hydralazine–protein conjugates will stimulate the production of antibodies in rabbits and antibodies in human sera from patients with the LE syndrome will recognize and agglutinate rabbit red blood cells to which

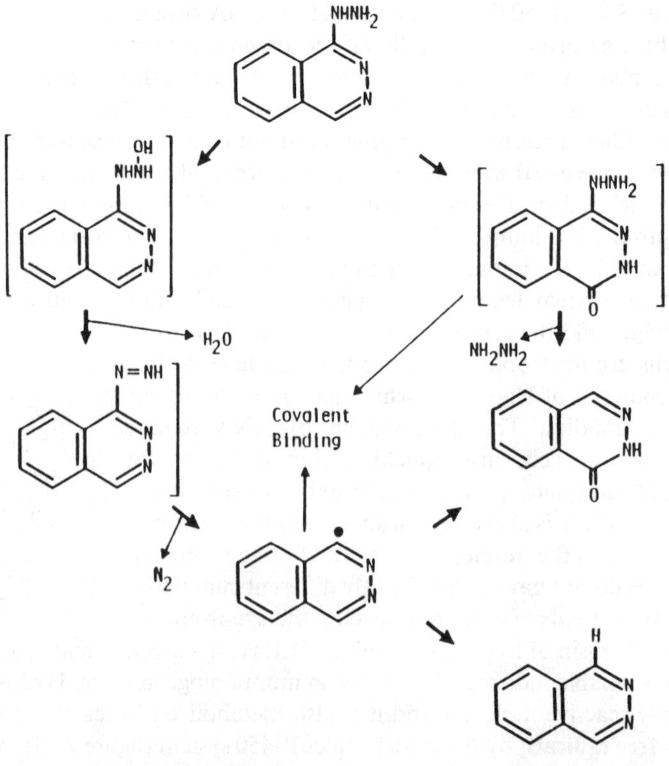

FIGURE 7.55. Possible routes for the metabolic activation of hydralazine. The oxidation of the hydrazine group may also involve the formation of a nitrogen centred radical which could also give rise to phthalazine with loss of nitrogen.

hydralazine has been chemically attached. Hydralazine will abolish this reaction *in vitro* indicating that it is the hapten or is similar to it.

Thus interaction of hydralazine or a metabolite with macromolecules may underlie the immune response. An alternative or additional hypothesis involves inhibition of the **complement system**. The complement system helps remove immune complexes by solubilization, but if it is inhibited deposition and accumulation of such complexes would be increased. Hydralazine and some of its metabolites interfere with part of the complement system, inhibiting the covalent binding of complement C4 by reaction with the thioester of activated C4. However, the concentrations required are high relative to the normal therapeutic concentration. More recently it has been shown that hydralazine inhibits DNA methylation in the T-cell. The inhibition of DNA methyl transferase may initiate immune reactions via activation of genes as a result of this interference with DNA methylation. However, although the mechanism of hydralazine-induced LE is not yet understood, it is an important example of drug-induced toxicity for two reasons:

(a) It illustrates the role and possibly the requirement for various predisposing factors in the development of an adverse drug reaction in a human population. An understanding of this should allow reduction of such adverse drug reactions by improved surveillance and prescribing.

(b) It reveals the difficulties of testing for this type of reaction in experimental animals when the various predisposing factors may not be present. However, the LE syndrome does occur in certain strains of mice and acetylation rates do vary between strains of laboratory animals. Using such specific models might therefore allow improved prediction.

Multi-organ toxicity

Ethylene glycol

This substance is a liquid used in anti-freeze, paints, polishes and cosmetics. As it has a sweet taste and is readily available it has been used as a poor man's alcohol, but it may also be ingested accidentally and for suicidal purposes. Diethylene glycol was once used as a vehicle for the drug sulphanilamide and when used for this it caused some 76 deaths.

The minimum lethal dose of ethylene glycol is about 100 ml and after ingestion death may occur within 24 h from damage to the CNS or more slowly (8–12 days) from renal failure.

There seem to be three recognizable *clinical stages*:

(a) Within 30 min and lasting for perhaps 12 h, there is intoxication, nausea, vomiting, coma, convulsions, nystagmus, papilloedema, depressed reflexes, myoclonic jerks and tetanic contractions. Permanent optic atrophy may occur.

(b) Between 12 and 24 h there is tachypnoea, tachycardia, hypertension, pulmonary oedema and congestive cardiac failure.

(c) Between 24 and 72 h the kidneys become damaged giving rise to flank pain and acute renal tubular necrosis.

The clinical biochemical features reflect the biochemical and physiological effects. Thus, there is reduced plasma bicarbonate, low plasma calcium and raised potassium. Crystals, blood and protein may all be detected in the urine (*crystalluria, haematuria* and *proteinuria*, respectively), and the urine may have a low specific gravity.

The mechanism of toxicity of ethylene glycol involves metabolism, but unlike previous examples this does not involve metabolic activation to a reactive metabolite. Thus, ethylene glycol is metabolized by several oxidation steps eventually to yield oxalic acid (figure 7.56). The first step is catalysed by the enzyme **alcohol dehydrogenase** and herein lies the key to treatment of poisoning. The result of each of the metabolic steps is the production of NADH. The imbalance in the level of this in the body is adjusted by oxidation to NAD coupled to the production of lactate. There is thus an increase in the level of lactate and *lactic acidosis* may result. Also, the intermediate metabolites of ethylene glycol have metabolic effects such as the inhibition of oxidative phosphorylation, glucose metabolism, Krebs' cycle, protein synthesis, RNA synthesis and DNA replication for example.

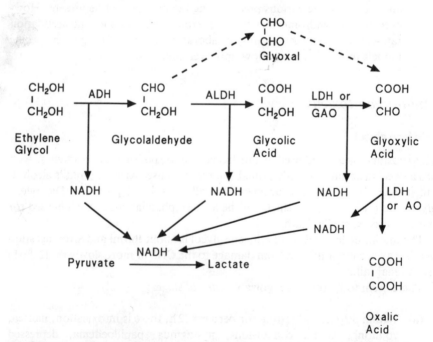

FIGURE 7.56. The metabolism of ethylene glycol. The NADH produced is utilized in the production of lactate. ADH: alcohol dehydrogenase; ALDH: aldehyde dehydrogenase; LDH: lactate dehydrogenase; GAO: glycolic acid oxidase; AO: aldehyde oxidase.

The consequences of this are as follows:

   (i) acidosis due to lactate, oxalate and the other acidic metabolites; this
       results in metabolic distress and physiological changes
  (ii) loss of calcium as calcium oxalate
 (iii) deposition of crystals of **calcium oxalate** in the renal tubules and brain
  (iv) inhibition of various metabolic pathways leading to accumulation of
       organic acids
   (v) impairment of cerebral function by oxalate and damage by crystals;
       also some of the aldehyde metabolites may impair cerebral function
  (vi) damage to renal tubules by oxalate crystals leading to necrosis.

Thus the pathological damage includes cerebral oedema, haemorrhage and
deposition of calcium oxalate crystals. The lungs show oedema, and occasionally
calcium oxalate crystals and degenerative myocardial changes may also occur.
There is degeneration of proximal tubular epithelium with calcium oxalate
crystals and fat droplets detectable in tubular epithelial cells. The degeneration of
distal tubules may also be seen.

Ethylene glycol is more toxic to humans than animals, and in general the
susceptible species are those which metabolize the compound to **oxalic acid**,
although this is quantitatively a minor route. The treatment of poisoning with
ethylene glycol reflects the mechanism and biochemical effects. Thus, after
standard procedures such as gastric lavage to reduce absorption and supportive
therapy for shock and respiratory distress, patients are treated with the following:

   (a) ethanol; this competes with ethylene glycol for alcohol dehydrogenase, but
       as it is a better substrate the first step in ethylene glycol metabolism is
       blocked – animal studies have shown that this doubles the LD_{50}
   (b) intravenous sodium bicarbonate; this corrects the acidosis – animal studies
       have shown that this increases the LD_{50} by around four times
   (c) calcium gluconate; this corrects the hypocalcaemia
   (d) dialysis to remove ethylene glycol.

Thus the treatment of poisoning with ethylene glycol is a logical result of
understanding the biochemistry of the toxicity.

Methanol

Methanol is widely used as a solvent and as a denaturing agent for ethanol and is
also found in antifreeze. Mass poisonings have occurred due to ingestion in
alcoholic drinks made with contaminated ethanol as well as from accidental
exposure. Inhalation and skin absorption may cause toxicity. In humans about
10 ml can cause blindness and 30 ml is potentially fatal, but there is variation in
the lethal dose.

The clinical features are an initial mild inebriation and drowsiness followed,
after a delay of 8–36 h, by nausea, vomiting, abdominal pain, dizziness,

headaches and possibly coma. Visual disturbances may start within 6 h and include blurred vision, diminished visual acuity and dilated pupils which are unreactive to light. These effects indicate that blindness may occur and changes to the optic disc can be seen with an opthalmoscope. Severe depression of the CNS may occur at later stages due to accumulation of metabolites.

The mechanism underlying methanol poisoning again involves metabolism, with **alcohol dehydrogenase** catalysing the first step in the pathway (figure 7.57), although the involvement of other enzymes such as **catalase** and a **microsomal ethanol oxidation system** has been proposed in some species. The second step, oxidation of formaldehyde to formic acid, is catalysed by several enzymes: hepatic and erythrocyte aldehyde dehydrogenase; the folate pathway; formaldehyde dehydrogenase (figure 7.57). The products of metabolism, formaldehyde and formic acid, are both toxic. However formaldehyde has a very short half-life (about 1 min), whereas formic acid accumulates. Formic acid is further metabolized to carbon dioxide or other products in a reaction which *in vivo* involves combination with **tetrahydrofolate** in a reaction catalysed by formyl tetrahydrofolate synthetase (figure 7.57). It seems that the hepatic concentration of tetrahydrofolate regulates the rate of metabolism and hence removal of formic acid and that this may be a factor in the accumulation of formate. Those species which are less sensitive to methanol poisoning, such as the rat, remove formate more rapidly than humans and other sensitive primates.

The result of formate accumulation is **metabolic acidosis**. However, at later stages the acidosis may also involve the accumulation of other anions such as lactate. This may be a result of inhibition of cytochrome oxidase and hence of mitochondrial respiration, tissue hypoxia due to reduced circulation of blood or an increase in the NADH/NAD ratio. The acidosis which results from methanol poisoning will result in more formic acid being in the non-ionized state and hence more readily able to enter the CNS. This will cause central depression and

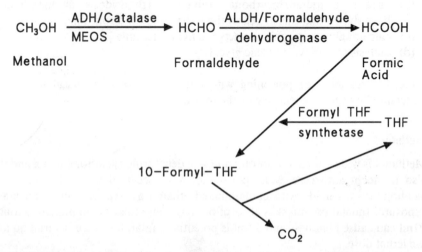

FIGURE 7.57. The metabolism of methanol. ADH: alcohol dehydrogenase; MEOS: microsomal ethanol oxidizing system; ALDH: aldehyde dehydrogenase; THF: tetrahydrofolate.

hypotension and increased lactate production. This situation is known as the **'circulus hypoxicus'**.

The other major toxic effect of methanol is the ocular toxicity. Although formaldehyde might be formed locally in the retina, this seems unlikely, whereas formate is known to cause experimental ocular toxicity. The mechanism suggested involves inhibition by formate of cytochrome oxidase in the optic nerve. As the optic nerve cells have few mitochondria they are very susceptible to this **'histotoxic hypoxia'**. The inhibition will result in a decrease in ATP and hence disruption of optic nerve function. Thus, stasis of axoplasmic flow, axonal swelling, optic disc oedema and loss of function occur. Studies have shown that formate alone will cause toxicity in the absence of acidosis, although this will exacerbate the toxicity. The liver, kidney and heart may show pathological changes and pulmonary oedema and alveolar epithelial damage can occur. In severe cases of methanol poisoning, death may occur due to respiratory and cardiac arrest.

The treatment of methanol poisoning involves firstly the administration of an **antidote**, ethanol, which blocks metabolism. **Ethanol** competes with methanol for alcohol dehydrogenase as the enzyme has a greater affinity for ethanol. Methanol metabolism can be reduced by as much as 90% by an equimolar dose of ethanol and the half-life becomes extended to 46 h. **4-Methylpyrazole**, which also binds to alcohol dehydrogenase, has been used successfully in monkeys to treat methanol poisoning, as has folic acid.

Secondly, i.v. sodium bicarbonate is given for correction of the metabolic acidosis. Haemodialysis may be used in very serious cases.

Multi-organ toxicity: metals

Cadmium

Cadmium is a metal which is widely used in industry in alloys, in plating, in batteries and in the pigments used in inks, paints, plastic, rubber and enamel. It is an extremely toxic substance and the major hazard is from inhalation of cadmium metal or cadmium oxide. Although it is present in food, significant oral ingestion is rare and absorption from the gut is poor (5–8%). However, various dietary and other factors may enhance absorption from the gastrointestinal tract. In contrast, up to 40% of an inhaled dose may be absorbed and hence its presence in cigarettes is a significant source of exposure.

Cadmium is bound to proteins and red blood cells in blood and transported in this form, but 50–75% of the body burden is located in the liver and kidneys. The half-life of cadmium in the body is between 7 and 30 years and it is excreted through the kidneys, particularly after they become damaged.

Cadium has many toxic effects, primarily causing **kidney damage**, as a result of chronic exposure, and **testicular damage** after acute exposure, although the latter does not seem to be a common feature in humans after occupational exposure to the metal. It is also hepatotoxic and affects vascular tissue and bone. After acute inhalation exposure, lung irritation and damage may occur along with other symptoms such as diarrhoea and malaise. Chronic inhalation exposure can result in progressive fibrosis of the lower airways leading to **emphysema**. This results from necrosis of alveolar macrophages and hence release of degradative enzymes which damage the basement membranes of the alveolus. These lung lesions may occur before kidney damage is observed. Cadmium can also cause disorders of calcium metabolism and the subsequent loss of calcium from the body leads to osteomalacia and brittle bones. In Japan this became known as **Itai-Itai ('Ouch-Ouch!') disease** when it occurred in women eating rice contaminated with cadmium. The raised urinary levels of **proline** and **hydroxyproline** associated with chronic cadmium toxicity may be due to this damage to the bones.

Kidney damage is a delayed effect even after single doses, being due to the accumulation of cadmium in the kidney, as a complex with the protein metallothionein. **Metallothionein** is a low molecular weight protein (6500 Da) containing about 30% cysteine, which is involved with the transport of metals, such as zinc, within the body. Due to its chemical similarity to zinc, cadmium exposure induces the production of this protein and 80–90% of cadmium is bound to it *in vivo*, probably through SH-groups on the protein. Thus, exposure to repeated small doses of cadmium will prevent the toxicity of large acute doses by increasing the amount of **metallothionein** available. The protein is thus serving a protective function. The cadmium–metallothionein complex is synthesized in the liver and transported to the kidney, filtered through the glomerulus and is reabsorbed by the proximal tubular cells, possibly by endocytosis. Within these cells the complex is taken up into lysosomes and degraded by proteases to release cadmium which may damage the cells or recombine with more metallothionein. The cellular damage caused by cadmium may be at least partly a result of its ability to bind to the sulphydryl groups of critical proteins and enzymes. There seems to be a critical level of cadmium in the kidney above which toxicity occurs. The damage to the kidney occurs in the first and second segments of the proximal tubule. This can be detected biochemically as glucose, amino acids and protein in urine. The proteins are predominantly of low molecular weight, such as β_2-**microglobulin**, which are not reabsorbed by the proximal tubules damaged by cadmium. Larger proteins in the urine indicate glomerular damage. The vasoconstriction caused by cadmium may affect renal function, and cadmium may cause fibrotic degeneration of renal blood vessels. The tubular cells degenerate and interstitial fibrosis can occur. The proximal tubular damage can progress in chronic cadmium toxicity to distal tubular dysfunction, loss of calcium in urine giving rise to renal calculi and osteomalacia.

The binding of cadmium to metallothionein decreases toxicity to the testes but increases the nephrotoxicity, possibly because the complex is preferentially, and more easily, taken up by the kidney than the free metal. Dosing animals with the cadmium–metallothionein complex leads to acute kidney damage, whereas exposure to single doses of cadmium itself does not.

The **testicular damage** occurs within a few hours of a single exposure to cadmium and results in necrosis, degeneration and complete loss of spermatozoa. The mechanism involves an effect on the vasculature of the testis. Cadmium reduces blood flow through the testis and ischaemic necrosis results from the lack of oxygen and nutrients reaching the tissue. In this case cadmium is probably acting mainly indirectly by affecting a physiological parameter. However, pre-treatment of animals with **zinc** reduces the testicular toxicity of cadmium by inducing the synthesis of metallothionein and hence reducing the free cadmium level.

The vasoconstriction which is caused by cadmium may underlie the hypertension observed in experimental animals. Cadmium is also carcinogenic in experimental animals causing tumours at the site of exposure. Also Leydig cell tumours occur in the testis of animals after acute doses of cadmium sufficient to cause testicular necrosis. This seems to be an indirect effect due to the reduced level of **testosterone** in the blood which follows testicular damage. This causes Leydig cell hyperplasia and tumours to occur.

Thus, cadmium causes multi-organ toxicity, and at least some of the toxic effects are due to it being a divalent metal similar to zinc and able to bind to sulphydryl groups.

Mercury

Mercury can exist in three forms, elemental, inorganic and organic, and all are toxic. However, the toxicity of the three forms of mercury are different, mainly as a result of differences in distribution. Some of these toxic properties have been known for centuries.

Elemental mercury (Hg°) may be absorbed by biological systems as a vapour. Despite being a liquid metal, mercury readily vaporizes at room temperature and in this form constitutes a particular hazard to those who use scientific instruments containing it for example.

Elemental mercury vapour is relatively lipid soluble and is readily absorbed from the lungs following inhalation and is oxidized in the red blood cells to Hg^{2+}. Elemental mercury may also be transported in red blood cells to other tissues such as the CNS. Elemental mercury readily passes across the **blood–brain barrier** into the CNS and also into the foetus. The metallic compound is only poorly absorbed from the gastrointestinal tract, however.

Inorganic mercury, existing as monovalent (mercurous) or divalent (mercuric) ions is relatively poorly absorbed from the gastrointestinal tract (7% in humans). After absorption inorganic mercury accumulates in the kidney.

Organic mercury is the most readily absorbed (90–95% from the gastrointestinal tract), and after absorption distributes especially to the brain, particularly the posterior cortex. All the forms of mercury will cross the placenta and gain access to the foetus, although elemental mercury and organic mercury show greater uptake. The concentrations in certain foetal tissues, such as red blood cells, are greater than in maternal tissue.

Mercury is eliminated from the body in the urine and faeces with the latter being the major route. Thus, with methyl mercury 90% is excreted into the faeces. Methyl mercury is secreted into the bile as a cysteine conjugate and undergoes extensive **enterohepatic recirculation**.

The half-life of mercury is long but there are two phases, the first being around 2 days, then the terminal phase which is around 20 days. However the half-life will depend on the form of mercury. Thus methyl mercury has a half-life of about 70 days whereas for inorganic mercury this is about 40 days.

Organic mercury compounds, especially **phenyl** and **methoxyethyl mercury** may also be biotransformed into inorganic mercury by cleavage of the carbon–mercury bond. Although such compounds are more readily absorbed than inorganic mercury compounds, the toxicity is similar.

Elemental mercury is oxidized *in vivo* to inorganic mercury, a biotransformation which is probably catalysed by catalase. It is selectively accumulated in the kidney and also by lysosomes. Inorganic mercury (Hg^{2+}) will induce the synthesis of metallothionein. Mercury binds to cellular components such as enzymes in various organelles, especially to proteins containing sulphydryl groups. Thus, cysteine and glutathione will react with mercury to produce soluble products which can be excreted into the bile.

Toxic effects

ELEMENTAL MERCURY VAPOUR

Although there may be toxic effects to the respiratory system from the inhalation of mercury vapour, the major toxic effect is to the CNS. This is especially true after chronic exposure. There are a variety of symptoms such as muscle tremors, personality changes, delirium, hallucination and gingivitis.

INORGANIC MERCURY

Mercuric chloride and other mercuric salts will, when ingested orally, cause immediate acute damage to the gastrointestinal tract. This may be manifested as bloody diarrhoea, ulceration and necrosis of the tract. After 24 h renal failure occurs which results from necrosis of the pars recta region of the proximal tubular epithelial cells. The epithelial cells show damage to the plasma membrane, endoplasmic reticulum, mitochondria and effects on the nucleus. The result of this damage is excretion of glucose (glycosuria), amino acids (aminoaciduria), appearance of proteins in the urine (proteinuria), and changes in various metabolites excreted into urine. After an initial diuresis there is a reduction in urine (oliguria), possibly developing into complete lack of urine (anuria). The effect on renal function can also be detected by determination of blood urea (BUN) which will be elevated in renal failure.

Chronic low-level exposure to inorganic mercury may lead to a glomerular disease which has an immunologic basis. This type of nephropathy is accompanied by proteinuria and may involve glomerular damage due to immune complexes. Also, chronic exposure can give rise to salivation and gingivitis and erethism which involves psychological effects such as nervousness and shyness.

Mercurous salts are less toxic than mercuric salts, probably as a result of lower solubility. Exposure of human subjects to **mercurous chloride (calomel)** may result in hypersensitivity reactions.

ORGANIC MERCURY

Mercury in this form, such as **methyl mercury**, is extremely toxic, mainly affecting the CNS. However, some organo mercury compounds such as phenyl and methoxyethyl mercury cause similar toxic effects to inorganic mercury. There have been a number of instances in which human exposure to methyl mercury has occurred, and consequently data is available on the toxic effects to man as well as experimental animals. Methyl mercury was responsible for the poisoning which occurred in Japan, known as **Minamata disease**. This resulted from industrial effluent containing inorganic mercury contaminating the water of Minamata Bay in Japan. The micro-organisms in the sediments at the bottom of the bay biotransformed the inorganic mercury ions into methyl and dimethyl mercury. As this form of mercury is lipid soluble it was able to enter the **food chain** and so become concentrated in fish as a result of their eating small organisms which had absorbed the methyl mercury. The local population who consumed the fish therefore became contaminated with methyl mercury. Another episode occurred in Iraq when seed grain treated with a methyl mercury fungicide was used to make bread. Over 6000 people were recorded as exposed and more than 500 died. The major features of methyl mercury poisoning are paresthesia, ataxia, dysarthria and deafness. There is a clear dose–response relationship for each of these toxic effects in exposed humans which has been derived from the poisoning episode in Iraq. Thus, there is a linear relationship between body burden and the frequency of cases in the exposed population. However, the occurrence of each of these

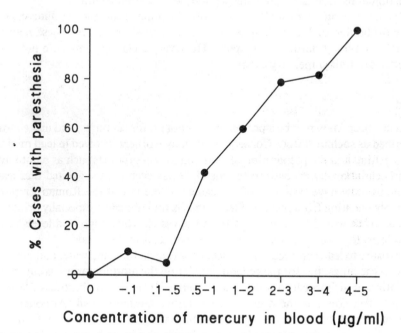

FIGURE 7.58. The dose-response relationship for methyl mercury in exposed humans. The concentration of mercury in the blood and the incidence of paresthesias are used as dose and response respectively. Data from Bakir *et al.*, *Science*, **181**, 230, 1973.

toxic effects shows a different profile and a different threshold (figure 7.58). The pathology involves degeneration and necrosis of nerve cells in the cerebral cortex, and particularly those areas dealing with vision. The blood–brain barrier is also disrupted.

Exposure of pregnant women to methyl mercury caused cerebral palsy and mental retardation in the offspring, despite lack of symptoms in the mothers.

Mechanism

Mercury is a reactive element and its toxicity is probably due to interaction with proteins. Mercury has a particular affinity for sulphydryl groups in proteins and consequently is an inhibitor of various enzymes such as membrane ATPase, which are sulphydryl dependent. It can also react with amino, phosphoryl and carboxyl groups. Brain pyruvate metabolism is known to be inhibited by mercury, as are lactate dehydrogenase and fatty acid synthetase. The accumulation of mercury in lysosomes increases the activity of lysomal acid phosphatase which may be a cause of toxicity as lysosomal damage releases various hydrolytic enzymes into the cell, which can then cause cellular damage. Mercury accumulates in the kidney and is believed to cause uncoupling of oxidative phsophorylation in the mitochondria of the kidney cells. Thus, a number of mitochondrial enzymes are inhibited by Hg^{2+}. These effects on the mitochondria will lead to a reduction of respiratory control in the renal cells and their functions such as solute reabsorption, will be compromised.

Mercury poisoning is usually treated with chelating agents such as **dimercaprol** or **penicillamine** or haemodialysis may be used in severe cases. These help to decrease the body burden of mercury. However, chelating agents are not very effective after alkyl mercury exposure.

Lead

Lead has been known to be a poisonous compound for centuries, and indeed was described as such in 300 BC. Consequently many workers involved in lead mining and smelting and the preparation of lead-containing products such as paint have been occupationally exposed to the metal. It has even been suggested that lead poisoning may have been one of the causes of the fall of the Roman Empire, probably resulting from the use of lead utensils for eating and especially drinking liquids which would leach the lead from the vessel. Thus, high lead levels have been detected in the skeletons of Romans dating from the period.

Exposure to lead can occur in a variety of ways; via food and water and inhaled through the lungs, but for the general population the most important is currently inhalation of airborne lead which mainly derives from the combustion of leaded petrol which contains the organic lead additive **tetraethyl lead**. Although the amount of lead in food may be greater than that in the air, absorption from the lungs is greater. Other sources in the environment are lead smelters, batteries, paints, lead water pipes and insecticides such as lead arsenate. Paint used to be a significant source of lead, and lead poisoning in children even recently may have been due to flakes of old lead paint still being present in slum housing areas. During the time of Prohibition in the U.S.A. earlier this century those drinking illicit Moonshine Whisky suffered renal damage as a result of lead poisoning which derived from the solder used to construct the stills. Lead poisoning has also resulted from the use of medicinal agents from certain countries. Industrial poisoning with lead became common in the Industrial Revolution with 1000 cases per year occurring in the U.K. alone at the turn of century, and exposure to lead still occurs in industry.

Exposure to lead may be to the metal, lead salts and organic lead.

Lead causes damage to a variety of organs and also causes significant biochemical effects. Thus the kidneys, testes, bones, **gastrointestinal tract** and the nervous system are all damaged by lead. The major biochemical effect is interference with haem synthesis giving rise to anaemia.

Acute exposure to inorganic lead causes renal damage, in particular damage to the proximal tubules. This is detectable biochemically as amino aciduria, and glycosuria. Lead adversely affects **reproductive** function in both males and females, and recent studies in men occupationally exposed to lead have indicated that testicular function is adversely affected by lead. Animal studies have shown that lead is gametotoxic.

After absorption lead enters the blood and 97% is taken up by red blood cells. Here lead has a half-life of 2–3 weeks during which there is some redistribution to liver and kidney, then excretion into bile or deposition in **bone**. After an initial, reversible, uptake into bone, lead in bone becomes incorporated into the hydroxyapatite crystalline structure. Because of this, past exposure to lead is possible to quantitate using X-ray analysis. It is also possible to detect lead exposure and possible poisoning from urine and blood analysis, and the amount in blood represents current exposure. However, as lead is taken up into the red blood cell, both the free blood lead level and that in the erythrocytes needs to be known.

'Normal' blood levels in adults not occupationally exposed in the U.S.A. are in the region of $0\cdot15$–$0\cdot7\,\mu g/ml$ with the average at $0\cdot3\,\mu g/ml$. The threshold for toxicity is $0\cdot8\,\mu g/ml$.

Interference with haem synthesis

Lead interferes with the synthesis of haem, and its inhibitory effects on the enzymes of this pathway can be readily detected in humans. Myoglobin synthesis and cytochromes P-450 may also be affected. The mechanism of the interference of haem synthesis by lead can be seen in figure 7.59. Lead inhibits haem synthesis at several points: aminolaevulinate synthetase, aminolaevulinate dehydrase, ferrochelatase, haem oxidase and coproporphyrinogen oxidase. Thus, there is excess protoporphyrin available, but the inhibition of ferrochelatase means that protoporphyrin rather than haem is inserted into the globin moiety and zinc replaces the iron. Lack of haem causes negative feedback control on aminolaevulinate synthetase leading to increased synthesis of aminolaevulinate. The final result of the inhibition of these various enzymes is a decrease in the level of

MITOCHONDRION

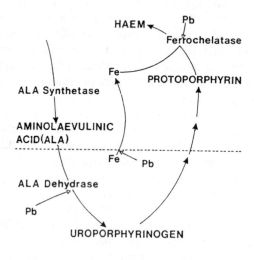

CYTOSOL

FIGURE 7.59. The synthesis of haem in the mammalian erythrocyte. The points of interference by lead are shown. From Timbrell, J. A., *Introduction to Toxicology*, Taylor and Francis, London, 1989.

haemoglobin and hence anaemia. The effects may be detected as coproporphyrin excretion in the urine, free erythrocyte protoporphyrin is increased and excess aminolaevulinate is excreted in urine. Significant inhibition of the enzyme amino-laevulinate dehydrase is detectable in volunteers with blood lead levels of $0 \cdot 4 \mu g/ml$ resulting from normal exposure in an urban environment. With extensive exposure to lead such as may occur following industrial exposure or accidental poisoning, morphological changes in red blood cells may be seen. Indeed the anaemia caused by lead is due to both shortened life time of red blood cells as well as decreased synthesis of haem. The reduced life time of the red blood cell is due to increased fragility of the red cell membrane.

Renal toxicity

Acute exposure to inorganic lead can cause reversible damage to the kidneys, manifested as tubular dysfunction. Chronic exposure to lead, however, causes permanent interstitial nephropathy which involves tubular cell atrophy, patho-logical changes in the vasculature and fibrosis. The most pronounced changes occur in the proximal tubules. Indeed lead–protein complexes are seen as inclusion bodies in tubular cells, and the mitochondria in such cells have been shown to be altered with impaired oxidative phosphorylation. Clearly this will influence the function of the proximal tubular cells in reabsorption and secretion of solutes and metabolites. Consequently one indication of renal dysfunction is amino aciduria, glycosuria and impairment of sodium reabsorption.

Neurotoxicity

The effects on the CNS are perhaps the most significant as far as humans are concerned, and children are especially vulnerable. The neurotoxicity is observed as **encephalopathy** and **peripheral neuropathy**. Encephalopathy is accompanied by various pathological changes including cerebral oedema, degeneration of neurones and necrosis of the cerebral cortex. It is observed particularly in children and can be manifested as ataxia, convulsions and coma. Cerebral palsy may also occur along with mental retardation and seizures. The mechanism of the neurotoxicity may involve direct effects of lead on neuronal transmission which are thought to occur at blood lead levels of $0 \cdot 3$–$0.5 \mu g/ml$ as assessed by behavioural studies, brain wave patterns detectable by electroencephalogram and CNS-evoked potentials. Inhibition of cholinergic function via interference with extracellular calcium may underlie these changes in neurotransmission. The functioning of other neurotransmitters such as dopamine and γ-aminobutyric acid are also affected by lead. There are also indirect effects via cerebral oedema and cellular hypoxia. Peripheral neuropathy involves the degeneration of peripheral, especially motor, nerves. Dysfunction of motor nerves is detectable as a decrease in nerve conduction velocity at blood lead levels of $0 \cdot 5$–$0 \cdot 7 \mu g/ml$.

Manifestations of lead toxicity

Biochemical changes such as increased aminolaevulinate excretion and inhibition of aminolaevulinate dehydrase may be detected in urine and blood, respectively, at blood lead levels of $0 \cdot 4$–$0 \cdot 6 \mu g/ml$. Anaemia is a late feature however. Neurotoxicity may be detectable at blood lead levels of $0 \cdot 8$–$1 \cdot 0 \mu g/ml$. At blood lead levels greater than $1 \cdot 2 \mu g/ml$ encephalopathy occurs. Peripheral nerve palsies are rare, and the foot and wrist drop which were once characteristic of occupational lead poisoning only occurs after excessive exposure and are now rarely seen. Similarly, seizures and impaired consciousness may result from involvement of the CNS. Bone changes are usually seen in children and are detected as bands at the growing ends of the bones and a change in bone shape.

Testicular dysfunction in humans occupationally exposed has been detected as a decrease in plasma **testosterone** and other changes indicative of altered testicular function.

Renal function changes are readily detectable as amino aciduria and glycosuria.

The treatment of lead poisoning involves the use of chelating agents to remove the lead from the soft tissues of the body. Thus, agents such as **sodium calcium edetate** may be used.

Organic lead is probably more toxic than inorganic lead as it is lipid soluble. For example, triethyl lead, which results from breakdown of tetraethyl lead, is readily absorbed through the skin and into the brain and will cause encephalopathy. Symptoms are delusions, hallucinations and ataxia and the effects are rapid. Organic lead, however, has no effect on haem synthesis.

Bibliography

General

BHATNAGAR, R. S. (1980) (editor) *Molecular Basis of Environmental Toxicity* (Ann Arbor: Ann Arbor Science Publications).

GILLETTE, J. R. and MITCHELL, J. R. (1975) Drug actions and interactions: Theoretical considerations. In *Handbook of Experimental Pharmacology*, Vol. 28, Part 3, *Concepts in Biochemical Pharmacology*, edited by J. R. Gillette and J. R. Mitchell (Berlin: Springer).

GILLETTE, J. R. and POHL, L. R. (1977) A prospective on covalent binding and toxicity. *J. Toxicol. Environ. Health*, **2**, 849.

HODGSON, E. and GUTHRIE, F. E. (1980) (editors) *Introduction to Biochemical Toxicology* (New York: Elsevier-North Holland).

HODGSON, E., BEND, J. R. and PHILPOT, R. M. (1971–1981) *Reviews in Biochemical Toxicology*, Vols. 1–12 (New York: Elsevier-North Holland).

JOLLOW, D. J., KOCSIS, J. J., SNYDER, R. and VANIO, H. (editors) (1977) *Biological Reactive Intermediates* (New York: Plenum Press).

KLAASSEN, C. D., AMDUR, M. O. and DOULL, J. (editors) (1986) *Casarett and Doull's Toxicology, The Basic Science of Poisons*, (New York: Macmillan).

KOCSIS, J. J., JOLLOW, D. J., WITMER, C. M., NELSON, J. O. and SNYDER, R. (editors) (1986) Biological reactive intermediates III. Mechanisms of action in animal models and human disease (New York: Plenum Press).

MITCHELL, J. R., POTTER, W. Z., HINSON, J. A., SNODGRASS, W. R., TIMBRELL, J. A. and GILLETTE, J. R. (1975) Toxic drug reactions. In *Handbook of Experimental Pharmacology*, Vol. 28, Part 3, *Concepts in Biochemical Pharmacology*, edited by J. R. Gillette and J. R. Mitchell (Berlin: Springer).

MONKS, T. J. and LAU, S. S. (1988) Reactive intermediates and their toxicological significance. *Toxicology*, **52**, 1.

NELSON, S. D. (1982) Metabolic activation and drug toxicity. *J. Medicinal Chem.*, **25**, 753.

PRATT, W. B. and TAYLOR, P. (editors) (1990) *Principles of Drug Action, The Basis of Pharmacology* (New York: Churchill Livingstone).

SLATER, T. F. (1972) *Free Radicals in Tissue Injury* (London: Pion).

SLATER, T. F. (editor) (1978) *Biochemical Mechanisms of Liver Injury* (London: Academic Press).

SMITH, D. A. (editor) (1977) Mechanisms of molecular and cellular toxicology. *J. Toxicol. Environ. Health*, **2**, 1229.

SNYDER, R., PARKE, D. V., KOCSIS, J., JOLLOW, D. J. and GIBSON, C. G. (editors) (1981) *Biological Reactive Intermediates 2: Chemical Mechanisms and Biological Effects* (New York: Plenum Press).

Symposium on Influence of Metabolic Activations and Inactivations on Toxic Effects. *Archs Toxicol.*, **39**, 1.

WITSCHI, H. and HASCHEK, W. M. (1980) Some problems correlating molecular mechanisms and cell damage. In *Molecular Basis of Environmental Toxicity*, edited by R. S. Bhatnagar (Ann Arbor: Ann Arbor Science Publications).

Chemical carcinogenesis

BROOKES, P. (1977) Role of covalent binding in carcinogenecity. In *Biological Reactive Intermediates*, edited by D. J. Jollow, J. J. Kocsis, R. Snyder and H. Vanio (New York: Plenum Press).

DIPPLE, A., MICHEJDA, C. J. and WEISBURGER, E. K. (1985) Metabolism of chemical carcinogens. *Pharmacol. Ther.*, **27**, 265.

GREIM, H., JUNG, R., KRAMER, M., MARQUARDT, H. and OESCH, F. (editors) (1982) *Biochemical Basis of Chemical Carcinogenesis* (New York: Raven Press).

HAGGERTY, H.G. and HOLSAPPLE, M.P. (1990) Role of metabolism in dimethyl-nitrosamine-induced immunosuppression: a review. *Toxicology, 63*, 1.

HATHWAY, D.E. (1984) *Molecular Aspects of Toxicology* (Royal Society of Chemistry: London).

HOLBROOK, D.J. (1980) Chemical carcinogenesis. In *Introduction to Biochemical Toxicology*, edited by E. Hodgson and F.E. Guthrie (New York: Elsevier-North Holland).

IOANNIDES, C., LUM, P.Y. and PARKE, D.V. (1984) Cytochrome P-448 and the activation of toxic chemicals and carcinogens. *Xenobiotica, 14*, 119.

LAWLEY, P.D. (1980) DNA as a target or alkylating carcinogens. In *Chemical Carcinogenesis*, edited by P. Brookes, *British Medical Bulletin, 36*, 19.

LEHR, R.E. and JERINA, D.M. (1977) Metabolic activation of polycyclic hydrocarbons. *Archs Toxicol., 39*, 1.

MAGEE, P.N. (1974) Activation and inactivation of chemical carcinogens in the mammal. In *Essays in Biochemistry*, edited by P.N. Campbell and F. Dickens, 10, 105.

MILLER, J.A. and MILLER, E.C. (1977) The concept of reactive electrophilic metabolites in chemical carcinogenesis: Recent results with aromatic amines, safrole and aflatoxin B_1. In *Biological Reactive Intermediates*, edited by D.J. Jollow, J.J. Kocsis, R. Snyder and H. Vanio (New York: Plenum Press).

MULDER, G.J., KROESE, E.D. and MEERMAN, J.H.N. (1988) The generation of reactive intermediates from xenobiotics by sulphate conjugation and their role in drug toxicity. In *Metabolism of Xenobiotics*, edited by J.W. Gorrod, H. Oelschlager and J. Caldwell (London: Taylor & Francis).

PEGG, A.E (1983) Alkylation and subsequent repair of DNA after exposure to di-methylnitrosamine and related carcinogens. In *Reviews in Biochemical Toxicology*, Vol. 5, edited by E. Hodgson, J.R. Bend and R.M. Philpot (New York: Elsevier-North Holland).

SAYER, J.M., WHALEN, D.L. and JERINA, D.M. (1989) Chemical strategies for the inactivation of bay-region diol-epoxides, ultimate carcinogens derived from polycyclic aromatic hydrocarbons. *Drug Metab. Rev., 20*, 155.

SEARLE, C.E. (editor) (1984) *Chemical Carcinogens*, 2nd edition (Washington D.C.: American Chemical Society).

SIMS, P. (1980) Metabolic activation of chemical carcinogens. In *Chemical Carcinogenesis*, edited by P. Brookes, *British Medical Bulletin, 36*, 11.

THORGIERSSON, S.S., GLOWINSKI, I.B. and McMANUS, M.E. (1983) Metabolism, mutagenicity and carcinogenicity of aromatic amines. In *Reviews in Biochemical Toxicology*, Vol. 5, edited by E. Hodgson, J.R. Bend and R.M. Philpot (New York: Elsevier-North Holland).

Direct toxic action: tissue lesions

ANDERS, M.W., ELFARRA, A.A. and LASH, L.H. (1987) Cellular effects of reactive intermediates: Nephrotoxicity of S-conjugates of amino acids. *Arch. Toxicol., 60*, 103.

BOYD, M.R. (1980) Biochemical mechanisms in chemical induced lung injury: Roles of metabolic activation. *CRC Crit. Rev. Toxicol., 7*, 103.

COHEN, G.M. (1990) Pulmonary metabolism of foreign compounds: Its role in metabolic activation. *Environ. Health Perspect., 85*, 31.

DEKANT, W., VAMVAKAS, S. and ANDERS, M.W. (1989) Bioactivation of nephrotoxic haloalkanes by glutathione conjugation. Formation of toxic and mutagenic inter-mediates by cysteine conjugate β-lyase. *Drug Metab. Rev., 20*, 43.

GILLETTE, J. R. (1973) Factors that effect the covalent binding and toxicity of drugs. In *Pharmacology and the Future of Man, Proc. 5th Int. Congr. Pharmacology, San Francisco, 1972*, Vol. 2, edited by T. A. Loomis (Basel: Karger).

HOOK, J. B. and HEWITT, W. R. (1986) Toxic responses of the kidney. In *Casarett and Doull's Toxicology, The Basic Science of Poisons*, edited by C. D. Klaassen, M. O. Amdur and J. Doull (New York: Macmillan).

LAURENT, G., KISHORE, B. K. and TULKENS, P. M. (1990) Aminoglycoside-induced renal phospholipidosis and nephrotoxicity. *Biochem. Pharmacol.*, **40**, 2383.

MAILMAN, R. B. (1980) Biochemical toxicology of the central nervous system. In *Introduction to Biochemical Toxicology*, edited by E. Hodgson and F. E. Guthrie (New York: Elsevier-North Holland).

MARET, G., TESTA, B., JENNER, P., EL TAYAR, N. and CARRUPT, P.-A. (1990) The MPTP story: MAO activates tetrahydropyridine derivatives to toxins causing parkinsonism. *Drug Metab. Rev.*, **22**, 291.

MITCHELL, J. R., POTTER, W. Z., HINSON, J. A., SNODGRASS, W. R., TIMBRELL, J. A. and GILLETTE, J. R. (1975) Toxic drug reactions. In *Handbook of Experimental Pharmacology*, Vol. 28, Part 3, *Concepts in Biochemical Pharmacology*, edited by J. R. Gillette and J. R. Mitchell (Berlin: Springer).

MONKS, T. J. and LAU, S. S. (1988) Reactive intermediates and their toxicological significance. *Toxicology,* **52**, 1.

POHL, L. R. (1979) Biochemical toxicology of chloroform. In *Reviews of Biochemical Toxicology*, Vol. 1, edited by J. Bend, R. M. Philpot and E. Hodgson (Amsterdam: Elsevier-North Holland).

PRESCOTT, L. F. (1983) Paracetamol overdosage: pharmacological considerations and clinical management. *Drugs,* **25**, 290.

RECKNAGEL, R. O. and GLENDE, E. A. (1973) Carbon tetrachloride hepatotoxicity: An example of lethal cleavage. *CRC Crit. Rev. Toxicol.*, **2**, 263.

RECKNAGEL, R. O., GLENDE, E. A. and HRUSZKEWYCZ, A. M. (1977) New data supporting an obligatory rule for lipid peroxidation in carbon tetrachloride-induced loss of aminopyrine demethylase, cytochrome P450 and glucose-6-phosphatase. In *Biological Reactive Intermediates*, edited by D. J. Jollow, J. J. Kocsis, R. Snyder and H. Vanio (New York: Plenum Press).

REYNOLDS, E. S. (1971) Liver endoplasmic reticulum: Target site of halocarbon metabolites. *Adv. Exp. Med. Biol.*, **84**, 117.

RUSH, G. F., SMITH, J. H., NEWTON, J. F. and HOOK, J. B. (1984) Chemically induced nephrotoxicity: role of metabolic activation. *CRC Crit. Rev. Toxicol.*, **13**, 99.

RUSH, G. F. and HOOK, J. B. (1986) The kidney as a target organ for toxicity. In *Target Organ Toxicity*, Vol. 2, edited by G. M. Cohen (Boca Raton, Florida: CRC Press).

SLATER, T. F. (1978) Biochemical studies on liver injury. In *Biochemical Mechanisms of Liver Injury*, edited by T. F. Slater (London: Academic Press).

SMITH, L. L. and NEMERY, B. (1986) The lung as a target organ for toxicity. In *Target Organ Toxicity*, Vol. 2, edited by G. M. Cohen (Boca Ration, Florida: CRC Press).

TARLOFF, J. B., GOLDSTEIN, R. S. and HOOK, J. B. (1990) Xenobiotic biotransformation by the kidney: pharmacological and toxicological aspects. In *Progress in Drug Metabolism*, Vol. 12, edited by G. G. Gibson (London: Taylor & Francis).

TIMBRELL, J. A. (1979) The role of metabolism in the hepatoxicity of isoniazid and iproniazid. *Drug. Metab. Rev.,* **10**, 125.

TIMBRELL, J. A. (1988) Acetylation and its toxicological significance. In *Metabolism of Xenobiotics*, edited by J. W. Gorrod, H. Oelschlager, and J. Caldwell (London: Taylor & Francis).

WALKER, R. J. and DUGGIN, G. G. Drug nephrotoxicity. *Annu. Rev. Pharmacol. Toxicol.*, **28**, 331.

Pharmacological, physiological and biochemical effects

COBURN, R. F. and FORMAN, H. J. (1987) Carbon monoxide toxicity. In *Handbook of Physiology, The Respiratory System,* Section 3, Vol. IV, edited by L. E. Fahri and S. M. Tenney (Bethesda, MD: American Physiology Society).

GOLDFRANK, L. R., FLOMENBAUM, N. E., LEWIN, N. A., WEISMAN, R. S. and HOWLAND, M. A. (editors) (1990) *Goldfrank's Toxicologic Emergencies,* 4th edition (Norwalk, Connecticut: Appleton & Lange).

GOSSEL, T. A. and BRICKER, J. D. (1984) *Principles of Clinical Toxicology* (New York: Raven Press).

HATHWAY, D. E. (1984) *Molecular Aspects of Toxicology* (Royal Society of Chemistry: London).

JOHNSON, M. K. (1982) The target for initiation of delayed neurotoxicity by organophosphorus esters: Biochemical studies and toxicological applications. In *Reviews of Biochemical Toxicology,* Vol. 4, edited by J. Bend, R. M. Philpot and E. Hodgson (Amsterdam: Elsevier-North Holland).

MAIN, A. R. (1980) Cholinesterase inhibitors. In *Introduction to Biochemical Toxicology,* edited by E. Hodgson and F. E. Guthrie (New York: Elsevier-North Holland).

MURPHY, S. D. (1986) Toxic effects of pesticides. In *Casarett and Doull's Toxicology,* edited by C. D. Klaassen, M. O. Admur and J. Doull (New York: Macmillan).

NEBERT, D. W. and WEBER, W. W. (1990) Pharmacogenetics. In *Principles of Drug Action, The Basis of Pharmacology,* edited by W. B. Pratt and P. Taylor (New York: Churchill Livingstone).

PENNEY, D. G. (1990) Acute carbon monoxide poisoning: animal model: A review. *Toxicology,* **62,** 123.

PENNY, D. G. (1988) A review: Hemodynamic response to carbon monoxide. *Environ. Health Perspect.,* **77,** 121.

SMITH, R. P. (1986) Toxic responses of the blood. In *Casarett and Doull's Toxicology, The Basic Science of Poisons,* edited by C. D. Klaassen, M. O. Amdur and J. Doull (New York: Macmillan).

VESELL, E. S. (1975) Genetically determined variations in drug disposition and response in man. In *Handbook of Experimental Pharmacology,* edited by J. R. Gillette and J. R. Mitchell (Berlin: Springer).

Lethal synthesis and incorporation

PETERS, R. A. (1963) *Biochemical Lesions and Lethal Synthesis* (Oxford: Pergamon Press).

SLATER, T. F. (1978) Biochemical studies in liver injury. In *Biochemical Mechanisms of Liver Injury,* edited by T. F. Slater (London: Academic Press).

ZIMMERMAN, H. J. (1976) Experimental hepatotoxicity. In *Handbook of Experimental Pharmacology,* Vol. 16, Part 5, *Experimental Production of Disease,* edited by O. Eichler (Berlin: Springer).

Teratogenesis

BROWN, L. P., FLINT, O. P., ORTON, T. C. and GIBSON, G. G. (1986) Chemical teratogenesis: Testing methods and the role of metabolism. *Drug Metab. Rev.,* **17,** 221.

GORDON, G. B., SPIELBERG, S. P., BLAKE, D. A. and BALASUBRAMANIAN, V. (1981) Thalidomide teratogenesis: Evidence for a toxic arene oxide metabolite. *Proc. Natn. Acad. Sci. USA,* **78,** 2545.

JUCHAU, M. R. (1990) Bioactivation in chemical teratogenesis. *Annu. Rev. Pharmacol. Toxicol.,* **29**, 165.

KEBERLE, H., LOUSTALOT, P., MALLER, P. K., FAIGLE, J. W. and SCHMID, K. (1965) Biochemical effects of drugs on the mammalian conceptus. *Ann. N. Y. Acad. Sci.,* **123**, 252.

LENZ, W. (1965) Epidemiology of congenital malformations. *Ann. N. Y. Acad. Sci.,* **123**, 128.

MANSON, J. M. (1986) Teratogens. In *Casarett and Doull's Toxicology, The Basic Science of Poisons,* edited by C. D. Klaassen, M. O. Amdur and J. Doull (New York: Macmillan).

MARTZ, F., FAILINGER, C. and BLAKE, D. A. (1977) Phenytoin teratogenesis: correlation between embryopathic effect and covalent binding of a putative arene oxide metabolite in gestational tissue. *J. Pharmac. Exp. Ther.,* **203**, 321.

RUDDON, R. W. (1990) Chemical teratogenesis. In *Principles of Drug Action, The Basis of Pharmacology,* edited by W. B. Pratt and P. Taylor (New York: Churchill Livingstone).

WILLIAMS, R. T. (1963) Teratogenic effects of thalidomide and related substances. *Lancet,* **i**, 723.

WILSON, J. G. (1965) Embryological considerations in teratology. *Ann. N. Y. Acad. Sci.,* **123**, 219.

WILSON, J. G. and FRASER, F. C. (1977) *Handbook of Teratology,* Vol. 2 (New York: Plenum Press).

Immunotoxicity

AMOS, H. E. (1979) Immunological aspects of practolol toxicity. *Int. J. Immunopharmac.,* **1**, 9.

PARK, B. K. and KITERINGHAM, N. (1990) Drug–protein conjugation and its immunological consequences. *Drug Metab. Rev.,* **22**, 87.

PERRY, H. M. (1973) Late toxicity to hydralazine resembling systemic lupus erythematosus or rheumatoid arthritis. *Am. J. Med.,* **54**, 58.

POHL, L. R., SATOH, H., CHRIST, D. D. and KENNA, J. G. (1988) The immunologic and metabolic basis of drug hypersensitivities. *Ann. Rev. Pharmacol. Toxicol.,* **28**, 367.

POHL, L. R., KENNA, J. G., SATOH, H., CHRIST, D. D. and MARTIN, J. L. (1989) Neoantigens associated with halothane hepatitis. *Drug Metab. Rev.,* **20**, 203.

PRATT, W. B. (1990) Drug allergy. In *Principles of Drug Action, The Basis of Pharmacology,* edited by W. B. Pratt and P. Taylor (New York: Churchill Livingstone).

TIMBRELL, J. A., FACCHINI, V., HARLAND, S. J. and MANSILLA-TINOCO, R. (1984) Hydralazine-induced lupus: is there a toxic metabolic pathway? *Eur. J. Clin. Pharmacol.,* **27**, 555.

UETRECHT, J. (1990) Drug metabolism by leukocytes and its role in drug-induced lupus and other idiosyncratic drug reactions. *CRC Crit. Rev. Toxicol.,* **20**, 213.

Multi-organ toxicity

DONALDSON, W. E. (1980) Trace element toxicity. In *Introduction to Biochemical Toxicology,* edited by E. Hodgson and F. E. Guthrie (New York: Elsevier-North Holland).

GOYER, R. A. (1986) Toxic effects of metals. In *Casarett and Doull's Toxicology, The Basic Science of Poisons,* edited by C. D. Klaassen, M. O. Amdur and J. Doull (New York: Macmillan).

JACOBSEN, D. and MCMARTIN, K. E. (1986) Methanol and ethylene glycol poisonings. Mechanism of toxicity, clinical course, diagnosis and treatment. *Med. Toxicol.,* **1,** 309.

FIELDER, R. J. and DALE, E. A. (1983) *Toxicity Review: Cadmium and its Compounds* (London: HMSO).

Glossary

α-carbon: first carbon after functional group.

acetylator status/phenotype: genetically determined difference in the acetylation of certain foreign compounds giving rise to rapid and slow acetylators.

acidaemia: decrease in blood pH.

acidosis: the condition where the pH of the tissues falls below acceptable limits.

aciduria: decrease in urinary pH.

acro-osteolysis: dissolution of the bone of the distal phalanges of the fingers and toes.

ACTH: adrenocorticotrophic hormone.

actin: cytoskeletal protein.

acyl: group such as acetyl, propionyl, etc.

acylation: addition of an acyl group.

ADCC: antibody dependent cell cytotoxicity.

adenocarcinoma: malignant epithelial tumour.

adenoma: benign epithelial tumour of glandular origin.

ADI: acceptable daily intake.

ADP-ribosylation: transfer of ADP-ribose from NAD^+ to a protein.

adrenergic: nerves responding to adrenaline.

β-adrenoceptor an autonomic receptor of which there are two types, β_1 and β_2.

aglycone: portion of molecule attached to glycoside as in a glucuronic acid conjugate.

AHH: aryl hydrocarbon hydroxylase.

Alkalosis: the condition where the pH of the tissues rises above acceptable limits.

Alkyl: group such as methyl or ethyl.

alkylation: addition of an alkyl group.

allosteric (change): alteration of protein conformation resulting in alteration in function.

allozymes: alternative electrophoretic forms of a protein coded by alternative alleles of a single gene.

allyl: the unsaturated group, $CH_2 = CH -$.

allylic: containing the allyl group.

ALT: alanine transaminase; alanine aminotransferase; (previously known as SGPT: serum glutamate pyruvate transaminase).

aminoaciduria: excretion of amino acids into the urine.

amphipathic (amphiphilic): molecules possessing both hydrophobic, non-polar and hydrophilic polar, moieties.

ANA: antinuclear antibodies.

anaphylactic (anaphylaxis): a Type I immunological reaction.

aneuploidy: increase or decrease in the normal number of chromosomes of an organism (karyotype).

antihypertensive: drug used for lowering blood pressure.

antiport: membrane carrier system in which two substances are transported in opposite directions.

antitubercular: drug used to treat tuberculosis.

anuria: cessation of urine production.

aplastic: absence of tissue such as bone marrow in *aplastic anaemia*.

apoprotein: protein component, e.g. of an enzyme minus the cofactor(s) and metal ions.

apoptosis: programmed cell death.

apurinic: loss of a purine moiety.

arteriole: small branch of an artery.

Arthus (reaction): Type III immediate hypersensitivity reaction.

aryl: aromatic moiety.

arylamine: aromatic amine.

arylated: addition of aromatic moiety.

arylhydroxamic acid: *N*-hydroxy aromatic acetylamine.

AST: aspartate transaminase (previously known as SGOT, serum glutamate oxalate transaminase).

astrocytes: cells found in the central nervous system.

AUC: area under the plasma concentration vs time curve.

autoantibodies: antibodies directed against 'self' tissues or constituents.

autoimmune: immune response in which antibodies are directed against the organism itself.

autoradiography: use of radiolabelled compounds to show distribution in a tissue, organ or even whole animal.

axoplasm: cytoplasm of an axon.

azo: N=N group.

basophil: a granulocyte (type of white blood cell) distinguishable by Leishman's stain as containing purple blue granules.

basophilic: cells which stain readily with basic dyes.

β-cells: insulin producing cells of the pancreas.

bioactivation: metabolism of a foreign substance to a chemically reactive metabolite.

bioavailability: proportion of a drug or foreign compound absorbed by the organism.

biosynthesis: synthesis within and by a living organism.

biotransformation: chemical change brought about by a biological system.

blebbing: appearance of blebs (protrusions) on outside of cells.

BNPP: *bis-p*-nitrophenyl phosphate.

bolus: a single dose.

bradycardia: slowing of heart beat.

bronchiole: small branch of the bronchial tree of the lungs.

bronchitis: inflammation of the bronchial system.

bronchoconstriction: constriction of the airways of the lungs.

bronchoscopy: visual examination of the bronchial system with a bronchoscope.

bronchospasm: spasms of constriction in the bronchi.

canalicular: relating to the canaliculi.

canaliculi: smallest vessels of the biliary network formed from adjoining hepatocytes.

carbanion: chemical moiety in which a carbon atom is negatively charged.

carbene: free radical with two unpaired electrons on a carbon atom.

carbinolamine: chemical moiety in which carbon to which amino group is attached is hydroxylated.

carbonium (ion): chemical moiety in which a carbon atom is positively charged.

carcinogenic: a substance able to cause cancer.

cardiac arrhythmias: abnormal beating rhythms of the heart.

cardiolipin: double phospholipid that contains four fatty acid chains. Found mainly in mitochondrial inner membrane.

cardiomyopathy: pathological changes to heart tissue.

cardiotoxic: toxic to heart tissue.

catabolized: metabolically broken down.

cDNAs: complementary DNA strands.

centrilobular: the region of the liver lobule surrounding the central vein.

chelation: specific entrapment of one compound, such as a metal, by another in a complex.

chemoreceptors: biological receptors modified for excitation by a chemical substance.

chiral: the presence of asymmetry in a molecule giving rise to isomers.

chloracne: a particular type of skin lesion caused particularly by halogenated hydrocarbons.

chloracnegenicity: ability to cause chloracne.

cholestasis: cessation of bile flow.

cholinergic: receptors which are stimulated by acetylcholine.

cis: when two groups in a chemical structure are on the same side of the molecule.

clastogenesis: occurrence of chromosomal breaks which results in a gain, loss or rearrangement of pieces of chromosomes.

clathrate: shape or appearance of a lattice.

co-administered: when two substances are given together.

co-oxidation: oxidation of two substrates in the same reaction.

co-planar: in the same plane.

co-transport: the transport of two substances together such as by a membrane active transport system.

CoA: coenzyme A.

codominance: full expression in a heterozygote of both alleles of a pair without either being influenced by the other.

COHb: carboxyhaemoglobin.

colo-rectal: colon and rectum portion of the gastrointestinal tract.

cooperativity: change in the conformation of a protein such as a haemoglobin or an enzyme leading to a change in binding of the substrate.
corneal: relating to the cornea of the eye.
corneum: horny layer of skin.
cristae: folds within the mitochondrial structure.
crystalluria: appearance of crystals in the urine.
cyanosis: condition where there is an excessive amount of reduced haemoglobin in the blood giving rise to a bluish coloration to skin and mucus membranes.
cytolytic: causing lysis of cells.
cytoskeletal: relating to the cytoskeleton.
cytoskeleton: network of protein fibrils within cells.
cytotropic: a class of antibodies which attach to tissue cells.

dalton: unit of molecular weight.
DBCP: dibromo-chloro-propane.
DCVC: dichloro-vinyl cysteine.
DDE: p,p'-Dichloro,-diphenyl dichloroethylene.
DDT: p,p'-Dichloro,-diphenyl trichloroethane.
deacetylation: removal of acetyl group.
dealkylation: removal of alkyl group.
deaminate: removal of amine group.
dechlorination: removal of chlorine group.
de-ethylation: removal of ethyl group.
dehalogenation: removal of halogen atom(s).
dehydrohalogenation: removal of halogen atom(s) and hydrogen, e.g. removal of HCl or HBr from a molecule.
demethylation: removal of methyl group.
denervated: a tissue deprived of a nerve supply.
depolymerize: the unravelling of a polymer.
derepression: remove repression of nucleic acid to allow transcription of gene.
desulphuration: removal of sulphur from a molecule.
deuterated: insertion of deuterium in place of hydrogen in a molecule.
diastereoisomers: stereoisomeric structures which are not enantiomers and thus not mirror images are *diastereoisomers*. They have different physical and chemical properties. A compound with two asymmetric centres may thus give rise to diastereoisomers.
dienes: unsaturated chemical structure in which there are two double bonds.
dimer: a macromolecule such as a protein may exist as a pair of sub-units or *dimers*.
distal: remote from the point of reference.
dopaminergic: receptors responsive to dopamine.
DR4: HLA type, DR4; human lymphocyte antigen.
ductule: small duct.
dyspnoea: laboured breathing.

ED$_{50}$: effective dose for 50% of the exposed population.

ED$_{99}$: effective dose for 99% of the exposed population.

electrophile: a chemical which is attracted to react with electron rich centre in another molecule.

electrophilic: having the property of an electrophile.

embryogenesis: development of the embryo.

embryolethal: causing death of embryo.

embryotoxic: toxic to embryo.

enantiomer: a compound with an asymmetric carbon atom yields a pair of isomers or *enantiomers* which are nonidentical mirror images.

encephalopathy: degenerative disease of the brain.

endocytosis: uptake or removal of substance from a cell by a process of invagination of membrane and formation of a vesicle.

endogenous: from inside an organism.

endometrium: mucus membrane of the uterus.

endonucleases: enzymes involved in DNA fragmentation.

endothelial/endothelium: layer of epithelial cells lining cavities of blood vessels and heart.

enzymic: involving an enzyme.

epinephrine: adrenaline.

epithelial/epithelium: tissue covering internal and external surfaces of mammalian body.

epitope: antigenic determinant.

exogenous: from outside the organism.

extracellular: outside the cell.

extrahepatic: outside the liver.

favism: syndrome of poisoning by Fava beans.

Fc: one of two segments of immunoglobulin molecule.

fenestrations: perforations.

FEV: forced expiratory volume.

fibroblast: connective tissue cell.

flavoprotein: protein in which the prosthetic group contains a flavin, e.g. FMN.

FMN: flavin mononucleotide.

folate: the cofactor folic acid and precursor of tetrahydrofolic acid.

gametotoxic: toxic to the male or female gametes.

genome: complete set of hereditary factors as in chromosomes.

glial cells: supporting cells, such as astrocytes, found in the central nervous system.

$\alpha 2\mu$-globulin: endogenous protein which may form complexes with exogenous chemicals which may accumulate in the kidney resulting in damage.

glomerulonephritis: inflammation of the capillary loops in the glomerulus.

glomerulus: a functional unit of the mammalian kidney consisting of a small bunch of capillaries projecting into a capsule (Bowman's capsule) which serves to collect the filtrate from the blood of those capillaries and direct it into the kidney tubule.

gluconeogenesis: synthesis of glucose.

glucosides: conjugates with glucose.

glucuronidation: addition of glucuronic acid to form a conjugate.

glucuronide: glucuronic acid conjugate.

glycolipid: molecule containing lipid and carbohydrate.

glycoprotein: molecule containing protein and carbohydrate.

glycoside: carbohydrate conjugate.

glycosidic: linkage between xenobiotic and carbohydrate moiety in conjugate.

glycosuria: glucose in urine.

granulocyte: leucocyte containing granules.

granulocytopoenia: deficiency of granulocytes.

GS-: glutathione moiety.

GSH: reduced glutathione (γ-glutamyl-cysteinyl-glycine).

GSSG: oxidized glutathione.

haem: iron protoporphyrin moiety as found in haemoglobin.

haemangiosarcoma: a particular type of tumour of the blood vessels.

haematuria: appearance of blood in urine.

haemodialysis: passage of blood through a dialysis machine in order to remove a toxic compound after an overdose.

haemolysis: breakdown/destruction of red blood cells and liberation of haemoglobin.

haemoperfusion: passage of blood through a resin or charcoal column to remove toxic compounds.

haemoprotein: protein containing haem.

haloalkanes: alkanes containing one or more halogen atoms.

Hb: haemoglobin as in Hb_4O_2 which is four subunits with one molecule of oxygen bound.

HCBD: hexachlorobutadiene.

heparin: endogenous compound which stops clotting.

hepatocarcinogen: substance causing tumours of the liver.

hepatocellular: relating to the hepatocytes.

hepatomegaly: increase in liver size.

hepatotoxin: substance causing liver injury.

heterocyclic: a ring structure containing a mixture of atoms.

heterozygote: possessing different alleles for a particular character.

histones: basic proteins associated with nucleic acids.

homocytotropic: having affinity for cells from the same species. *See* **cytotropic**.

homogenate: mixture resulting from homogenization of a tissue.

homolytic: equal cleavage of a chemical bond.

homozygote: possessing the same alleles for a particular character.

humoral: immune response which involves the production of specific antibodies.

hydrolase: hydrolytic enzyme.

hydroperoxide: chemical compound containing the group $-OOH$.

hydrophilic: having an affinity with water. A substance which associates with water.

hydrophobic: a substance which does not tend to associate with water.

hydropic degeneration: pathological process involving the accumulation of water in cells as a result of alteration of ion transport.

hydroxyalkenals: hydroxylated aliphatic aldehyde products of lipid peroxidation, e.g. 4-hydroxynonenal.

hydroxylases: oxidative enzymes which add a hydroxyl group.

hyper-: prefix meaning increased or raised.

hyperbaric: higher than normal atmospheric pressure.

hyperplasia: increase in the number of cells in a tissue or organ.

hyperpnoea: increased breathing.

hyperpyrexia: raised body temperature.

hypersensitivity: increased sensitivity to immunogenic compounds.

hyperthermia: abnormally raised temperature.

hypo-thyroidism: decreased activity of the thyroid gland.

hypocalcaemia: low blood calcium level.

hypolipidaemia: low blood lipid.

hypophysectomy: removal of pituitary gland.

hypothalamus: region of the brain lying immediately above the pituitary responsible for coordinating and controlling the autonomic nervous system.

hypoxia: low oxygen level.

hypoxic: tissue suffering a low oxygen level.

imine: the group $H-N=R'$.

immunogen: substance able to elicit a specific immune response.

immunogenic: description of an immunogen.

immunoglobulin (Ig): one of five classes of antibody protein involved in immune responses: IgA; IgD; IgE; IgG and IgM.

immunohistochemical/-histochemistry: use of specific antibody labelling to detect particular substances in tissues.

immunosuppression: reduction in the function of the immune system.

immunotoxic: toxic to or toxicity involving the immune system.

infarction: loss of blood supply to a tissue due to obstruction and subsequent damage.

initiation: first stage in chemical carcinogenesis in which ultimate carcinogen reacts with genetic material and causes a heritable change.

inulin: soluble polysaccharide.

invagination: infolding of biological membrane.

isoelectric point: pH at which there is no net charge on a molecule such as an amino acid and therefore there is no movement on electrophoresis.

isoenzyme; isozyme: one of several forms of an enzyme where the different forms usually catalyse similar but different reactions.

karyolysis: loss of the nucleus during necrosis.

karyorrhexis: fragmentation of nucleus to form basophilic granules.

kDa: kilodaltons; units of molecular weight.

kinase: enzyme catalysing transfer of high energy group from a donor such as ATP.

kinins: polypeptides such as bradykinin found in the plasma and involved in the inflammatory response.

L-DOPA: dihydroxyphenylalanine, a drug used to treat Parkinson's disease.

lacrimation: promoting tear formation.

lactoperoxidase: peroxidase enzyme found in mammary glands.

lavage: washing with fluid such as isotonic saline as in gastric lavage.

LD_{50}; (LC_{50}): lethal dose (or concentration) for 50% of the exposed population.

LD_1: lethal dose for 1% of the exposed population.

leucocytopoenia: deficiency of leucocytes.

leucopenia: *see* leucocytopoenia.

LH: luteinizing hormone.

ligand: substance which binds specifically.

liganded: bound to ligand.

ligandin: binding protein identical with glutathione-*S*-transferase B.

ligase: enzyme catalysing the joining together of two molecules and involving breakdown of ATP or other energy rich molecule.

ligation: tying off a vessel or duct.

lipases: enzymes which degrade lipids.

lipoidal: lipid like.

lipophilic: lipid liking. A substance which associates with lipid.

lipophilicity: a term used to describe the ability of a substance to dissolve in or associate with lipid and therefore living tissue.

lipoprotein: macromolecule which is a combination of lipid and protein. Involved in transport of lipids out of liver cells and in blood.

lithocholate: secondary bile acid.

lobule: a unit of structure in an organ such as the liver.

LOOH: lipid hydroperoxide.

lyase: enzyme which adds groups to a double bond or removes groups to leave a double bond.

lymphocyte: type of white blood cell produced in the thymus and bone marrow. Two types, T and B.

lymphoid tissue: tissue involved in the immune system which produces lymphocytes.

lymphokines: soluble factors which are associated with T lymphocytes and cause physiological changes such as increased vascular permeability.

lysis: breakage of a chemical bond or breakdown of a tissue or macromolecule.

macrophage: a phagocytic type of white blood cell.

megamitochondria: exceptionally large mitochondria.

MEOS: microsomal ethanol oxidizing system.

mercapto-: $-SH$ group.

metallothionein: metal binding protein.

methylation: addition of a methyl group.

Michaelis-Menten kinetics: kinetics describing processes such as the majority of enzyme mediated reactions in which the initial reaction rate at low substrate concentrations is first order but at higher substrate concentrations becomes saturated and zero order. Can also apply to excretion for some compounds.

microbodies: peroxisomes.

microflora: microorganisms such as bacteria.

midzonal: the region between the periportal and centrilobular regions of the liver.

miosis: constriction of the pupils.

mispairing: alteration of the pairing of the bases in DNA.

mitogens: stimulants of the immune system.

μm: micrometres.

MNNG: *N*-methyl-*N'*-nitro-*N*-nitrosoguanidine.

monomer: single subunit of a compound such as a protein.

monomorphic: with one form.

mucosa: mucus membrane.

muscarinic receptors: receptors for acetylcholine found in smooth muscle, heart and exocrine glands.

mutagen: substance which causes a heritable change in DNA.

mutagenesis: process in which a heritable change in DNA is produced.

myalgia: muscle pain.

myelin: concentric layers of plasma membrane wound round a nerve cell process.

myeloid: pertaining to bone marrow.

myelotoxicity: toxicity to the bone marrow.

myocardial: relating to heart muscle.

myoclonic: relating to myoclonus, a shocklike muscle contraction.

myoglobin: protein similar to haemoglobin.

myosin: protein found in most vertebrate cells and always when actin is present.

N-dealkylation: removal of an alkyl group from an amine.

NAPQI: *N*-acetyl-*p*-benzoquinone imine.

nephron: functional unit of the kidney.

nephropathy: pathological damage to the nephrons of the kidney.

nephrotoxicity: toxicity to the kidney.

neuronal: relating to neurones (nerve cells).

neuropathy: damage to the nervous system.

neurotoxic: toxic to the nervous system.

neurotransmission: passage of nerve impulses.

neurotransmitter: endogenous substance involved in the transmission of nerve impulses such as adrenaline.

neutrophil: phagocytic granulocyte (white blood cell).

nitroso: the group $-N{\rightarrow}O$.

nitroxide: the group $\equiv N{\rightarrow}O$.

nitrenium: the group $-N^{+}-H$.

nm: nanometres.

nomogram: graph such as blood concentration of a compound vs time used for the estimation of the toxicity expected from the compound as in overdose cases.
non-vascularized: lacking in blood vessels.
nucleoside: base-sugar (e.g. adenosine).
nucleotide: base-sugar-phosphate (e.g. adenosine triphosphate).

oliguria: concentrated urine.
ontogenesis: development of an organism.
oocyte: an egg cell; may be primary (diploid) or secondary (haploid).
opacification: becoming opaque.
ophthalmoscope: instrument for examining the retina of the eye.
organogenesis: formation of organs and limbs in the embryo during the gestation period.
osteomalacia: increase in uncalcified matrix in the bone resulting in increased fragility.
β-oxidation: degradative metabolism of fatty acids leading to the production of acetylCoA.
oxirane: epoxide ring.
oxygenase: enzyme catalysing oxidation reaction in which molecular oxygen is utilized.

pachytene: one of the stages in meiosis and also of development of the mammalian spermatocyte.
pampiniform plexus: collection of vessels delivering blood to and draining blood from the mammalian testis.
papilloedema: oedema of the optic papilla.
parenchyma: tissue of a homogeneous cell type, such as the liver which is composed mainly of hepatocytes.
parenteral: routes of exposure to a compound other than via the gastrointestinal tract.
paraesthesia: tingling sensations in the fingers and toes.
PBBs: polybrominated biophenyl's.
PCBs: polychlorinated biphenyl's.
pentose: carbohydrate with five carbon atoms.
peptidase: enzyme which cleaves peptide bonds.
peptide: molecule made up of amino acids joined by peptide bonds (–CO–NH–).
percutaneous: through the skin.
perfusion: passage of blood or other fluid through a tissue.
perinatal: the period between the end of pregnancy (7th month in humans) and the first week of life.
periportal: the region of the liver lobule surrounding the portal tract.
peroxidized: oxidized by the addition of $-O_2$.
peroxisome: intracellular organelle which carries out fatty acid oxidation and other oxidative transformations.
peroxy: the chemical group –O–O–.
phagocytic: a cell which is able to engulf foreign particles.

pharmacodynamic (toxicity): relating to the *action* of a compound with a receptor or enzyme for example.

pharmacokinetic (toxicity): relating to the *concentration* of a compound at a target site.

photosensitization: sensitization of skin to light.

physico-chemical characteristics: characteristics of a molecule such as lipid solubility, size, polarity.

pinocytosis: uptake of a solution by a cell.

polymerase: enzyme which catalyses polymerization.

polypeptide: a chain of amino acids.

polysaccharide: a chain of carbohydrate residues.

porphyrin: ring composed of four pyrrole rings.

post-transcriptional: event occurring after transcription of DNA code to mRNA.

portal tract (triad): the group of vessels seen in sections of liver tissue consisting of a bile ductule, hepatic arteriole and portal venule.

potentiation: when an effect due to two substances with different modes of action is greater than expected from the effects of the individual components.

PPi: diphosphate.

ppm: parts per million.

prostaglandins: endogenous chemical mediators derived from unsaturated fatty acids such as arachidonic acid.

proteases: enzymes which degrade proteins.

proteinuria: presence of protein in the urine.

protonated: molecule to which a proton is added.

pseudosubstrate: substance which mimics the normal substrate for an enzyme.

pyknosis: change occurring in a cell during necrosis in which there is shrinkage of the nucleus and increased staining intensity.

pyrolysis: breakdown of a substance or mixture by heat/burning.

racemate: mixture of isomers; racemic mixture.

Raynaud's syndrome: severe pallor of the fingers and toes.

redox: reduction and oxidation cycle.

reticuloendothelial: system of phagocytic cells found in the liver, bone marrow, spleen and lymph nodes.

rhinitis: inflammation of the nasal passages.

ribosomes: particles found on the smooth endoplasmic reticulum involved with protein synthesis.

rodenticide: pesticide used specifically for killing rodents.

rRNA: ribosomal RNA.

RT: room temperature.

semiquinone: quinone reduced by the addition of one electron.

sensitization: sensitization of the immune system as a result of exposure to an antigen.

SER: smooth endoplasmic reticulum.

sinusoid: blood filled space which may be a specialized capillary such as found in the liver in which the basement membrane is modified to allow for efficient passage.

sp2: orbital.

sp3: orbital.

spectrin: cytoskeletal protein associated with the cytoplasmic side of the membrane.

spermatocytes: cells in the testes which develop into sperm.

splenomegaly: enlarged spleen.

stasis: stoppage or slowing down of flow.

steatosis: fatty infiltration in organ or tissue.

steric factors: factors relating to the shape of a molecule.

sulphation: addition of a sulphate group.

superoxide: $O_2 \cdot ^-$; oxygen radical.

symport: membrane carrier system in which two substances are transported in the same direction.

synergism: when an effect due to two substances with similar modes of action is greater than expected from the effects of the individual components.

synergist: a compound which causes synergism with another compound.

t$\frac{1}{2}$: half-life.

tachypnoea: increased breathing rate.

TCDD: 2,3,7,8-tetrachlorodibenzo-p-dioxin; dioxin.

TD$_{50}$: toxic dose for 50% of the exposed population.

teratogen: compound which causes abnormal development of the embryo or foetus.

teratogenesis: development of abnormal embryo or foetus.

thiol: –SH group.

thiolate: –S⁻.

thiyl: –S·; sulphur free radical.

thrombocytopoenia: low level of platelets in the blood.

thymocytes: cells found in the thymus.

thyroidectomy: removal of the thyroid gland.

tinnitus: 'ringing' in the ears.

TLV: threshold limit value.

α-TH: α-tocopherol; vitamin E.

α-TQ: α-tocopherol quinone.

toxication: increase in toxicity.

toxicokinetics: study of the kinetics of toxic substances.

trans: when two groups in a chemical structure are on the opposite side of the molecule.

transversion: transformation of a purine base into a pyrimidine or *vice versa*.

trifluoroacyl: the group CF_3–CO–.

triglycerides: lipids in which glycerol is esterified with three fatty acids.

trihydric: molecule having three alcoholic hydroxyl groups.

trimodal: frequency distribution which divides into three groups.

trisomy: possession of an extra chromosome.
tumorigenic: able to cause tumours.

UDP: uridine diphosphate.
UDPGA: uridine diphosphate glucuronic acid; glucuronic acid donor.
ultrafiltrate: fluid formed in renal tubule from blood passing through the glomerulus/Bowman's capsule in the kidney.
uncouplers: compounds which uncouple oxidative phosphorylation from ATP production in the mitochondria.
uniport: membrane carrier system in which one substance is transported.
urticaria: appearance of wheels on the skin; may occur as part of an immunological reaction.
uterine: part of uterus.

vacuolated: cell showing vacuoles.
vacuoles: spaces inside cells.
vasculature: blood vessels in a tissue.
vasculitis: inflammation of the vascular system.
vasoconstriction: constriction of blood vessels.
vasodilator: compound which dilates the blood vessels.
VCM: vinyl chloride monomer.
veno-occlusive: occlusion of blood vessels.
venule: very small tributary of a vein.
VLDL: very low density lipoprotein.

wheals: raised patches on the skin; may be part of an immune reaction.

xenobiotics: substances foreign to living systems.

Index